D0971523

The Dark Side of Man

ALSO BY MICHAEL P. GHIGLIERI

Canyon

East of the Mountains of the Moon: Chimpanzee Society in the African Rain Forest

The Chimpanzees of Kibale Forest: A Field Study of Ecology and Social Structure

The DARK SIDE of MAN

Tracing the Origins of Male Violence

Michael P. Ghiglieri

A Cornelia and Michael Bessie Book

HELIX BOOKS

PERSEUS BOOKS
Reading, Massachusetts

The author is grateful for permission to reprint the following copyrighted material:

page 1: "Compelled by the urge to pass on genetic material" from *Female Choices: Sexual Behavior of Female Primates,* by Meredith F. Small, copyright 1993 by Cornell University. Reprinted by permission of Cornell University Press.

page 54: "Is it not unreasonable to anticipate that our understanding of the human mind would be aided greatly by knowing the purpose for which it was designed" from *Adaptation and Natural Selection: A Critique of Some Current Evolutionary Thought,* by George C. Williams. Copyright 1966 by Princeton University Press. Reprinted by permission of Princeton University Press.

pages 54–56: "'I've had it,' said Tom, 'When do we head back to camp?'" from *Lucy: The Beginnings of Humankind,* by Donald C. Johanson and Maitland A. Edey. Copyright 1981 by Donald C. Johanson and Maitland A. Edey. Reprinted by permission of Simon and Schuster.

page 156: "If they had had firearms and been taught to use them, I suspect they would have used them to kill" from *The Chimpanzees of Gombe,* by Jane Goodall. Copyright 1986 by the President and Fellows of Harvard College. Reprinted by permission of Harvard University Press.

page 205: "It seems, then, that the attacks are an expression of the hatred" from *Through a Window,* by Jane Goodall. Copyright 1990 by Soko Publications, Ltd. Reprinted by permission of Houghton Mifflin Company.

• • •

ISBN 0-7382-0076-X

Library of Congress Catalog Card Number: 99-60682

Perseus Books is a member of the Perseus Books Group

A Cornelia and Michael Bessie Book

Jacket design by Adrian Morgan
Text design by Mark Corsey
Set in 11-point Apollo by Eclipse Publishing Services

123456789—0302010099
First printing, March 1999

Find Helix Books on the World Wide Web at http://www.aw.com/gb/

With thanks to Charles Darwin,
the intellectual father of modern biology

In a world full of hatred, repression, terrorism, small wars, and preparation for immense wars, human conflict is a subject that deserves the most careful and searching inquiry.

— David A. Hamburg, *Science*

CONTENTS

TO THE READER

YOU CAN EXPECT at least two things from this book. The first is the facts on men's violence. You'll see which men assault or kill other people (and where and when they do it) globally and in the United States. The second is the most reasonable, accurate, and sound evolutionary explanations for why men are as violent as they are. This book is about men and women and sex and violence, but it also explores the basic underlying biological and cultural forces that urge people to act as they do. Do not, however, expect political correctness from this book. Politics and an open-minded scientific discussion of human nature mix about as well as oil and water.

What prompted this book was a dozen episodes of warlike murder among wild chimpanzees during the 1970s and 1980s. This humanlike violence among male great apes seemed so significant in the ultraviolent world we humans have created—and it felt so eerie—that it haunted me.

Fortunately, I was in a good position to do something about my feelings. Two years before the first observed ape war in 1974, I had used GI Bill benefits, earned during my stint as an army platoon sergeant, to study gorilla behavior for a master's degree. That first chimp war (described in Chapter 6) linked me up with the grand master of chimpanzee research, Jane Goodall. Terrorist kidnappings led by Laurent Kabila (who recently conquered the Congo) and aimed at Goodall halted all expatriate research at Goodall's Gombe Stream site. Two years later, however, I launched a Ph.D. study of the behavior of wild chimps in Uganda. By 1980, the facts of primate violence had yanked the rug out from under a legion of social scientists who had been certain that humans held the monopoly on rape, murder, and war. Our nearest evolutionary cousins, it turns out, routinely commit violence in the wild.

Now we know that, instead of being an unfortunate by-product of civilization, human violence has much deeper roots. The purpose of this book is to reveal precisely how deeply entrenched in the male psyche the triggers of men's violence lie—and why they are there at all.

The first three chapters of *The Dark Side of Man* explore the differences between men and women, human emotions, and the evolution of human behavior. Chapters 4 through 7 form the core of the book. These chapters explore rape, murder, war, and genocide. They focus on the violence of men, relying on hard statistics, examples, and evolutionary interpretations to do so. Each chapter also includes a brief natural history tour of similar violence by one species of great ape. Chapter 8 offers an antidote to the levels of male violence we have allowed to flourish among us.

Nearly every chapter of *The Dark Side of Man* begins (and often ends) with what appears to be a fictional story of violence: rape, murder, or warfare. Each of these stories is completely factual. The people in each are or were real people. I learned their stories via interviews with the survivors. I have included these stories because they embody the focus of this book: the resolution of conflict by violence.

Readers often wonder why an author is impelled to write a particular book. The origin of this book requires a confession. This project started for me long before the Gombe ape wars, when I was only five years old. To save on baby-sitter money one night, my parents took me with them to the movies. We saw *The Day the Earth Stood Still*.

In this 1951 classic, Michael Rennie plays an alien, Klaatu, sent to Earth to deliver an ultimatum to *Homo sapiens:* either join the Galactic Federation and surrender all their atomic weapons to the aliens' robot police force, or the Federation—via Klaatu's immensely powerful robot, Gort—will scorch the Earth to a cinder. Our addiction to war and doomsday weapons, says Klaatu, threatens the galaxy. He has been sent to nip this addiction in the bud before we invent space travel and wreck the universe. To show that he is serious, Klaatu neutralizes electricity on the planet for an hour. Atomic disarmament versus certain death for the planet should not have been a hard choice for any logical being. But for the political leaders in this movie—all of whom are men—it is. When American soldiers machine-gun Klaatu in the back, killing him, Gort retrieves his body and revives him. Klaatu stops Gort from barbecuing the Earth. But despite this brush with annihilation, the human politicians still will not agree to dismantle their atomic weapons.

Even at age five, I was struck by the stupidity of these men. Only many years later, after I had been drafted into the U.S. Army in 1966, did I realize that the movie reflected the mind-sets of real politicians and generals and it depicted the violence of real men all too well.

Ever since, I have been intrigued by why men are so violent. Later, my experiences as a platoon sergeant, a field scientist working among African apes, an international wilderness guide among remote and primitive tribesmen, and a martial arts student deepened my quest for understanding.

I fully realize the risk involved in writing a book that promises to reveal as much as this one does. Michael Crichton explains just how I feel in his acknowledgments in *Five Patients,* where he quotes an influential professor who taught one of Crichton's mentors, Alan Gregg:

> Whenever you say anything explicitly to anyone, you may also say something else implicitly, namely, that you think you are the guy to say it. Such sentiments trouble all but the most egotistical writers; the others recognize that their sense of enfranchisement is a gift of the people around them, whom they can only hope not to disappoint.

ACKNOWLEDGMENTS

IN WRITING THIS BOOK, I owe a debt to the hundreds of scientists and other professionals who slogged through muddy rain forests to spy on wild apes, slinked down wet alleys in pursuit of killers, worked to understand what makes felons felonious, lay awake in uniform in killing zones waiting for ambushes, groped with the repercussions of doomsday weapons, probed the hidden mysteries of the human mind, sifted the sands of time for evidence of life before us, or otherwise sacrificed part of their lives in the quest to understand the complex phenomenon of *Homo sapiens*. Beyond these, I acknowledge that my greatest intellectual debt is owed, as it is by all biologists, to Charles Darwin.

Most of the pioneers to whom I owe the greatest debts for blazing the trail are mentioned in the text. But several deserve special mention because their work was indispensable to our understanding. At the risk of offending all too many very deserving scientists through my forced omission of them here, I include my special thanks to Napoleon A. Chagnon, Martin Daly, Richard Dawkins, Irenäus Eibl-Eibesfeldt, Jane Goodall, William D. Hamilton, Sarah Blaffer Hrdy, Melvin Konner, Louis S. B. Leakey, John Maynard Smith, Desmond Morris, George B. Schaller, Thomas T. Struhsaker, Donald Symons, Nancy and Randy Thornhill, Lionel Tiger, Robert L. Trivers, James Watson, George C. Williams, Edward O. Wilson, Margo Wilson, and Solly Zuckerman. I also thank my mentor Dale Marvellini for talking me into my first study on apes, lowland gorillas, and I owe my mentor, Peter S. Rodman, a debt of gratitude for helping me orienteer my way to the shoulders of these giants.

I also owe several individuals thanks for helping directly on this project with source materials or feedback: Samuel Craighead Alexander, Ron Blanchard, John R. Brownlee, Gary Buckley, Florence Carlstrom, Napoleon Chagnon, Ed Chen, John Farella, Elaine Gammil,

Pepi Granat, Dustin Hurlbut, Daniel D. James, Kirk Jensen, Donald Johanson, Dana Kline, William J. Klingenberg, Ray Martinez, Alfonso Sakeva, Casey Simpson, Joe Skorupa, Randy and Nancy Thornhill, Scott Thybony, Evan Widling, and Margaret Zahn. I owe my wife, Connie Ghiglieri, a debt of thanks for hanging in there with me on draft after draft of *The Dark Side of Man*—for her forbearance when I ceased to be a husband or a father during my quest to get it right and also for her feedback on chapter after chapter. I owe my kids a few thousand hours of quality time lost to rewriting drafts of this book. And I owe my friend Robert Hoffman a debt for his many late hours of performing computer alchemy to save what I had written from evil word processing programs. My friend Meg Mitchell slogged through a semi-polished draft for a hundred or so hours and pointed out at least a thousand places I might improve. I owe her a great debt for helping to sift the wheat from the chaff. Finally, I owe my most serious debts to veteran editor Simon Michael Bessie for having the faith to champion this book and to Amanda Cook, editor at Perseus Books, for identifying what worked and what didn't and for suggesting what else would.

Any errors, omissions, or troublesome opinions or conclusions that might remain in this book are strictly my responsibility.

• • •

PART ONE

Roots

Sexual selection apparently has acted on man, both on the male and the female side, causing the two sexes to differ in body and mind.

— Charles Darwin, 1871

Compelled by the urge to pass on genetic material to the next generation, the sexes must often cooperate in mating and parenting, but each sex cooperates only under duress because females and males operate under different reproductive rules set down in opposing directions eons ago. Like an open wound that never heals, the conflict between males and females will never be resolved because the evolutionary interests of the two sexes are forever locked in opposing position. There is no right or wrong here, no sex better than the other, just two types of individuals trying to win in the game of reproductive success. The ground rules of the battle include cooperation, conflict, and exploitation, and both sexes use these tactics equally.

— Meredith F. Small, 1993

CHAPTER ONE

Born to Be Bad?

There were dead people along the road, so my friends gave up trying to go to Fort Portal," Tedi explained. She had just walked eight miles through the primeval rain forest, a hike no Mutoro woman in her right mind would make due to its dangers. But her husband, Otim, was my game ranger, and she had come to warn us. The dead people, Acholi and Langi—both East Nile tribes—were victims of Idi Amin Dada's latest (February 1977) genocidal pogrom.

A Kakwa from the West Nile, Amin was determined to erase from Uganda all Langi and Acholi people, hereditary enemies from the east. President Milton Obote, whom Amin had deposed six years earlier, was a Langi. And Obote's tribesmen, all conveniently bearing surnames beginning with the letter *O,* could never be anything but enemies. For Amin, the only good enemy was a dead enemy. Our problem was that Otim was a Langi.

Only a dozen miles away from our tiny encampment at Ngogo, in the center of Kibale Forest, blood was flowing. Amin had ordered his army of West Nile rogues, illiterates, and sadists to eliminate all Langi and Acholi occupying all official positions, from postal clerks and schoolteachers to district commissioners. His gangs of armed thugs were breaking into schoolrooms, offices, businesses, private homes, and even bush huts to drag off innocent Langi people. Once these people were outside, the thugs hacked off their heads with pangas (machetes). Thousands of Acholi were simply machine-gunned en masse. A dozen miles from our enclave in the jungle, entire rooms full of living Acholi and Langi prisoners were wrapped together with wire, doused with kerosene, and set ablaze.

That same night, Amin's censored Voice of Uganda radio network broadcast allegations that an invasion force from Tanzania (where Obote still lived in exile) had violated Uganda's borders and was being assisted by Langi and Acholi rebels under Obote's leadership. By definition, all Langi and Acholi were state enemies. Amin's dreaded gestapo, the State Research Bureau (SRB), was now carrying out mass executions of them. Outside this jungle that concealed us, Otim would die.

I worked out a plan for Otim's escape into the forest should we hear Land Rovers approaching. Engine noise two miles away would be audible, giving us plenty of time to make him vanish. With only a little luck, I reckoned, Otim would live through this pogrom. And when Amin cooled off again, Otim could seek sanctuary in Tanzania or Kenya.

Meanwhile, the Voice of Uganda reported a new twist in the alleged Tanzanian invasion. The rebel force was being assisted by American mercenaries. All Americans in the country (there were forty-six of us, mostly missionaries and pilots) were ordered by Life President Idi Amin Dada to assemble immediately at Entebbe International Airport for expulsion—after signing over all our personal property to the government of Uganda.

Being only three months into my research on wild chimpanzees but already having made major breakthroughs, I was now about to lose out again to African politics. I perused my maps. The nearest escape route would skirt the southern fringe of the Mountains of the Moon into Zaire. To avoid becoming one of Amin's three thousand murder victims this week, I would have to avoid all roads and villages, travel at night, and bivouac in thick foliage. But I had not seen much of this terrain yet, and the trip would take several days even if everything went well. Either way, I would carry all of my data sheets.

Should Otim come along? Or should we both sit tight a bit longer? One thing we had going for us was that Ngogo was incorrectly located by several miles on government topo maps. Even so, there were villagers out there who knew where we were. And some of them were poachers whom we had chased out of the nature reserve—men with grudges.

I decided to stay put a bit longer. Maybe we'd survive Amin's genocide against rival tribes (including Americans) long enough for me to gain some insights from the wild chimpanzees on the origins of such barbaric genocidal behavior.

• • •

Differences Between Girls and Boys

My wife, Connie, and I have a daughter and a son eighteen months apart. By age three, our daughter, Crystal, spent hours building beds for her dolls, stuffed animals, ponies, and even her dinosaurs and a Halloween bat. She tucked them lovingly into their blankets and propped their heads on pillows. She put bandages on their imaginary wounds. She talked and sang to them and brushed their manes. She arranged them all in a cozy circle around a tablecloth and fed them with a spoon at polite, orderly picnics. She nurtured her menagerie as if she had studied child psychology. And she scolded them if they acted naughty. She also demanded of her mother and me—on pain of a tantrum—that she wear dresses, even in two feet of snow. She *had* to look feminine. Today, eight years later, Crystal no longer insists on dresses, but she is fixated on horses and is even more nurturing and security-conscious.

At three years old, her brother, Cliff, shot the same stuffed animals with toy guns, stabbed them with a rubber knife, hacked them with his plastic sword, and toppled them off the staircase in death plunges. His dinosaurs did nothing but attack and kill each other and gorge themselves on the bloody carcasses. Cliff never spoon-fed, hugged, bandaged, consoled, instructed, or scolded anything. His clothes simply had to be comfortable. Today he's a Boy Scout, wants to be a river guide and a U.S. Air Force pilot, and plays computer and video games like an addict. The goal of these games? To kill the bad guys.

As the years pass, the gender gap continues to widen.

So a brother and sister are different, you might say. Big deal. Two kids prove nothing.

Two friends of ours, one a social psychologist and the other a psychoanalyst/social anthropologist, have a girl and a boy of similar ages. To avoid programming their four-year-old son with violent tendencies, the mother allowed him no toy weapons. Much to her dismay, he relentlessly prowled the woods around their house in search of sticks shaped like guns, which he used to shoot things. Her daughter never did this. Instead, she acted like a girl. This tormented the mother. But this also proves nothing.

Scientific or not, most of us know that men and women, or boys and girls, are different. Parents know it. Teachers know it. Husbands usually suspect it about their wives. Wives are certain about their husbands. It is no secret, despite the politically correct insistence that men and women are equal in every sense. Of course, women and men are equal in their value as people and in their legal rights. But otherwise men and women really are different—so different, so early, that infant boys and girls behave as if they were programmed for widely divergent roles. As if men were born to be bad.

We do know that children everywhere identify as masculine or feminine before they are two years old and that they insist on copying their own gender. Not even surgical sex changes on male babies eighteen months old can reverse the pattern to female. Children fine-tune their behavior by watching older people. By age two, girls copy or imitate their mothers (or, if they lack a mother to copy, they imitate other mothers), boys imitate their fathers. Significantly, this division seems to prime boys for violence. It happens everywhere.

All this is clear from an astoundingly detailed global study by German ethologist Irenäus Eibl-Eibesfeldt. It reveals that older boys worldwide mostly play contest games of pursuit and scuffling, do experiments, and commonly practice fighting—despite their being punished far more often than girls for being aggressive. Girls, meanwhile, play more sedate and even solitary games that often focus on security. Even more intriguing, children are likely to imitate behaviors that they see as appropriate to their own sex regardless of the sex of the actor. American experiments in child development, for example, reveal that a girl will copy "feminine" behavior seen in a man before she will imitate aggressive or bullying "masculine" behavior seen in a woman.

The human urge to adopt an "appropriate" gender is so powerful that it succeeds even without gender roles to copy. The Israeli kibbutz experiment provides an inadvertent test of this. This Israeli system tried to create monogenderal roles by rearing children communally. But most of the children, who had no family role models to copy, invented their own families. The kibbutz also failed to eradicate stereotypical roles, even during play. Girls grew up focusing not only on female role models but also on maternal role models. The fact that gender among kibbutz children emerged as the most powerful and unchangeable root of their identity, despite communal rearing, hints at the depth of the human instinct to lock into the "right" gender.

Yes, parents do reinforce this natural process, often unconsciously, and they usually do so from the moment their children are born. Mothers and fathers, however, nurture their sons and daughters differently. Mothers, for example, soothe and comfort their infant girls more than their infant boys, but they burp, rock, arouse, stimulate, stress, look at, talk to, and even smile at their infant boys more. Mothers also hold their infant boys closer.

Still, it is children's self-directed sex differences so early in life that spotlight the most fundamental question of human behavior and men's violence: Are the psyches of men and women intrinsically different in design? And if so, how and why? Are men somehow born to be bad? Or do they start out innocent and then get corrupted?

The Evolution of Sex Differences

ADMITTEDLY, THESE QUESTIONS about the basic designs of the human male and female psyches are politically radioactive, but the answers to them are matters of life and death. Finding these answers demands that we sweep the table clean of several widely held but now untenable ideas about human behavior. As the nineteenth-century humorist Artemus Ward observed, "It ain't so much the things we don't know that get us in trouble. It's the things we know that ain't so."

That *Homo sapiens* is still locked in an identity crisis is ironic. How hard could it be for us to figure ourselves out as a species? Why can't we simply stand in front of the mirror of science, so to speak, and look at ourselves with an objective eye? The answer is that the old nature versus nurture debate fogs the mirror. For example, the claim by Franz Boas, Friedrich Engels, John Locke, Karl Marx, Margaret Mead, and B. F. Skinner that humans are born as malleable blank slates that become pure products of cultural indoctrination has consistently sabotaged the exploration of human nature by denying that we have a psyche equipped with instincts. It is society, the modern-day protégés of these philosophers and social scientists still insist, that creates the mental software that rules people's behavior. Meanwhile, the rival claim by many biologists that we humans carry a legacy of instincts from our primeval past—a human nature—disturbs some of us so much that the heat we feel during such discussions stops us from seeing the light.

To break the cycle of dogma here, biologists insist, we first must admit that humans are a biological phenomenon. Like all other mammals, we must eat, breathe oxygen, excrete, and seek warmth—in short, survive. If our DNA is to make it to the next generation, we must also mate and rear our offspring successfully. Just how biological are we? For the answer, ask any medical doctor how she was trained. She will tell you that she was inundated with every known aspect of human biology. Boring though this may seem to many of us, it is a good idea. If medical doctors instead focused mostly on sociology and political theory for their medical training, we would likely feel far more nervous than we already do when that ice-cold stethoscope stings our chest.

OK, so we're biological. But is human behavior truly influenced in a substantial way by our biology—by our genes? We know that behavior in other animals is. Robert Plomin found that many behaviors can be enhanced, created, or eliminated through selective breeding. More to the point, Plomin discovered that inheritance plays a role in human behavior. For example, at least one hundred different gene effects, most of them very rare, are now known to lower IQ. The specific genes that influence behavior are numerous "needles in the haystack" of the DNA molecule. "Just 15 years ago," Plomin concludes, "the idea of genetic influence on complex human behavior was anathema to many behavioral scientists. Now, however, the role of inheritance in behavior has become widely accepted, even for sensitive domains such as IQ."

Another study found that the IQs of 245 adopted children more closely matched those of their biological parents than those of their adoptive parents, who created the children's environment. Among identical twins from the Minnesota Study of Twins Reared Apart, 50 to 70 percent of variance in IQ was associated with genetics. Even more amazing, note Thomas J. Bouchard, Jr., and his colleagues, authors of the study, "on multiple measures of personality and temperament, occupational and leisure time interests, and social attitudes, monozygotic twins reared apart are about as similar as monozygotic twins reared together!"

Human behaviors now known or suspected to be based on genetics include amount of alcohol consumed, autism, language disability, panic disorder, eating disorders, antisocial personality disorder, and Tourette's syndrome. Even the tendency to divorce a spouse seems significantly influenced (52 percent heritability) by one's genes. Recent research by psychologist Jim Stevenson suggests that personality traits, especially "nice" traits, are genetically linked. Working with twins, Stevenson found that more than half of the variance associated with "prosocial" behavior was linked to genes. Only about 20 percent of "antisocial" behavior was.

In short, much of our behavior is substantially influenced by our genes, but much also is influenced by our environment. "Any analysis of the causes of human nature that tends to ignore either the genes or environmental factors," concludes physical anthropologist Melvin Konner, "may safely be discarded."

The basic premise of this book is that we are understandable both from a biological perspective and in an environmental context. Nature equipped each of us with a complex brain ruled by chemical neurotransmitters that spur in us instinctive emotional responses to situations, which in turn influence our behavior. This may not be a comfortable way to look at ourselves, but biology tells us that this is the only accurate way and, more to the point, that it is the only way that offers us any real hope of understanding our behavior, including our use of violence.

The counterargument that humans are impossible to understand because our culture shapes our behavior far more than biology does is often little more than a monkey wrench tossed into the discussion to grind it to a halt before it treads on politically incorrect turf. To deflect this monkey wrench and to test ideas about human violence, this book will look not just at human beings but also at our nearest living relatives, the great apes, none of which have been immersed in human culture. Indeed, a close look at orangutans, gorillas, or chimpanzees is as startling as walking in front of a mirror, knowing the reflection is of ourselves but seeing someone else's face.

The great apes offer us more than just an eerie glimpse of the basic behavioral software from which humanity emerged. They also provide us with insights into the origins of human violence—insights that help make it possible to understand the human male psyche.

This glimpse of our evolutionary past did not come easily. It took more than thirty-five years of field research by hundreds of scientists to reveal how and why the great apes socialize with each other. We now know that *each individual ape socializes—cooperatively or aggressively—based on its own decision on how to enter the reproductive arena, an arena that demands being social in some way.* The lives these apes lead are shaped by instinctive social "rules" that are violent, sexist, and xenophobic. Understanding how these rules work offers us the only informed approach to understanding the roots of human violence. The upcoming chapters on rape, murder, and war each offer a short natural history tour of similar violence by great apes. First, however, we must look at the evolutionary process that made the priorities of males and females so utterly different and made men so violent.

If not for the insights of a medical school drop-out who became a naturalist on a five-year voyage around the world, we still might not have a clue as to why the sexes differ. The reasons are revealed first in Charles Darwin's 1859 blockbuster, *On the Origin of Species by Means of Natural Selection, or the Preservation of Favoured Races in the Struggle for Life,* a book that sold out the same day it hit the book shops, and for good reason. In it Darwin redefined the "Hand of God."

Although Darwin is now a household name, many of us are hazy on what he actually said. So here is Darwin's concise definition of evolution's prime architect:

> As many more individuals of each species are born than can possibly survive; and as, consequently, there is a frequently recurring struggle for existence, it follows that any being, if it vary however slightly in any manner profitable to itself, under the complex and sometimes varying conditions of life, will have a better chance of surviving, and thus be naturally selected. . . . This preservation of favourable variations and the rejection of injurious variations, I call Natural Selection.

Evolution, we now know, works by natural selection "editing" new alleles (alternate forms of a gene that arise through mutation) via each allele's effects on reproductive success. Indeed, the modern, neo-Darwinian definition of evolution is simply *a change in the frequency of alleles in a population from one generation to the next*. Field research on hundreds of species of wild plants and animals has documented natural selection in action. Jonathan Weiner's *The Beak of the Finch* offers a fascinating view of this process as seen during long-term work on Darwin's finches in the Galápagos Islands. Neo-Darwinian theory is so well established that, regardless of the biochemistry of heredity, we know that natural selection must be happening on every planet in the universe on which life exists. How important is this? "For a biologist," notes Nobel Prize–winning immunologist Sir Peter Medawar, "the alternative to thinking in evolutionary terms is not to think at all."

A dozen years after redefining the "Hand of God," Darwin boosted our understanding of why males and females are so different—and why males are so violent—by explaining a special form of natural selection he called *sexual selection*. This process enhances traits within one sex that help its members win against sexual rivals. It works in both sexes, and it works in two ways. Among males, one way is the "pretty male strategy." Males winning via this strategy sire more offspring because females choose these males more often as mates based on traits the females prefer. The second way sexual selection works is the "macho male strategy," in which some males breed more than others because they defeat rival males and exclude them from breeding. (Females, depending on their species, also can compete via either of these strategies. But females also use a third strategy, called the "supermom strategy," in which they compete based on the efficiency of their reproductive equipment.) It is macho male sexual selection that leads to war, rape, and most murder in nature.

The explanation of how this process occurs was refined by biologist Robert L. Trivers, who defined the concept of *parental investment* as "any investment by the parent in an individual offspring that increases the offspring's chance of survival (and hence reproductive success) at the cost of the parent's ability to invest in other offspring." Among all mammals, for example, reproduction is limited by the physiology of females, who have no choice but to invest more than males in offspring by nursing them during their months or years of infancy.

How does this lead to violence? Among human hunter-gatherers, mothers must invest four to five years exclusively in each child. These women cannot hope to raise more than three or four children who survive to adulthood. But during this approximately twenty-year period, men, whose bodies are not required as an exclusive nurturing machine bonded for years to one infant, can sire a hundred offspring, or a thousand. Physiologically, a man can fertilize a different woman every day or two. And some men try to.

As long as some of these men's "extra" children survive, sexual selection will favor male genes that increase their chances of convincing "extra" women to mate with them. Indeed, it seems that all male mammals are governed by the same rule: he who mates the most wins. The problem every male faces in mating with yet more females, however, is that males and females normally exist in nearly equal numbers, and there are few "extra" females.

Enter violence. The sex that must donate more parental investment per offspring becomes the limiting factor for the genetic fitness of the other sex, which, since it donates less investment, is free to wander. This leads to huge differences in the reproductive strategies of the sexes. As we see in many nature videos, male mammals use their "free time" to compete violently for very limited opportunities to mate again and again.

In nonmammalian species, the sex that competes for extra mates most violently may be the female sex. Among cassowaries in Australia's tropical rain forests, for example, females are the aggressive (macho) sex. To defend or expand their territories and to evict all female competitors, these six-foot-tall birds, armed with a three-inch claw/spike that can disembowel a dingo, battle each other using brutal kicks. A victorious female then mates with as many males as she can find, one after the other. She leaves each with a clutch of her eggs. The males, who are one-third smaller than the females, dutifully incubate the eggs, driving off predators and sometimes going up to fifty days without food to protect the clutch. When the eggs hatch, the male leads his brood of tiny hatchlings around the rain forest and demonstrates survival skills. The lesson? Sexual selection is an equal opportunity process, but it cannot work unless fueled by disparities in parental investment.

Among mammals, females always bear a heavier burden in terms of parental investment. And the bigger the gap between what female mammals must invest and what males must invest, the more intensely the males compete. If we knew only this about mammals, we would expect males to be violent toward one another.

Though far from being politically correct, *sexual selection favors the genes of males who sire more offspring no matter how they have done it*. Via the "pretty male strategy," sexual selection reinforces males' allure to females—as epitomized by the most brilliant plumage on male birds of paradise or the longest, most extravagant tails on male widow birds, both of which lead to the highest reproductive success for males. Via the "macho male strategy," sexual selection reinforces greater size, strength, speed, weapons, prowess in combat, intelligence, sneakiness, strategic sense, and even proclivity to cooperate with male kin in coordinated combat.

"All's fair in love and war" was penned by sexual selection, whose core logic commands individuals to "beget more children no matter how." Sexual selection reinforces an endless arms race of *sexual dimorphism,* in which males end up "this way" and females end up "that way."

What does this have to do with men? Unlike gorgeous male birds, most male primates are drab. In surveying three hundred published reports of mating behavior among higher primates, physical anthropologist Meredith F. Small found no example of females preferring specific types of males. Instead, females simply mate with those males who are victorious in dominance fights with other males—mainly because these are the only males remaining on their feet in the mating arena. In short, the "pretty male strategy" is close to meaningless to polygynous male primates, who instead use the "macho male strategy" to the hilt. It also seems meaningless to females, because their only option is the male who possesses the guts, savvy, and strength to be physically present. Among male primates, not only does "might make right," but superiority in combat is the only sure road to reproductive success.

How closely nature follows the "might makes right" axiom has even been quantified. For example, a team led by Tim Clutton-Brock, Fionna Guiness, and Steven Albon spent twelve years measuring the lifetime reproductive success of red deer on Scotland's Isle of Rhum. Stags are twice as big as hinds and brandish their huge antlers not as ornaments but as weapons. Without larger than average body size and larger antler size, a stag cannot forcibly exclude other stags and thus mate with multiple hinds. Combat is so intense that it prunes the reproductive life span of stags to a breeding prime only half as long as a hind's. Young and old stags are excluded from breeding by stags in their prime. Indeed, losing in combat eliminates some stags from breeding throughout their lives, while winning makes others into macho studs. The lifetime range for stags was zero calves sired by losers in combat but up to thirty calves sired by winners. Meanwhile, no hind bore more than twelve calves.

The lesson here is that the single most useless—and dangerous—approach one could take in trying to explain human violence is to look only at nurture, while ignoring how sexual selection has sculpted the evolutionary software programming the differences between men and women.

What differences?

The Sexual Blueprints

"NEXT TO BEING HUMAN, your most obvious characteristic is being male or female," notes psychologist Herant Katchadourian. "Virtually every society expects men and women to behave differently both in occupational and relational settings—particularly in sexual interactions." This is sensitive territory. Even to mention sex differences these days is to stumble into a political minefield, because putative differences have been widely misused to justify unfair double standards in gender roles, sexual habits, opportunities, worth, and freedom. Since the 1960s, many social scientists have, in an effort to preserve their ideology, redefined *gender* (sex roles) to mean behaviors resulting from socialization and redefined *sex* as a physical trait due to chromosomes.

Although it is true that sex is biological and that gender consists mostly of behaviors that children learn, gender emerges from both sex-specific instinct and socialization. The violence that "genderization" fosters is an instinctive strategy that men learn they can get away with when other strategies fail. Everything we do—eating, defecating, mating, parenting, and defending ourselves (all of which are entirely biological in origin)—is shaped by both biology and social learning. The biology of behavior is always mitigated to some degree by nurture. It is this degree, on a case-by-case basis, that makes our quest to understand men's violence so fascinating. To get a better view of why men differ from women, let's do an experiment.

Try to imagine men and women being identical in all respects: behavior, psychology, sexual orientation, size, physiology, athletic ability—everything except their genitals. Are you visualizing this? Does it compute?

Probably not. This is because the sexes are so different that most of us cannot even imagine them being identical. Sex, above all other traits, is The Cornerstone of human behavior. But sex is defined biologically by the types of gametes, or sex cells, an individual produces. Females produce large eggs containing DNA plus nutrients for the developing embryo. Males produce tiny motile sperm containing little more than DNA. The mission of these tiny swimmers is to search, find, and fertilize.

Beyond these gametes, sexuality itself is merely a reproductive strategy that allows two individuals to mix their genes and thus maximize the odds that some of their offspring will be born better adapted to flourish—or at least better able to outfox parasites—in a changing world. But this strategy of being sexual carries a price tag for both sexes: first, in the additional cost of the mating effort required to compete for, attract, and support a quality mate, and second, in the parental investment each sex must make to raise offspring. *The differences in reproductive effort faced by men and not women—and vice versa—are precisely what molded both human nature and gender and what fueled sexual selection to design men and women to be born destined to diverge in their behaviors.* "Females and males," explains Meredith F. Small, "are apples and oranges thrown into the same basket."

Ironically, both sexes start out almost identical. The basic blueprint of all mammals is a female one, and it stays female unless it is changed by masculine hormones. Hormones are encoded by genes. Each person's genetic makeup consists of twenty-three pairs of chromosomes, twenty-two of which contain DNA possessed by both sexes. The last pair are the sex chromosomes, X and Y. Females have two X chromosomes; males have one X and one Y. Among our hundred thousand genes arrayed on these forty-six chromosomes, the message "become male" is written by just one.

This "master sex switch," which instructs XY people to become males, is a chain of 140,000 nucleotide bases called *interval 1A2*. Interval 1A2 is a mere five-hundredths of the Y chromosome—only 0.2 percent—but it holds *the* master gene, *testis-determining factor*. Identified in 1994, this critical gene encodes for SRY, a DNA-binding protein.

The acid test of interval 1A2 and gene SRY was done by nature. People with an X and a Y chromosome lacking interval 1A2 are sex reversed; they are women. In contrast, XX people, whom we'd expect to be women but who possess the anomaly of interval 1A2 attached accidentally to one of their X chromosomes, are men.

In normal XY human embryos, interval 1A2 and gene SRY trigger the testes to develop at nine weeks and then a penis and scrotum at twelve weeks. The testes then secrete Müllerian-inhibiting hormone, causing the rudimentary female ducts to degenerate. The testes also secrete testosterone. By contrast, normal XX embryos develop ovaries by the tenth week. These secrete the female hormones, estrogen and progesterone. Four weeks later, the clitoris and labia develop.

Interestingly, both sexes secrete nearly the same hormones but in different proportions. These two hormonal "recipes" induce either maleness or femaleness. These recipes are so critical that two or three days after a male is born, his testosterone levels spike and stimulate the brain and hypothalamus to produce a male pituitary gland that secretes male gonadotrophic hormones. Injecting female hormones into any male mammal at birth will make its brain permanently nonmale. Such males will, among other things, fail to recognize females as mating partners.

The human brain is, by default, female until male sex hormones change it. Psychiatrist Richard Pillard suspects that Müllerian-inhibiting hormone helps defeminize the brain. So does testosterone. Abnormally high levels of testosterone in infant girls (due to a single enzyme disorder precluding cortisol production) cause them to become aggressive tomboys who rarely want to settle down by marrying one man. Some of the women experienced a malelike libido so intense that, upon finally receiving cortisone treatments to reverse it, they admitted liking the loss of their constant edge of sexual urgency and finally feeling like "normal" women. In contrast, low levels of testosterone in utero predispose men to homosexuality.

There is no question that human embryos sit on the gender fence until hormones topple them to one side or the other. Research shows that male sexual orientation is genetic, not environmental. A gene has even been found—among the several hundred genes in interval Xq28 (on the X chromosome)—that permanently switches men's sexual orientation from women to men.

Testosterone is so powerful that it has become a cliché used to explain male idiocy. Yet testosterone's reputation for making males act like males is well deserved. Testosterone reduces fear, increases aggression, and speeds the glucose supply to muscles. At adolescence, male levels of testosterone increase twenty- to thirtyfold and spur a growth spurt in the young male's trunk, shoulders, heart muscles, lungs, eyes, facial bones, and overall height. Even the number of red blood cells suddenly increases. Men average seventy-seven pounds of muscle mass, compared to fifty-one pounds in women. This disparity is even greater than it seems: male muscles are 30 to 40 percent stronger biochemically, pound for pound, than women's and are quicker to neutralize chemical wastes such as lactic acid.

In women, meanwhile, estrogen from the ovaries stimulates widening of the hips and causes the onset of menarche and the maturation of the uterus. Genes on the X chromosome actually limit women's muscle mass so that women need only two-thirds the calories that men do in basal metabolism. When these growth spurts end, men are so much stronger in athletics that even in our politically correct society, the sexes do not compete as equals except in horseback riding and shooting. *The biochemical evidence is unrelenting: men are designed by nature for higher performance in aggressive, physically demanding action.*

But testosterone does even more. Primatologist Robert M. Sapolsky found that aggression by a male is what most perpetuates more aggression in that same male—and that testosterone is the key player. Clashes between baboons in Kenya's Masai Mara, for example, are chronic and stressful. Testosterone levels in most males plunge immediately when they are stressed. In dominant males, however, testosterone levels *rise* during the first hour of stress. Sapolsky found that dominant males have the ability to inhibit the production of cortisol (the human stress hormone that demasculinized those women with male libidos discussed earlier) and thus keep high levels of testosterone. This ability lies in each male's personality. Males who keep their cool—and their testosterone—when challenged by rivals do three things: they recognize whether a rival is neutral or threatening; they attack all who threaten;

and, even if they lose a fight, they grab an innocent scapegoat and punish him or her soundly. The top males fight aggression with aggression and fire with fire in a self-feeding cycle that keeps them jacked up on testosterone and hyperaggressive.

"Attitude counts," Sapolsky concludes, so much so that perception of external events "can alter physiology at least as profoundly as the external events themselves." That attitude can determine reality is an important lesson. But the more important lesson here is that *male primates are designed—and primed by testosterone—to create their own reality through an aggressive attitude.*

In humans, this becomes obvious early. Three- to five-year-old boys are far more aggressive (fighting and threatening) than girls. They also share food altruistically less often than do girls. By age nine, boys create peer hierarchies in which the most aggressive boys predictably get what they want first. Girls are also aggressive, but in a very different way: they generally use prosocial aggression to enforce rules. Common among girls is the threat, "If you don't stop doing that, I'll tell." Far rarer are fistfights and raw physical intimidation.

The divergent behaviors of girls versus boys and women versus men pose another question: Are the brains of men and women different? And if so, do men have "more violent" brains?

Differences Between Women and Men

THE HUMAN BRAIN, male or female, is nature's magnum opus. It contains 100 billion neurons, each interconnected by up to 100,000 dendrites. The brain is structurally and functionally organized into discrete units, or modules, that create our mental states and cognitive thoughts. The brain is also a use-it-or-lose-it organ. The act of learning is one of the major events that stimulates the brain's dendritic connections and terminal branches to multiply. Not using one's brain is as devastating to mental potential as not using one's body is to athletic potential.

Are the brains of men and women truly different? Intelligence tests offer clues. Before 1972, women performed better on tests of verbal skills. Since then, men have consistently outscored women in mathematical calculating, conceptual ability, spatial orientation (especially in mapping their own environments, even at six years old), and the ability to separate a figure from its background. "Much independent evidence suggests that male hormones [androgens], both in utero and at puberty, elevate spatial ability," conclude Steven Gaulin and Harold

Hoffman. These authors hint that men's spatial ability stems from the need to defend their territory.

At least to some extent, hormones are again the culprits. Changes in women's cognitive performance, for example, follow their cyclical levels of estrogen and progesterone. Women tested in the 1980s performed fine motor tasks best when their sex hormone levels were high and did significantly better in spatial reasoning tasks when those levels were low. Also intriguing is evidence that women commit significantly more crimes just before their menstrual periods.

These differences emerge not just in psychological testing but in professional life. The top ten American mathematics departments were staffed in the 1990s by 303 tenured men and four tenured women. Further, on Scholastic Aptitude Tests (SATs), nearly all the most mathematically gifted students are boys. Oddly, boys scoring above 700 (the 99th percentile) were found to have five times more immune system disorders, such as asthma, than normal boys. This suggests, too, that hormonal biochemistry may supercharge or limit certain brain functions.

In contrast, no evidence so far confirms the argument that reinforcement of sex roles at home, in school, in sports, or at work enhances men's abilities to perceive space and to solve insight problems more than women's. Instead, it is clear that the brains of men and women do differ—just as male and female brains in all mammals differ. In humans, they differ specifically in the numbers and sizes of neurons; in dendritic spines and length of branching; in synaptic numbers, types, and organization; in regional nuclear volume; in volume of neural structures (thalamus, anterior hypothalamus, and corpus callosum); and in the locations of verbal processing. Although these differences are real, none of them tells us why men are more violent than women. What they do tell us is that the organ that initiates behavior is visibly different in men and women.

Brain research aside, the commonly perceived differences—physical, mental, and emotional—between men and women have fueled a double standard. In 1912, for example, the *Titanic* sank with lifeboats for less than half her 2,208 passengers and crew. The rule was "women and children first." The mere 705 survivors included at least 90 percent of all women on board. Some men in first class also escaped, but 92 percent of the men in second class drowned. Whether this reveals that men in power value women over other men, or worse, that men in power take any opportunity to eliminate other men—or whether it reveals that both are true—remains an open question. But in reality, men died and women lived due to the sexual double standard in which women are viewed as prime property more valuable than men.

The opposite side of the double standard is the sexist devaluation of women. As recently as 1990, for example, American women earned salaries only 72 percent as much as those paid to men with equal training. Why? Industrial giant J. Paul Getty offers one perspective in his book *As I See It*. During World War II, Getty supervised the 5,500 employees of his Spartan Aircraft Corporation, one-third of whom were women. He writes:

> One of the more surprising discoveries I made was that women were completely honest and straightforward about their capacities and limitations. Asked to perform some task beyond their ability and experience, they would openly admit they could not do it and ask to be taught or shown. Not so the men. They could not bring themselves to confess ignorance or ineptitude. Instead they would usually claim full understanding and try to bluff their way through—making very costly mistakes and blunders in the process.
>
> Roles reversed when it came to taking criticism. Male workers accepted criticism of their work matter-of-factly, taking no personal offense. Women almost invariably reacted to any critical remark about their work as though it was an all-out attack on them as individuals. Their eyes would fill with tears or they would burst out crying or flee into the women's restroom. Afterwards, they were likely to sulk for hours—or days—or even quit entirely.

Although Getty never says whether he considered men or women to be better employees, he leaves no doubt that he saw them as different kinds of employees.

Unfortunately, many men do insist that women are inferior, and they often refuse to value them as equals on the job. In *The Descent of Woman,* feminist Elaine Morgan concedes a key difference between men and women that moves us closer to understanding violence: "When all the factors of prejudice and self-interest have been discounted, the fact remains that on average women put less of themselves into their work than men do, simply because they are childbearers and wives as well as workers."

Here Morgan focuses on a fundamental chasm between men's and women's priorities and psychologies that goes well beyond the workplace and into basic reproductive biology. Indeed, this chasm reflects one of the deepest but most poorly recognized roots of men's violence. To see how and why this is so, we must first ask a key question concerning human sexual behavior: what do most women and men really want from each other?

I Love You Because . . .

WHILE ANSWERS to the question of what men and women want from each other could fill this book—and indeed have filled many books aimed at helping befuddled men (which includes all men at one time or another)—women have already stated what they want. And in just a handful of words.

"Whatever women say in public about their willingness to share the burden of making a living, in private I hear something entirely different," says veteran marriage counselor and clinical psychologist Willard F. Harley, Jr., who, by 1986, had spent twenty years interviewing fifteen thousand troubled couples. "Married women tell me they resent working, if their working is an absolute necessity. . . . A husband's failure to provide sufficient income for housing, clothing, food, transportation, and the other basics of life commonly causes marital strain in our society," Harley adds. "No matter how successful a career woman might be, she usually wants her husband to earn enough money to allow her to feel supported and to feel cared for."

Harley suggests that although women may want several things from their husbands, most women place a very high priority on material security. Anthropology agrees with him. Biological anthropologist Laura Betzig, for example, found that more women worldwide prefer to marry an economically successful man who already has a wife rather than a man who is single but poor (most human societies are polygynous). Betzig also found that wealthier men worldwide marry more wives and have many more lovers than poorer men.

Anthropologists Kim Hill and Hillard Kaplan likewise found that Ache women of the Paraguayan rain forest are most attracted to the best hunters. This is true even if the women can manage only an adulterous relationship with these men. Ache hunters supply 87 percent of all calories consumed. Hunters with shotguns are the most successful. Not only do they raise their catch from 910 calories to 2,360 calories per hour of hunting; women seek them most often as mates or lovers. In contrast, mediocre Ache hunters rarely can find a woman willing to marry them. For Ache men, shotguns are the equivalent of a six-figure salary among North American men; both groups have access to sex with more partners than do their less wealthy peers.

What about men and women globally? For six years, psychologist David M. Buss and fifty colleagues explored preferences among ten thousand people in thirty-seven countries from Africa to North America. Both sexes, Buss found, prefer a partner who is loving, stable, dependable, and pleasant. But these traits do not tell us much. They alone do not satisfy most people's criteria for a mate except perhaps a man who wants a "Stepford wife." For example, the most common trait people wrote in (Buss did not list it on his original questionnaire) was "a sense of humor."

On top of these qualities, Buss found that men are universally attracted to young, attractive, and "spunky" women. (Revealingly, America's $8-billion-per-year porn business also demands young women in their late teens or early twenties, yet this industry uses actors well into their forties.) "In each of the 37 cultures," Buss reports, " men valued physical attractiveness and good looks in a mate more than did their female counterparts. These sex differences were not limited to cultures that are saturated with visual media, Westernized cultures or racial, ethnic, religious, or political groups. Worldwide, men place a premium on physical appearance."

In contrast, although many women in Buss's study said that they were attracted to physically strong men, "in 36 out of the 37 cultures, women placed significantly greater value on financial prospects than did men. . . . Women desire *social status* and *ambition-industriousness* in a long-term mate more than their male counterparts do."

Lonely hearts ads by American women reveal that they too most often seek mates with resources and status. These ads also reveal that heterosexual women are three times more likely to seek socioeconomic resources in a mate than are lesbians. Even more revealing, Willard F. Harley, Jr., found that among the women he interviewed, attractive women often said that they found ugly but economically successful men who treated them well physically attractive. To these women the men's blemishes vanished.

Sexual "attractiveness" is so sex specific and, for women, so intermeshed with "promises" of ability to provide resources, that even researchers are surprised. Psychologist John Marshall Townsend studied the mate preferences of 1,180 American men and women from varying backgrounds. He presented people with an array of photos of models dressed either in high-status garb or in Burger King uniforms. The women in this study found an ugly man wearing a blazer and a Rolex watch equally acceptable as a dating partner as a handsome man wearing a Burger King uniform. These women's bias toward wealthy men increased even more when they were asked to consider the ugly man as a potential husband or father of their future children. Townsend concluded that women who say that they focus on "love" and "commitment" in a man are actually more concerned with the man's ability to invest in them financially. When a woman says "I have to respect him," Townsend writes, what she really means is "I have to respect his socioeconomic status."

Meanwhile, the men in Townsend's study vastly preferred pretty women in Burger King uniforms to unattractive women in expensive high-status garb.

Before we condemn these men as hopelessly superficial, let's look at some recent research by Judith H. Langlois and Lori A. Roggmann. These researchers digitized the facial features of ninety-six female students, averaged them into a composite picture, and then presented all ninety-six photos plus the composite to male students. The men rated the composite "average" face in the top five most attractive faces in the study. Only four real women scored higher. By contrast, the men gave low ratings to faces that were extreme in any measure. Men, it appears, are most attracted to women whose looks are symmetrical and, in a biological sense, truly average.

An extension of this test clarifies men's attraction to average looks versus "beauty." When researchers presented men with three composites of female faces, the most popular turned out not to be a true average of real faces but instead a composite of composites, with digitized, deliberately exaggerated "desirable" (and often childlike) features—full lips, high cheekbones, small chin, widely spaced eyes—that the men had focused on in their earlier top fifteen real choices. The upshot is that although men prefer biologically average women, they also can be seduced by unreal, hyperattractive beauty embodied most in childlike female traits.

Even more striking evidence exists of men's tendency to focus on the physical. In a survey of college students asking for characteristics they found most attractive in the opposite sex, women students named intelligence and a sense of humor as the top traits they sought in men. Men said they focused most on a woman's breasts.

"It turns out," concludes David M. Buss after reviewing his thirty-seven cultures, "that a woman's physical appearance is the most powerful predictor of the occupational status of the man she marries. A woman's appearance is more significant than her intelligence, her level of education or even her original socioeconomic status in determining the mate she will marry."

Thus, in contrast to women, whose mating priorities revolve around security and certain behavioral qualities in men, men are apparently seeking genetically sound breeding stock, and they are using physical, not behavioral, cues to find it. The bottom line? According to Buss, the bottom line is "men possessing what women want—the ability to provide resources—are best able to mate according to their preferences."

Are women justified in focusing first on men's success? Children of good Ache hunters do have a higher survival rate than those of mediocre hunters. The same holds true for professional parents in England; their adolescent children averaged two inches taller and also matured earlier than those of unskilled laborers, regardless of family size. American wives of wealthier men have more and healthier children than middle-class wives do. In short, women seem to know what they are doing—at least where their children's health is concerned—when they seek men with money. "A successful man is one who makes more money than his wife can spend," actress Lana Turner once wryly pointed out. "A successful woman is one who can find such a man."

What all this implies is that the priorities of American men and women differ drastically. Consider, for example, what 15,000 husbands admitted to Willard F. Harley, Jr., about their top five priorities in a wife. These men wanted sexual fulfillment, recreational companionship, an attractive spouse, domestic support, and admiration. Meanwhile, women's top five priorities in a husband were affection, conversation, honesty and openness, financial support, and family commitment. (Another 4,500 women interviewed by feminist Shere Hite matched Harley's findings.) In *His Needs, Her Needs* Harley concludes, "His needs are not hers."

No argument there, but the first three priorities that women listed are not as intimate and cozy as they may seem. Affection, conversation, and honesty and openness are not merely sensitive and pleasant; they also are the strongest reassurances by a husband that his financial support and family commitment are predictable and secure. A lack of affection and open communication, on the other hand, is a glaring signal that a husband may be wandering.

The divergent priorities of men and women tumble into a notoriously widespread failure of the sexes to speak the same language. In *You Just Don't Understand: Women and Men in Conversation,* linguist Deborah Tannen notes that from childhood, women use language to seek confirmation and to reinforce intimacy. Men use it to guard their independence and to negotiate status. The sexes' purposes for language diverge so much, notes Tannen, that messages assumed by both parties but never spoken far exceed those that are spoken. Men and women are normally at such cross-purposes verbally, she adds, that they often walk away from the same conversation with entirely different impressions and opinions of what was said.

While the paradox of men and women who use the same language but remain mutually unintelligible is no laughing matter, for men and women, neither is laughing itself. Psychologist Robert R. Provine found that less than 20 percent of conversational laughter is "a response to anything resembling a formal effort at humor." Instead, laughter seems to work more as a contagious social lubricant. When we hear someone laugh, we laugh, and generally we feel better. But, as might be guessed, men and women differ in how they "use" laughter.

In conversations, women laugh far more than men. "Female speakers laugh 127 percent more than their male audience," Provine reports. "In contrast, male speakers laugh about 7 percent less than their female audience. Neither males nor females laugh as much to female speakers as they do to male speakers." The trend for women to laugh more than twice as much as men during a conversation between them seems to be cross-cultural. And laughter itself tends to be disarming, even ingratiating. This difference between men and women poses several questions, not the least of which are why is this so and, perhaps more important, is there something wrong here?

Playing by the Same Rules but in Two Different Games

A DECADE AGO, only 7 percent of American Ph.D.'s in engineering, 10 percent in computer sciences, and 16 percent in the physical sciences and math were awarded to women. Recently, the prestigious U.S. National Academy of Sciences selected sixty new members, only five of whom were women. And, whereas 68 percent of men teaching science in U.S. universities hold tenure, only 36 percent of female science professors do. Moreover, women today hold only 3 percent of the top executive jobs in U.S. companies. Is this gulf between what men and women have achieved in these areas due to boys having been encouraged and girls not?

Yes, say feminist Irene Frieze and four women psychologist colleagues, who believe that these differences result from entrenched sexism aimed at "training women for nonachievement." They note that "the majority of women, even professional women, currently tend to put family concerns first. This means that women generally are not as productive nor as successful as men."

Perhaps. But before we accept the conclusion that women are more concerned about their families than their careers only because society forces or tricks them into it, we should ask, *Do women decide for other reasons that their natural and most rewarding talents lie outside the realms of mathematics, hard science, and capitalistic business?*

In fact, most women realize that a professional career will conflict with raising children. One study found that the mother-toddler bond was weaker in working mothers than in nonworking mothers. Toddlers with working mothers also felt and acted measurably more negative (bratty, uncooperative, disobedient) by the time they started kindergarten. In addition, Frieze and her colleagues present data suggesting that most professional women consider raising a family to be the top criterion for "success." Hence a mother who puts her preschool children before her career may not be making a less "successful" or less "productive" decision than one who chooses to work or chooses not to be a mother at all. Strangely, however, Frieze and her colleagues ignore what American professional women actually say they want most (a family) and instead base their definition of "successful" and "productive" on the criteria of Western men. These writers are not alone in this illogic, nor are their views extreme.

Feminist Germaine Greer says that women's lack of "success" is due to men's having "castrated" them and forced them to become vapid, self-sacrificing sex objects. Moreover, she notes, "the 'normal' sex roles that we learn to play from our infancy are no more natural than the antics of a transvestite." The only real success women can know, Greer insists, is by beating men at their own game. She says that the game women now play is sick: "The intimacy between mother and child is not sustaining and healthy." To Greer, marriages are disasters, and nuclear families are unhealthy for children, who would be better off raised en masse by trained women, much as Friedrich Engels prescribed in his communist manifesto.

Though interesting reading, these explanations for sex differences are wrong. Most of us agree that no matter what a woman's priorities may be, double standards and sexism are targets deserving of reform. But the more we know about these targets, the more elusive they become, especially for women scientists trying to sort out the effects of biology and those of socialization. "These women [scientists] are doing a balancing act of formidable proportions," writes physical anthropologist Melvin Konner in *The Tangled Wing: Biological Constraints on the Human Spirit*.

> They continue to struggle, in private and in public, for equal rights and equal treatment for people of both sexes; at the same time they uncover and report evidence that the sexes are irremediably different—that after sexism is wholly stripped away, after differences in training have gone the way of the whalebone corset, there will still be something different, something that is grounded in biology.

That "something different" seems to be that most women are born programmed to want to raise a family far more than they are driven to wrestle in the political arena. But raising a family in America in the *Ozzie and Harriet* mode is impossible for most people now that only one job in five pays enough to support a family of four. Hence most married mothers with young children must work, and to do this they often must compete with male workers. The dilemma produced by working for a living *and* raising offspring is an age-old problem faced by most female social primates. But the eye-opener in relation to men's violence is how most human females throughout history have solved this problem: by marrying males able and willing to support and protect them.

Yet even in hunter-gatherer societies, husbands rarely earn enough to support a woman and their offspring fully. Thus most mothers also have to work, usually by gathering plant foods for the family larder. In modern society, working mothers find it extremely demanding to perform all the tasks of marriage, motherhood, managing a household, and succeeding in an economic career simultaneously. Not surprisingly, this leads to marital conflict: over a woman's dashed expectations when she finds she must work (and do everything else) despite her husband's income, or over a husband's dashed expectations when his wife cannot (or will not) perform all the roles demanded of her. Such conflicts, usually over money, contribute to divorce in half of all marriages throughout the world.

If being successful as a wife, mother, and career woman seems like a hopeless catch-22, good. It should. In truth, mastering this challenge is possible only for an exceptionally adept woman. Meanwhile, the less spectacular majority give working women a bad rap, and they add fuel to the perceived "male conspiracy" to uphold the sexist professional double standard.

The true cause of the professional double standard is not a male conspiracy. It is simple competition emerging from men's instinctive reproductive strategies. As we have seen, men make themselves attractive to women—and boost their odds of raising a family—by economically outcompeting everyone who crosses their path, male or female, whether by hunting elephants more cleverly or by inside trading. Conversely, a woman who pursues an economic career not only absents herself from raising a family but also makes herself less attractive to men interested in marrying a future mother who will raise their children carefully and not be financially independent. On top of all this, the working mother also poses additional direct competition to working men, who, in typical male fashion, do what they can to squash that competition.

None of this is an American, or even a First World, syndrome. People in all known cultures encourage men more than women to perform in economics, politics, and war. That ten times more men than women worldwide are politicians is no coincidence. Nor is it by chance alone that people worldwide encourage women far more than men to be good nurturers. All societies do this.

These sex differences are basic biology for mammals, which have been shaped by natural selection for maximum individual reproductive success. No matter how well a man might protect, console, teach, or play with an infant, he cannot nurse it, hence women's instincts to perform most nurturing. But for reasons that will become all too clear later in this book, unmarried women (outside of socialistic government programs) have less success than married women in raising their offspring. Anthropology reveals that the most reproductively successful women have the assistance of a husband who sustains and protects both mother and child. Germaine Greer's claims notwithstanding, no other arrangement has ever improved on, or even matched, the nuclear family and its extensions in maximizing a woman's reproductive success. One clue to why this is so is children themselves. Most children are resilient enough to bounce back from near starvation or disease. But emotional distress due to poor nurturing depresses secretion of growth stimulating hormone to the point where unloved children raised communally not only fail to thrive but actually stop growing. "Communally reared children," admits feminist Alice Rossi, "far from being liberated, are often neglected, joyless creatures."

In short, women are urged by their psyches to seek reproductive success via very different strategies than men. And although we may convince men and women to play by the same rules, they are still playing very different games.

Are Men Born to Be Bad?

DESPITE ALL SCIENTIFIC RESEARCH, untangling the biology of human behavior is no easy task. Nor is the quest to understand humans helped by our own social behavior in the United States. America breeds alternative social values—often with junk food–like ingredients—and pop psychologies with the life spans of mayflies. Americans have subjected themselves to experiment after experiment: utopias, communes, free love, Skinner boxes, single parents, welfare, socialism, God as business, even mass suicide to hitch a ride on an imagined alien spaceship riding shotgun behind comet Hale-Bopp. For many people, TV has become a virtual alternative to having a life. When the sitcom *Seinfeld* ran its final episode, for example, viewers admitted that their lives would be fundamentally changed by no longer being able to tune in. Our quest to identify human nature also is not helped by the grim reality that even our most basic social institutions confuse many college professors.

For example, the fact that husbands worldwide work to support their wives and children perplexed anthropologist Margaret Mead. "What's distinctive about the human family," Mead wrote in 1949 in *Male and Female,* "lies in the nurturing behaviour of the human male, who among human beings everywhere helps provide food for women and children." Mead's confusion? "We have no indication that man the animal, man unpatterned by social learning, would do anything of the sort. . . . Male sexuality seems originally focused to no goal beyond immediate discharge; it is society that provides the male with a desire for children."

More than any other anthropologist, Mead indoctrinated most of modern America with the idea that human nature does not exist, except inasmuch as we are all social learners. Mead gained her credibility by parlaying her three months of interviewing fifty Samoan girls maturing from childhood to womanhood into *Coming of Age in Samoa: A Psychological Study of Primitive Youth for Western Civilization.* This 1928 book was used in more anthropology classes than any other book in history. Why? Because in it, free love flourished in a guiltless, pacifistic society, where violence existed only in occasional stylized war, almost as an afterthought. Samoans, Mead told us, lived in a societal paradise. Mead's message? We could, too. The "right" cultural upbringing could free us of the evils of violence, sexism, sexual guilt, dysfunction, and jealousy engendered by Western civilization. Inadvertently perhaps, Mead kicked off America's era of social junk food. She literally derailed any chance of our understanding male violence until field observations of the great apes woke us up. Her ideas, however, still shape education and politics in America despite the known fact that every major assertion she made about Samoan sex and violence is, and was then, false—but only partly because some of the girls Mead interviewed were simply having fun by telling her fibs about their exaggerated sex lives.

In fact, Samoans in 1925–1926 commonly raped girls. Brothers assiduously guarded the highly prized virginity of their sisters. Sexual jealousy led to mutilation and murder. Samoan men killed in warfare, often in staggering numbers. By contrast, New York City was more idyllic.

In short, because Mead ignored biology in favor of her own wishful thinking—and made things worse by spending only twelve weeks on the job, by neither living with nor interviewing Samoan adults, and by not even learning to speak Samoan very well—many of her major conclusions on human behavior were on a par with the flat-earth hypothesis.

For example, male nurturing is not, as Mead claimed, unique to men. Most male birds and many male social mammals are paragons of child support. A male African hornbill walls up his female partner inside a tree hole for months on end, from the time she lays her eggs until the nestlings are grown, to protect them all from predators. The male works unceasingly, day after day, to gather prey, which he feeds to the female through a slit left open in the mud wall they built. He also feeds their growing nestlings. If the male dies, the entire family may die. Males of other species work just as hard for their families. None was taught to do so by its "society." Nor do men nurture their families merely by imitating other husbands and fathers (although this certainly helps). The compulsion men feel to invest in their children is universal;

it is yet another instinct, sculpted by sexual selection, that is deeply rooted in the male psyche.

Ironically, it is physical anthropology (anathema to many cultural and social anthropologists such as Mead) that provides much of the evidence for human instincts. All hunter-gatherers divide labor by sex: women forage for plants (carbohydrates), while men hunt or fish (protein). Our marital patterns today are not all that different: Just replace the meat or fish that "man the hunter" risks his life to wrest from the wilds with the money that "man the worker" risks an ulcer and death on the freeway to bring home from the job. Then replace the plants that "woman the gatherer" brings home with the additional cash that "woman the part-time worker" earns. A key point here is that, in the normal mating pattern of humans over the ages, men have been expected to bring in more or better resources (protein being superior to carbohydrates, for example) than women. This is how most men attract, and hold, most women. Even male chimps hunt for meat most often when a sexually receptive female is nearby. "If women didn't exist," financial giant Aristotle Onassis noted, "all the money in the world wouldn't have any meaning."

After studying ethnographies worldwide, physical anthropologist Donald Symons concluded that the human psyche is genetically programmed to learn a sexual division of labor and roles, one that profits both men and women. But "hunting, fighting, and that elusive activity, 'politics,'" adds Symons, "[are] highly competitive, largely male domains." Hunting, fighting, and politics, of course, are the primary arenas in which men wrest from other men control of resources critical to attracting and supporting women. And men often do this violently, by robbing, murdering, and waging war—by no-holds-barred mayhem.

But the origin of men's violence is not a question of nature versus nurture. Instead, nurture is genetically programmed by nature. As this book will show, *women and men are designed by nature to be different in sex and in gender—the most basic ruling elements of the human psyche and self-identity—and they are also instinctively designed to learn appropriate and competitive gender roles culturally through parental nurturing as adaptations to help them win in all forms of reproductive competition with other people of the same sex. Men's violence emerges as a reproductive strategy shaped by each facet of this process: nature, sex, nurture, and gender.*

Gender roles allow us to survive, compete, mate, and raise our own children. Even the great apes share this need and drive to be programmed with gender behaviors; nongenderized apes either fail to mate or kill their infants by improper care. Gender is, in fact, our best example of how nature and nurture work together to shape us. Hence it is no coincidence that men and women are designed differently in body and mind to best execute gender-specific roles. Failing to identify with and perform the appropriate gender behaviors may result in natural selection culling one's "fail-to-genderize" genes. Indeed, it is certain that during most of human existence, failing to be violent enough seriously reduced a man's reproductive success.

Now we return to the big question: are men born to be lethally violent? The answer is yes. Aggression is programmed by our DNA. A Dutch team even identified a gene for hyperaggression in men. But even normal men are born killers. Melvin Konner surveyed 122 societies and found that weapon making was done by men, and never by women, in all of them. In another study of 75 societies, men in all of them even dreamed more violent dreams than women. Konner concludes that "men are more violent than women."

The statistics on homicide confirm this conclusion. And as we will see, although socialization does help men to choose their weapons, socialization is not what causes men to use those weapons to kill so much more often than women. What does cause men to kill, rape, rob, and wage war is something far more basic—and is something utterly alien to most women.

Yes, men *are* born to be bad—but not always bad, rarely gratuitously bad, and rarely bad in cold blood. Instead, most male mayhem erupts from a cascade of emotions far more primitive than those of the first cave man.

CHAPTER TWO

The Puppet Masters

"Another damned roadblock," I groaned.

To quell the rebellion now brewing (January 1981), President Milton Obote had ordered roadblocks manned by Tanzanian troops every thirty miles or so along this dirt road linking eastern and western Uganda. As if any rebels would be stupid enough to roar along this main road, I thought. When I had been here in Uganda previously, from 1976 to 1978, during the reign of Life President Idi Amin Dada, Uganda had been a hellhole. But now, on my second sojourn, under martial law and with all the roads blocked, it was even worse.

More than a year earlier, in 1979, Amin had dressed his Ugandan troops as Tanzanian soldiers and staged a mock raid into his own nation. Crying foul play by the Tanzanians, Amin then ordered his troops to don their Ugandan uniforms and invade Tanzania in "retaliation." Tanzania's president, Julius Nyerere, was so disgusted by Amin's buffoonery that he counterattacked, shoving Amin's troops back into Uganda and pursuing them relentlessly past Uganda's biggest city, Kampala.

His posturing as a military genius suddenly very hollow, Amin fled to Libya with his Swiss bankbooks and the treasure he had pillaged during his life presidency, which had lasted eight years. Meanwhile, Amin's troops fled willy-nilly from the more disciplined Tanzanian soldiers. To help them blend into the local peasantry, Amin's men traded their automatic G-3 rifles for food and civilian clothes. Nyerere's army had accomplished a first: it had invaded another postcolonial African nation, usurped its government, and taken it over. Many Ugandans celebrated their deliverance from Amin's hideous reign of torture and genocide. But not for long.

Nyerere eventually reinstated as president of Uganda fellow socialist Milton Obote, whose presidency had been usurped by Amin in

1971. Tribal politics resurged immediately. To consolidate his presidency against rebel factions of other tribes, Obote needed troops, but Uganda no longer had its own army. So Obote borrowed Nyerere's Tanzanian troops. Hence this damned roadblock I was facing. The biggest problem was that Tanzania was a poor country that had spent its last shilling to invade Uganda, also a poor country. It had no money to also occupy Uganda to help Obote to solidify his position. Nyerere, however, devised a simple solution to this problem: he ordered his Tanzanian troops to stay in Uganda indefinitely and to live off the land.

I pushed the brake pedal of my Land Rover. Over the years, I had been arrested twice as an American mercenary and detained yet another time because I looked too much like an Israeli commando (the Entebbe Raid that so humiliated Amin had occurred only months earlier; see Chapter 8). On yet another occasion, one of Amin's henchmen, the district commissioner of Toro, ordered his troops to intercept me before I left town. Their orders were to cut off my head. As you might have guessed, my foot was heavier on the accelerator.

Yet one more reason why I hated these road blocks. If only my rain forest was in another country. *Any* other country—except Zaire.

"Chai," hissed the Tanzanian soldier through the open window. The muzzle of his Chinese AK-47 submachine gun wobbled inches from my face. In traditional Swahili, *chai* means "tea." The kind you drink. But this was not the kind of *chai* this bandit was demanding. In post-Amin, Tanzanian-occupied Uganda, *chai* meant "bribe"—a bribe to drive the same two hundred miles of rutted road that I had been driving for years. *Another* bribe. At each roadblock, I was growing more annoyed at this demand. I decided, once again, that I was not going to pay anything.

"Hakuna chai" (I have no gift), I answered, trying to squelch my irritation over being subjected to highway robbery every hour on the hour. The guard's face pinched with greed. He was the predator; I the prey—or so he thought. He was merely a carbon copy of the soldiers at the last two roadblocks who had been beating the hapless Ugandans who had no money to pay. I found myself wishing again for my .45. I did not have it because, under this socialist dictatorship, it was illegal for a white person to possess a pistol. Customs officials were pawing through every bag with a fine-tooth comb.

"Chai," the uniformed goon hissed again in my face, *"au wewe siwezi kuendesha gari"* (or you cannot drive on). Saliva flecked his lips. His eyes were as empty as a cobra's.

Righteous anger and aggression swirled in me like a tornado. I became a testosterone cocktail. I wanted to treat this highway robber to a scenic vista down the barrel of my own weapon and then to . . .

• • •

NO MATTER HOW MUCH we may view ourselves as creatures of intellect, we are equally creatures of instinct and emotion, of passions hot and cold, of love and hate and fear and friendship—of dictates of the law of the jungle. And although our emotions are as primitive as a duck-billed platypus—and most of us know it—we still *feel* our way through our relationships with people. And we pay a great deal of attention to how we feel. Yet what most reveals the functions of these whirlwinds of emotions acting as our puppet masters are the ways in which men and women, even when faced with the same situation, do not feel the same.

What is it that makes us "feel" at all? What exactly are these feelings that pull the strings of sex and violence in men as puppeteers control their marionettes? And why do emotions possess such power?

One answer is that our emotions are the law of the jungle transcribed into brain chemicals. But, while true, this answer is very incomplete. To find a better one, we must trek through the dark, twisted terrain of the human psyche, where emotions prowl like tigers.

Compared to anything else on Earth, the three-pound natural computer we each own is nature's masterpiece of order, logic, and light. Of our nearly 100 billion brain cells, linked by 20 quadrillion synapses, 10,000 "think" for every one that controls a muscle or taps a sensory nerve. Our cerebral cortex integrates an impressive 70 percent of these "thinkers" to generate intellectual feats seen nowhere else in the known universe. But underlying this 70 percent of the brain capable of rational thought, insight, inspiration, and creativity, there lurks a far more ancient and powerful neural complex: the spiderlike limbic system, that dark land where tigers prowl. Here are built pleasure, pain, and all the emotions making the human condition one of endless tension.

The limbic system comprises the 20 percent of the human brain sandwiched above the old "reptilian" brain stem, which controls the autonomic processes of heartbeat, respiration, and the like, and below the "new" brain, which generates rational thought. The limbic system is also referred to as the old "mammalian" brain because all mammals have one.

The limbic system is a complex of neural structures, each with a set of functions. The hippocampus, for instance, is an essential mediator in memory and in helping to generate emotions. The amygdala mediates our sense of smell. More important, it allows us to read other people's emotions and to feel fear. The thalamus filters all other sensory and motor information from muscles and nerves, then relays it "upstairs" to the cerebrum. The cherry-size hypothalamus is truly immense in its power. Lying beneath the thalamus and receiving the richest supply of blood in the body, the hypothalamus dictates our emotional and physiological responses to outside challenges. Ruled by more than thirty of the brain's regulatory hormones, the hypothalamus governs heat, sweat, pleasure, pain, thirst, hunger, sexual desire, behavior, and fulfillment, as well as flight, aggression, and rage.

Exploration of the brain has revealed much about hypothalamic function. The septum connecting the front of the hypothalamus, for example, is a pleasure center. Rats with electrodes implanted there shock themselves ten thousand times an hour, giving up food, water, and even sex to do so, until exhausted. The hypothalamus also regulates hormone production by the pituitary gland; this is how it rules our libido and sexual impulses.

The hypothalamus makes men and women behave differently. Dutch researcher Dick Swaab found that the sexually dimorphic nucleus of the hypothalamus is two and a half times larger in men than in women. Similarly, neurobiologist Laura Allen found that the INAH-2 and INAH-3 interstitial nuclei of the anterior hypothalamus are significantly larger in men than in women. Neurobiologist Simon LeVay found not only that men's INAH-3 nucleus is bigger and shaped differently than women's but also that the INAH-3 nucleus in most homosexual men is the same size and shape as that of women. These findings suggest that *sexuality, including sexual orientation and emotions, are dictated by tangible and measurable sex differences in the morphology of the hypothalamus.*

The subtleties of the old mammalian brain allow us to feel myriad nuances of emotion, but only a few of these are basic, universal, pure, raw, and unadulterated. The best book dissecting them is Melvin Konner's *The Tangled Wing* (mentioned in Chapter 1). The following pages summarize (with apologies in advance to Konner for my own extrapolations) how these basic, instinctual emotions foster men's violence.

Rage, Jealousy, and Fear: Passions of Violence

ONE PRIMARY EMOTION is *rage,* an overwhelming, out-of-control response to someone taking something from us—property, a loved one, our self-esteem. Rage is produced by a cascade of endocrine hormones, the best known of which are norepinephrine (for aggression) and epinephrine (for fear). Serotonin also is emerging as a key player in rage. Excessively violent men and suicidal men show abnormally low levels of serotonin, a brain chemical also implicated in chronic depression. As yet no one knows why. We do know that serotonin levels increase in people who are well nurtured by loved ones or friends but plummet in people poorly nurtured.

The triggers of rage vary. We all understand the rage that erupts when a loved one is willfully hurt or when something precious is stolen from us. But rage also may erupt in response to a trivial or even an imagined injury.

Imagine that you are standing in line. It is a long, slow-moving line. One snaking across hot asphalt in the scorching sun. This line and the waiting you are doing are a major pain. Beads of sweat sting your eyes. The guy behind you is exhaling an inescapable cloud of bad breath. And you have been waiting a lot longer than you think is reasonable. You also know that if you are too far back in the line, the doors into the cool airiness of the building ahead will close and lock for the day, and you will miss out on a very important opportunity.

Are you in that line? Good.

You've been in it for an hour now. And, at this point, you'd rather be anywhere else on Earth.

Someone cuts in front of you.

He's a jerk. He does not even mumble an apology. He thinks he can just get away with it. After you have been waiting patiently in a muggy toxic cloud, he simply gives you a "You're a pushover" look: "You're nothing, I can step on you."

How do you feel?

More important, what would you like to do to this guy?

Most men would like to do something extreme. A few would even consider homicide. And remember, you are only imagining this guy.

Intriguingly, the old mammalian brain cannot discriminate between justified and utterly unjustifiable rage (say, from an imagined scenario). The spate of hormones that fuel rage are secreted automatically when we conceive of an insult or injury. Whether it be someone who cuts in line or someone who steals our parking place or our mate—or some ignorant bush soldier with an AK-47 and a dedicated intent to rob you at an unnecessary roadblock in the rolling green scrubland of Uganda—is irrelevant. Rage is automatic. What is not automatic is what we do about it.

Oddly, recent brain imaging of people recalling episodes of anger revealed a burst of activity in the septum adjacent to the hypothalamus in women but not in men. The difference is so glaring that we cannot help but ask, Are anger and rage the same emotion for women and men?

We already know that *jealousy* is not the same for men and women. Jealousy is what we feel when a sexual partner (even a potential or imagined one) is stolen. It can erupt into rage, often lethal rage. Psychological studies reveal that neither the rage nor the jealousy men feel is the same as what women feel. Although jealousy is equally severe for both sexes, the specific threat perceived by them differs. Most men in a study by David M. Buss, Randy Larsen, Drew Westen, and Jennifer Semmelroth admitted that sexual infidelity by their partners would disturb them far more than her emotional infidelity. Physiological tests backed this up. Men experienced measurable rage—they sweated, their heart rates spiked, and their brows wrinkled—at the mere suggestion that their women were sexually involved with other men because of the sex act itself, even if the women's emotional ties to the men, were weak. By contrast, 85 percent of the women tested admitted that emotional infidelity would upset them more than illicit sex. These women worried most about their men's emotional reattachment from them to the other women.

These diverging emotional reactions are due to a fundamental difference in the biological consequences of an unfaithful spouse to a man versus those to a woman. A man, for example, may be uncertain of the paternity of his unfaithful wife's children, whom he, the cuckold, will be expected to support as his own. In contrast, a woman remains certain that she is the mother of her children; what she is uncertain about is the likely diversion of her man's resources away from her and her children to the other woman and that woman's future children. Although jealousy is an extremely powerful emotion in both sexes, among men it is the fast lane on the highway to homicide (see Chapter 5).

WHEREAS JEALOUSY is murderous, *fear* is our most crippling emotion. Fear is the reaction to an expectation, realistic or not, that a thing will hurt us. The almond-size amygdala is the limbic system's fear center. Surgically removing it can leave a person smiling beatifically even while standing in the path of a platoon of murderous Hell's Angels. Because of the amygdala, we feel fear: innate or learned, rational or goofy. The amygdala does not discriminate; it simply reacts by sending fear commands to every other part of the brain. It even overrides our rational prefrontal cortex, which knows that we can overcome any fear—of snakes or heights or deep water—through de-learning it by taking action.

Predictably, it is the hypothalamus—our emotional command post for personal defense—that mediates the instinctual "fight or flight" response triggered by fear. The hypothalamus does this by spurring the pituitary to release a cataract of norepinephrine and adrenocorticotrophic hormones. These substances open bronchial tubes, elevate heart rate and blood pressure, contract muscles, shunt blood from the skin to internal organs and muscles, and increase the blood sugar supply to muscles. They also dilate our pupils for sharper vision and erect our hair shafts to both enhance our sense of touch and make us look bigger. All of these physiological responses are ancient biological adaptations that instantly prep the body for top performance in combat or escape. These automatic reactions to fear that turn us into temporary supermen and superwomen were designed from a jungle blueprint carried down through the millennia. We see them in all mammals, epitomized in the proverbial cornered rat.

Joy, Lust, Love, and Grief: Windows into Human Mating Strategies

THE OPPOSITE OF FEAR is the emotion people seek most relentlessly: *joy,* the fulfillment of a desire. The epicenter of joy, pleasure, happiness, or a general warm glow is in the ancient hypothalamus, and the epicenter of sexuality is in the limbic system surrounding the hypothalamus. What triggers the limbic system to type out J-O-Y hormonally is an event that reassures us that all is well with our ruling priorities: a warm smile from our mates, play with our healthy children, positive recognition by our peers, getting a raise or promotion, seeing a good performance by our children, or, even more basic, the birth of our children or grandchildren. The most powerful events triggering joy are ones in which we are successful reproductively, socially, financially, or in love.

SEX IS A COMMON TRIGGER for joy. But *lust* is what actually leads to sexual behavior. Lust is one of the most popular human emotions—as both a personal experience and a vicarious one. As we have already seen, men's libidos depend on high testosterone levels. Women's libidos depend on estradiol, on dopamine concentrations, and on the pituitary, hypothalamus, and other parts of the limbic system and the temporal lobe nearby. But testosterone affects women, too. In one study, women with higher levels of testosterone in their blood during ovulation reported more frequent sex. Women with higher levels of testosterone also were less depressed, enjoyed sex more, and formed relationships more easily.

Hence when someone says, "Sex is all in the mind," they are half right. Most of sex *is* in the brain, but it is in the old mammalian brain, not the conscious mind of the cerebrum. Sexuality swims through the brain as a cocktail of hormones.

Just how strongly it swims is surprising. In the United States, for example, the video rental business opens a window to what really turns people on. In the 1990s, the video rental business for all that Hollywood sex, violence, and comedy (and occasionally drama) was huge. But above that, in 1996, for example, 25,000 outlets rented hard-core pornography videos 665 *million* times! Proceeds from customers renting these videos and paying for other sex products (magazines, peep shows, cable programming, sex toys, live sex acts) grossed $8 billion, more than the entire Hollywood movie industry, videos included. Americans spent nearly $1 billion just on telephone sex. America's—and the world's—appetite for pornography is so insatiable that the United States produces, on average, nearly one new porn video (for a few thousand dollars each) almost every hour of the day and night, around the clock, from January through November.

"The birds and the bees" has become a cliché for explaining human lust to young people. But such an analogy is inappropriate, because, unlike nearly every other species, humans have no breeding season. Human females, unique among mammals, do not cycle into "heat" and suddenly become lustful for only a few days each month. Instead, women (and men) can be receptive to sex almost anytime. Even so, human sexual cycles do form, and often these cycles synchronize socially.

Women can even be hormonally "regularized" by a man. Actually, the man himself is not necessary, only his stench. Experiments "distilling" men's odors from T-shirts showed that the presence of the pheromones secreted in a man's armpit sweat alone can trigger and regularize a woman's menstrual cycle.

While this may take a large chunk of the romance out of romance, and may well also launch a whole new perfume industry (we can visualize the ads around Christmas time: "*Armpit,* a man's most natural gift to a woman"), the important point remains that a woman can have satisfying sex with a man on nearly every day of the month.

The evolutionary explanation for a woman's "anytime, anyplace" sexuality is that it is in a woman's best interest to be *the* one with whom her man mates most—or, better yet, exclusively. And the best way to accomplish this is by being sexually willing most of the time. Even so, potentially omnipresent lust as a bonding emotion often backfires when a woman is seduced by the wrong man (i.e., any extra one). This leads to adultery, which usually erupts into jealousy and rage, which in turn often leads to homicide.

The concatenation of lust, jealousy, and rage is why our sexual relationships generate such emotional heat. It is no literary accident that words such as *sultry, torrid, steamy,* and *hot* so aptly describe our love affairs. But what is hot for men may be stone cold for women. In fact, the differences between men's and women's mating strategies not only shove men and women into divergent roles, but they catapult them into a war between the sexes.

Years of research have revealed that the natural mating strategy of men focuses on three priorities: first, a man strives to be the sole mate of one or more women; next, the man guards these women from other men; and finally, he tries to mate with still more women. So it is no surprise that two of the Ten Commandments of Exodus forbid men to steal copulations from women who do not "belong" to them: "You shall commit no adultery" and "You shall not covet your neighbor's wife" (these were, of course, written by Hebrew men, not women). In India, men guard against infidelity by their wives even to the point of publicly killing wives they suspect of adultery. The catch is that men by nature seek sex with extra women, both via polygynous marriage and outside marriage altogether.

In feudal times, for example, the European custom of droit du seigneur (or jus primae noctis—"right of the first night") allowed the lord to have first intercourse with any new bride in his realm. Such lords impregnated an estimated one out of five new brides. "Much more than women," Donald Symons concludes, "men are predisposed to desire a variety of sex partners for the sake of variety."

Symons did not deduce this from Wilt "the Stilt" Chamberlain's admission that he had had sex with 1.2 new women per day for more than forty years (close to 20,000 women). Rather, Symons gleaned it from a sweep of global anthropology. But American libidos do match the world's. One survey, for example, showed that among American married couples, only 15 percent of wives have had an affair, while 24.5 percent of husbands have. These figures may or may not be accurate, but the more important point is that men do not seek multiple women merely for sexual variety. The results of their lust go well beyond those fleeting first moments of "variety" with a new lover. Planned Parenthood reckons that 25 percent of U.S. babies born are illegitimate. Indeed, American births outside marriage are soaring so high, note researchers Ronald B. Rindfuss, S. Phillip Morgan, and Gary Swicegood, that "there is no necessary relationship between marriage and becoming a parent." In short, philandering males sow infants, not wild oats.

The reproductive urge of lust is so powerful a puppet master that men are willing to pay dearly to satisfy it with "extra" women. Anthropologist Monique Borgerhoff Mulder found that Kipsigis men in Kenya, for example, pay immense bride prices for second and third wives. These prices are based on the wives' youth and virginity and also on their labor value and family connections. Are they worth the price? Borgerhoff Mulder found that Kipsigis men who married twice or more sired more than twice as many children as those who did not. Men in some cultures, in fact, often continue to choose and strive for polygyny even when it adversely impacts the health of their wives.

Indeed, a man's quest for "enough" sex can be expensive to the man himself, as Willard F. Harley, Jr., explains:

> One of the strangest studies in human behavior is the married man who is sexually attracted to another woman. He often seems possessed. I have known bank presidents, successful politicians, pastors of flourishing churches, leaders in every walk of life who have thrown away careers and let their life achievements go down the drain for a special sexual relationship. They explain to me in no uncertain terms that without this relationship everything else in life seems

> meaningless to them. . . . To the typical man, sex is like air or water. He doesn't have any "options."

So much for men. What about women?

"Because women in 99.5 percent of cultures around the world marry only one man at once," notes Helen Fisher, "it is fair to conclude that monandry, one spouse, is the overwhelmingly predominant marriage pattern for the human female." Hence women's natural mating strategy seems to be to marry one good man and convince him to invest heavily, or better yet exclusively, in her children. Only by acting monogamous, however, can a woman convince her mate of his paternity and thus gain his support.

"It is a woman's business to get married as soon as possible," playwright George Bernard Shaw observed, "and a man's to keep unmarried as long as he can." Although this may sound trite, sexist, or unfair, there is no question that it accurately describes what most people do. In the 1990s, unmarried American men under twenty-five years of age outnumbered unmarried women of the same age by a ratio of 1.2 to 1. These divergent sexual strategies are so entrenched in the human psyche that even homosexuals retain them. "Gay men and straight men also seem to display an identical strong drive for multiple sexual partners," notes journalist Chandler Burr; "lesbians and straight women seem to be alike in favoring fewer sexual partners."

These divergent strategies, despite individual exceptions we may know personally, also beget the infamous sexual double standard in which a man's lust is deemed more important than, or at least different from, a woman's lust. "The double standard implies that the world is divided into two classes of women," feminists Patricia Faunce and Susan Phipps-Yonas note, "good women and bad women, virgins and nonvirgins, procreative women and pleasure-seeking women." The "self-affirmed woman," they add, is free to ignore the double standard.

Of course she is, but at a cost. The double standard punishes girls for promiscuity but turns a blind eye toward, or even encourages, this behavior in boys. According to Shere Hite's survey of 2,500 college males, 92 percent of them said that the double standard is unfair to women. But only 35 percent said that they could seriously consider as a mate a woman friend who had had sex with ten to twenty men during one year, whereas 95 percent said that they would have no trouble taking as a serious friend a male who had had sex with twenty women. Two-thirds of the men agreed that the fairest solution is to allow women to be as promiscuous as men. But, they added, "of course the woman they marry would probably not be one of those women who had chosen to have sex with that many men."

Hite argues that this attitude is unfair and that the sexual double standard was created by Christianity. It was not. Considerations of fairness aside, the sexual double standard is as natural as the rotation of the Earth around the sun. ("Natural" merely means that it is a product of sexual selection.) The psychology of the double standard is so natural, and so insistent, that it is a major part of what shapes the violent jealousy and rage that we see when lust spurs mates to wander.

The sexual double standard evolved because promiscuous sex in a woman makes paternity uncertain. Men, as we shall see, generally invest little in children they suspect are not theirs. Such an investment is too expensive. Thus infants born of a known promiscuous mother usually lack a father's support. Hence, promiscuous women—at least those living outside socialist societies—see fewer of their children survive than do faithfully monogamous women. This is why most people in most cultures insist on female fidelity.

"In about two-thirds of the international sample," reports David Buss from his survey of mating preferences in thirty-seven cultures, "men desire chastity more than women do. . . . In no cultures do women desire virginity in a male more than men. In other words, where there is a difference between the sexes, it is always the case that men place a greater value on chastity. . . . [F]idelity is the characteristic most valued by men in a long-term mate."

Among the Dogon of the Sahel, the groom's relatives even insist that a bride be menstruating during the marriage ceremony as assurance that she is not already pregnant by another man. For an example closer to home, a survey of three hundred middle-class Los Angeles women revealed a strongly negative correlation between promiscuity and wealth. The wealthier the women, the fewer sexual partners they had had and the more children they bore.

Interestingly, 84 percent of the single women Shere Hite surveyed agreed that their relationships should be monogamous. Hite notes that "77 percent of single women in relationships are monogamous, a higher figure than among married women (although, of course, relationships tend to be shorter)." Even so, she adds, women often fail at monogamy. She reports that "70 percent of women married more than five years are having sex outside of marriage, although almost all believe in monogamy." (This figure clashes with the University of

Chicago's National Opinion Research Center's survey that only 15 percent of married women have ever had an affair.) Intriguingly, only 19 percent of Hite's "cheating" women fell in love with their lovers. But 89 percent of them kept their affairs secret.

Once again, this is because attitudes worldwide about adulterous women are condemnatory. In a study of 116 societies, 65 percent were found to be more permissive of male adultery than of female adultery. None was more accepting of adulterous women. In another study of 104 societies, nearly half deemed adultery by a wife to be grounds for divorce or far worse punishment. In no culture, however, was adultery by a husband grounds for divorce.

All of this not only is instructive about lust, but it also opens a window to our understanding of both jealousy and rage, as well as of men's violence prompted by sexual infidelity. Even more, it explains why people who cheat on their spouses usually lie about it.

Although humans are the world's consummate liars, we did not invent lying, nor did spoken language suddenly make lying possible. Body language is the most convincing way to lie. And other primates, especially apes, know it. And lust is the primary emotion allowing them to lie. I watched a female gorilla give a male false messages of her willingness to copulate, only to rob him of a rare item once she had him fooled. Chimps do this, too, as do female baboons. Primatologists Richard Byrne and Andrew Whiten cataloged 253 episodes of primate deception. The typical ones were of subordinate (female) baboons boldly and deliberately deceiving dominant males with offers of sex, which the females "used" to steal meat from the males. Whiten and Byrne conclude that tactical deception—"an individual's capacity to use an 'honest act' from his normal repertoire in a different context, such that even familiar individuals are misled"—is a deeply embedded instinct in social primates.

How often do humans lie? Psychologist Bella DePaulo and a team of researchers tracked the lies of seventy-seven college students and seventy local townspeople for a week. Although "little white lies" were more prevalent than self-serving ones, lying was rampant. Students lied twice daily, townspeople once. Students lied to their mothers during 46 percent of their conversations and to strangers during 77 percent. Students also lied to their acquaintances during 48 percent of their conversations and to their best friends during 28 percent. More to the point, both students and townspeople lied to their romantic partners during about one-third of their conversations.

Although it may be surprising how rampantly humans lie, it is astounding how early they start. Toddlers begin lying—to escape punishment or gain rewards they do not deserve—even before they develop the skills to be convincing. And although by age ten most children judge lying to be immoral, most become convincing liars anyway. Human lying is so common and so sophisticated, concludes biologist Richard Alexander, that *Homo sapiens* must be designed by instinct to outflank each other by lying.

Research on preschoolers supports Alexander's conclusion. Psychologist Michael Lewis instructed 33 three-year-olds not to peek at a new toy left on a table until the experimenter returned five minutes later, at which time the toy would be theirs. Of 15 boys and 18 girls, all but 1 boy and 3 girls peeked. The real eye-opener was which children admitted to peeking. Eleven confessed, 9 of them boys. Another 11 lied about peeking; 8 of them were girls. In short, girls at three years old were significantly more likely than boys to lie. What's more, videotapes showed that these little liars were so convincing that 60 university students who viewed the tapes could not tell who was lying and who was telling the truth. "Deception is an adaptive behavior," concludes Lewis, "that takes root early in life as part of a child's emerging moral code."

Why might women need a greater talent to lie than men?

As with so much of human behavior, the "need" to lie is tied to lust and fear, men's violence, and sexual strategies in general. And the answer reveals further sex differences in the emotions that rule our behavior—differences that often prompt men to try to control women. Behavioral researchers John Tooby and Leda Cosmides analyzed what happens when one person threatens another. First, all threats are coercive; they promise the use of force to take what is demanded unless the demand is met. But threats can be of three types: they can be true (compliance avoids being hurt); they can be bluffs (no punishment is forthcoming no matter what); or they can be double crosses (compliance leads both to losing what is demanded *and* to being injured). Tooby and Cosmides found that women and men differ both in how well they identify threats and in how they react to them.

In this study, men accurately identified threats as true, bluff, or double cross 70 percent of the time. Women showed only 48 percent accuracy. Why the difference? Because men often bluff other men, against whom carrying out a true threat with physical punishment is dangerous. Hence men are forced to be good at telling a bluff from a true threat. By contrast, men rarely bluff women. Instead, men use true threats against women because punishing a woman poses little danger to the men. Hence women can rarely afford to call any male threat as a bluff. Doing so would get them hurt too often.

This leads to three consequences. First, it is less important for women to distinguish between threats. Second, women treat most threats as true ones. Third—and this is where Tooby and Cosmides hit pay dirt—women are forced more often than men to use a sneaky strategy of *appearing* to comply with threats. In short, women need to lie more convincingly and more frequently to protect themselves from threatening men. This unequal logic of threat among men and women helps explain how people in lust often use deception to sidestep the double standard—and also lie to defuse the jealousy, rage, and violence of their spouses. As we will see, however, when lying fails, murder often follows.

This connection between deception and lust is more than academic. The hypothalamus prompts lust so often that it is nearly an omnipresent state in men. The report *Sex in America,* for example, notes that 54 percent of men think about sex at least once to several times a day, compared to 19 percent of women. Women, however, are fascinated by illicit sex. For instance, 94 percent of sex scenes on TV soap operas involve people not married to one another, and 87 percent of acts of intercourse during prime time also are extramarital.

In contrast to women, the mere glimpse of the genitalia of the opposite sex will trigger sexual arousal in men. Often it takes far less: a bare leg; a soft, interested voice; a scent; a memory; nearly anything feminine—especially *young* and feminine. This has led to yet another double standard, one based on men's preference for very young women (as mentioned in Chapter 1), dictating an impossible feminine quest to look young forever. This has led to the yearly $1.7 billion cosmetic surgery industry in America, which helps women lie about their age (and other imperfections) in an effort to inspire lust in men.

But if women are prone to lie to create their own best reality by fueling one man's lust or by sidestepping another's, men are equally guilty of using force to harness women's lust. In all cultures, most men satisfy their lust with more than one woman. And in most cultures, men may marry more than one wife, even if each marriage is separated by a divorce. Indeed, men compete for sex so much more than women do, notes Donald Symons, that "in preliterate societies competition over women probably is the single most important cause of violence."

To test this idea in part, anthropologist Laura Betzig analyzed 104 societies ruled by despots. Overall, she found the evidence "overwhelming that rich and powerful men do enjoy the greatest degree of polygyny . . . [and] most privileged access to more fertile, more attractive, wives." In most despotic societies, Betzig found, the king, khan, pharaoh, caesar, emperor, chief, or sultan typically kept more than 100 wives. These despots punished trespassers into their harems by ghastly tortures. Some enslaved the offender's women relatives. Many other despots castrated, amputated, impaled, or crucified offenders. Still others burned offenders at the stake, dismembered them alive, allowed elephants to trample them, or threw them to ravenous predators.

These savage strategies allowed despots to amass harems that were virtual cities of beautiful women. Late Roman emperors, for example, had hundreds of concubines. Azande kings of the upper Nile had 500 wives. Inca rulers conscripted 700 of the most beautiful girls in the Andes as wives or concubines (at age eight to ensure their virginity). The king of Dahomey not only had first pick of all women captured in war, but he also could take any women in his domain. His harem held thousands more than he could impregnate. Merely by withholding these women from other men, he outbred his rivals.

The champion of despotic breeders was Moulay Ismail the Bloodthirsty, a seventeenth-century sharifian emperor of Morocco. He sired 888 children.

Betzig concluded that men seek women above all else. Nixon aide John Dean stated the male perspective of lust most plainly when he admitted, "Power is an aphrodisiac."

Extreme measures in enforcing the sexual double standard have a long history. All wives and concubines in the harems of despots were (and are) guarded in seclusion, often by eunuchs. Furthermore, in twenty-three countries from Africa to Indonesia, women today are also often mutilated by circumcision of their clitorides and/or by having their vaginas sewn closed (infibulation) to prevent adultery and cuckoldry. The most extreme tactic to prevent cuckoldry yet is to take one's harem through the gates of death. When a Cahokia chief along the Mississippi River died, for example, his court sacrificed fifty women ages eighteen to twenty-three as brides to be buried with him. Chimu chiefs in Peru sacrificed two hundred to three hundred young women to be buried with them.

All this convinced Betzig not only that men seek women above all else but also that Lord Acton was right: "Absolute power corrupts absolutely."

THE BRIGHTER SIDE of lust is when it is combined with joy, which may lead to erotic *love*. Admittedly, *love* is an illusive term. It is often an ephemeral emotion as well. Nor is all love erotic. The most unconditional love is that of a parent for its child. Among all higher primates, in fact, motherly love is exemplary. But as cerebral or lofty as love may seem, it is hormones and the hypothalamus that prompt primate parents to care.

The hormone oxytocin, for example, floods new human mothers as soon as they breast-feed. Oxytocin acts as a tranquilizer, blunting pain. It also acts as a neurotransmitter that triggers tender maternal devotion. Some mammals, such as sheep, cannot even recognize their own newborns without oxytocin.

Oxytocin also appears to urge lovers to bond. It supercharges the vagus nerve between the brain and the sex organs. Of course, more than just this one hormone is responsible for men and women in every known culture on the planet falling in love. Unfortunately, the biochemistry that makes this happen is less romantic than Romeo and Juliet's not-even-death-do-us-part bond, even though it *is* what prompts one to care for someone else's welfare more than one's own.

Psychiatrist Michael Liebowitz hypothesizes that love is spurred by a flood of phenylethylamine mixed with other brain neurotransmitters. Phenylethylamine does two things: it speeds the transmission of impulses from one brain neuron to the next (from the limbic system to the neocortex, for instance), and it acts as a natural amphetamine, shifting the brain into overdrive. No wonder lovers can stay awake all night and still wander around the next day in euphoria. And no wonder some lovers become addicted to love.

Although it has become cliché to say that love does not last, the reality is that love downshifts. After two years or so with one partner, both the feeling of infatuation and the accompanying flood of phenylethylamine in the brain recede. If passion is replaced by the warmth of attachment, Liebowitz contends, the new chemistry responsible for bonding is based on endorphins, the natural opiates of the brain. Endorphins calm the mind and reduce pain and anxiety. Now lovers can finally get a night's sleep.

IN CONTRAST to love and joy, *grief* is the emotion of acute anguish, sorrow, or empathy over losing a loved one. It is the syndrome of true helplessness. A child's loss of a parent, for example, can cause depression all the way into adulthood. Grief leads not just to mental depression but also to a cascade of hormonal changes, including excessive cortisol secretion, which causes disorders such as malaise and loss of appetite.

Grief is one of the hardest emotions to overcome, because the conscious mind must do endless battle against each stage in the syndrome of grief—shock, denial, anger, bargaining, and affectlessness. This battle is a strange one, because grief is not ruled by the conscious cerebrum but by the unapproachable and illogical limbic system. I have known men, for example, who, dumped by their wives, became imprisoned for years by their own grief despite their conscious knowledge that they would eventually build a new life. Each one's cerebrum might say a hundred times a day that he will find a new love, stronger perhaps than the first. But, meanwhile, the amygdala in the old mammalian brain relentlessly spurs the limbic system to secrete cortisol, which whispers incessantly, "She's gone; you've lost her. She's gone; you've lost her. She's gone. . . ."

It is hardly surprising that grief differs in men and women. Grieving women tend to cry and punish themselves. Men tend to be angrier and more aggressive toward others. These latter emotions may actually stimulate the body to overcome the cortisol-driven malaise created by grief. They also may spur mass violence.

Gluttony

ONE OF OUR STRANGEST emotions is *gluttony*. In America, the land of the ever fatter, greedy overeating fuels a multibillion-dollar diet industry. It even kills many of those who fail to rein in their gluttony. In contrast to gluttony, however, hunger is a vital sensation governed by the hypothalamus. It tells us that we must eat food or suffer unpleasant consequences. But only a fine line exists between hunger and gluttony.

The wild chimps I studied in Kibale Forest, Uganda, for example, impressed me daily with their ability to eat dozens and dozens of wild figs in a single-minded, gut-straining obsession. But this was not gluttony. Despite their virtual orgies of fruit gobbling at those rare fig trees, these chimps were—as are all wild chimps—drastically underweight compared to apes cared for in captivity. To survive in the feast-or-famine world of nature, they needed to be "gluttons" whenever nature was bountiful. As if slaves to the first law of the jungle—to get it while you can—these apes were driven to pig out because next week could well bring famine.

Homo sapiens evolved in a world of disintegrating forests that was even more uncertain—but also a world in which ample nourishment was crucial to survive and win success against competitors. That nutrition counts, by the way, is not merely an idea that sells cornflakes; it rules our old mammalian brain.

Imagine a primeval woman designed to stay lean during times of plenty. How would she do during lean times? Her lack of fat reserves might kill her *and* the infant she is nursing or bearing. Indeed, we now know that fat reserves are the most important factor in fertility for hunter-gatherer women. We also know that women who train hard athletically can lower their body fat to a point at which their menstrual cycles cease. Physiologist Rose Frisch found that a reduction in women's muscle-to-fat ratio (in body weight) from 2.5:1 to 4:1 (a 15 percent weight loss) signals the hypothalamus to stop secreting gonadotrophin-releasing hormone. This critical hormone stimulates the pituitary gland to release the follicle-stimulating hormone and luteinizing hormone that are vital to estrogen release and ovulation. Nor is this infertility always temporary. In one study, 30 percent of women who cut back after habitual heavy exercise had so disrupted their hypothalamus that they remained permanently sterile. Clearly, nature deals harshly with women lacking a psyche that urges them, via the limbic reward of pleasure, to "overeat" whenever possible.

Gluttony owes its bad rap to the invention of farming, which replaced feast and famine with three square (and overly fatty) meals a day for the duration. But the old limbic system was designed in the wild. It does not recognize this new reality and may continue to urge us to eat, eat, eat, and then eat some more. Meanwhile, the cerebrum, which knows that if we endure hunger now and then our lives will be longer and healthier, strives by counting calories and saying no to rich desserts to stop us from committing suicide with food. As usual, the cerebrum often loses its battle with the primeval limbic system. The old mammalian brain can scream its command to eat more so loudly that, as we will see in Chapter 6, hunger can spur male violence on the most massive scale.

Programmed Emotions Versus Free Will

WHAT IS THE UPSHOT of these storms of emotions we are programmed to feel and to obey as if they were our puppet masters? The human cerebrum is unique in its cognitive power. Its capacity for logical analysis, technical expertise, insight, and inventiveness is light-years beyond that of all other primates. But when melded to a jungle legacy of testosterone-induced aggression, which is programmed by a superpowered hypothalamus tuned by millions of years of conflict for mates and territory, this incredible thinking machine becomes not only a very smart combatant but also a very aggressive one—so aggressive that emotions such as rage, jealousy, fear, lust, love, grief, and gluttony inspire men to kill.

In men, one chemical, testosterone, triggers these decisions to kill. Testosterone lowers the excitatory threshold for firing a bundle of nerve fibers called the stria terminalis, which connects the amygdala to the hypothalamus. Testosterone kick-starts the old mammalian brain, which in turn controls sex and aggression. And no matter how clichéd it has become to blame testosterone for the evil that men do, it is no joke. Testosterone is *the* natural chemical of male aggression. It is the parent of those emotional tigers and the most powerful string pulled by the master puppeteer.

But, one might object, as children we are taught to use our cerebrum to control the violent emotional impulses of our primeval limbic system. Granted, civilization is an endless university intended to teach self-control and to punish failure. But the soaring suicide rate among American teenage boys (in the mid-1990s, up more than 500 percent from 1950) tells us, among many other things, how unnatural civilization really is and how wildly it conflicts with the emotions prompted by severe storms of testosterone in the old mammalian brain.

This is partly because, as dangerous as our emotions can be when uncontrolled, they are not atavistic, excess baggage. Nor should they be excised like some useless appendix. Rather, emotions are absolutely vital biological compasses—chemically dictated urgings shaped by millennia—that point us in the direction of our self-interest. They tell us which way to jump, and how high, when our prospects to survive or reproduce wax or wane.

Civilization, by contrast, was rigged to protect us by forbidding those acts prompted by emotions that hurt people. Unfortunately for civilization, the human cerebrum is so clever that it easily finds ways to sidestep man-made laws and instead follow the puppet master's urgings of the old mammalian brain. So, despite civilization, the innate strategies programmed chemically into the male hypothalamus superimpose the law of the jungle on the behavior of men. Nor do men need to understand what fuels the dark side of the male psyche for such impulses to become violent acts. On the contrary, *I suspect that most violence erupts because men do not understand themselves.*

The big lesson here is this: to conquer violence, we must first understand it, and we must recognize the instinctive emotions that drive humans to commit violent acts. To pretend instead that men are not violent by nature is insane. "[The] continued pretense by some social scientists and philosophers that human beings are basically peaceable," warns Melvin Konner, "has so far evidently prevented little of human violence—which latter achievement would be the only possible justification for its benighted concealment of the truth."

Ironically, the biggest challenge in exploring the biology behind human behavior is not the difficult scientific research it demands. Rather, it is the difficulty of overcoming the fear of the results of such research. Many academicians throw up their hands and insist that if natural selection has indeed equipped us with violent emotions, our destiny is hopeless, because we cannot resist surrendering to those inescapable puppet masters. Many add that it would be better if we did not even talk about biological theory in connection with human behavior. It is far too dangerous, they say, because it offers license to be violent and to give up on our hopes of peace. Shere Hite gives a classic example of this denial:

> If the system we now have is "human nature," not an ideology, if the system as we now know it is something growing out of our very biological natures, and not a historical system which, once entrenched, becomes hard to dislodge—"reality" seems to mean living with an increasing amount of violence, massive inequity in global food distribution, in health and educational opportunities—and a grating friction in people's personal lives, plus destruction of the natural environment, not to mention destruction of each other in hurtful psychological games. If all this is true, then there is nothing to be done except for each person to retreat to her or his own mountaintop and hope for the best. But we do not have to believe this.

Unfortunately, this head-in-the-sand attitude permeates a lot of American social science dedicated to the idea that by twisting the dials of society, we can tune out all its ills. Such denial, however, serves only to foster a new "dark age." Biology tells us fairly that the human psyche, and its chemical programming of emotions, is a product of nature. It further reveals that the human psyche can be very violent. But by no means must we believe that a natural origin of violence means that men are doomed to rob, rape, murder, or wage war like so many genetic robots. To assume such would be a gross insult to the human intellect and spirit.

The violent behaviors emerging from the male psyche via the hypothalamus are the legacy of our ancestors. But their natural origin cannot be construed to mean that violence is admirable, justifiable, tolerable or, worst of all, deterministic to the point of being inevitable. The behavior of each of us is a product of our genes, our environment, and the choices we make. The truly important question here is, Are we smart enough to actively reduce the violence we live with? "Intelligence," notes brain researcher Robert J. Sternberg, "can be understood as a kind of mental self-government."

The larger lesson is this: if we ignore or deny, rather than strive to understand, the ultimate causes of violence and emotions and the legacy of rape, murder, and war lurking in the psyches of men, we will—as men have throughout history—doom ourselves to remain slaves to our own dark side.

We are now equipped with the tools to understand violence as behaviors shaped evolutionarily by men's and women's reproductive strategies, and to understand how and why men and women differ mentally and emotionally and in their priorities. These differences between men and women are not unique to humanity. We share them with our closest relatives. *Homo sapiens* did not invent violence but instead emerged from violent origins. In the next chapter, we will strap ourselves into a time machine to view these origins. Then, in the following chapters, we will tour the dark side of the human male psyche, beginning with rape.

Fasten your seat belts; this will be a rough ride.

• • •

I stared into the Tanzanian guard's grinning, avaricious face—and into the muzzle of his AK-47. There's the rub, I thought. Even if I did have a .45, I would have to be crazy to use it.

I struggled against my desire to shove his AK-47 where the sun doesn't shine and instead did what I had done so many times before. I told the guard that I understood he had a hard, lonely job. Then I praised his sense of duty, and I commiserated over his low pay. I encouraged him to keep up the good work, and I thanked him in advance for opening the gate.

He opened it as cheerfully as if it had been his own idea.

CHAPTER THREE

What Manner of Creature?

Is it not reasonable to anticipate that our understanding of the human mind would be aided greatly by knowing the purpose for which it was designed?

— George C. Williams, 1966

By late morning, Don Johanson and Tom Gray had surveyed the sun-blasted, erosion-dissected sediments for hours. The temperature now was soaring toward 110°F. The morning's search had yielded no real secrets: fossilized teeth and bones of tiny extinct horses, huge extinct pigs, an antelope, and a bit of monkey jaw. All of these things already had their counterparts in the team's collection back at base camp, and none of them was going to help answer the big question: where did *we* come from?

The day was stacking up to be a repeat of hundreds of others. It was one more pull on the arm of the early-man slot machine in hopes of finding the jackpot. Only this pull, today, had yielded all lemons and cherries. Try another pull tomorrow.

Don Johanson and Maitland Edey explain what happened next:

"I've had it," said Tom. "When do we head back to camp?"

"Right now. But let's go back this way and survey the bottom of that little gully over there."

The gully in question was just over the crest of the rise where we had been working all morning. It had been thoroughly checked out at least twice before by other workers, who had found nothing interesting. Nevertheless, conscious of the "lucky" feeling that had been with me since I woke, I decided to make that small final detour. There was virtually no bone in the gully. But as we turned to leave, I noticed something lying on the ground partway up the slope.

"That's a bit of hominid arm," I said.

"Can't be. It's too small. Has to be a monkey of some kind."

We knelt to examine it.

"Much too small," said Gray again.

I shook my head. "Hominid."

"What makes you so sure?" he said.

"That piece right next to your hand. That's hominid too."

"Jesus Christ," said Gray. He picked it up. It was the back of a small skull. A few feet away was part of a femur: a thighbone. "Jesus Christ," he said again. We stood up, and began to see other bits of bone on the slope: a couple of vertebrae, part of a pelvis—all of them hominid. An unbelievable, impermissible thought flickered through my mind. Suppose all these fitted together? Could they be parts of a single, extremely primitive skeleton? No such skeleton had ever been found—anywhere.

"Look at that," said Gray. "Ribs."

A single individual?

"I can't believe it," I said. "I just can't believe it."

"By God, you'd better believe it!" shouted Gray. "Here it is. Right here!" His voice went up into a howl. I joined him. In that 110-degree heat we began jumping up and down. With nobody to share our feelings, we hugged each other, sweaty and smelly, howling and hugging in the heat-shimmering gravel, the small brown remains of what now seemed almost certain to be parts of a single hominid skeleton lying all around us.

"We've got to stop jumping around," I finally said. "We may step on something. Also, we've got to make sure."

"Aren't you sure, for Christ's sake?"

"I mean, suppose we find two left legs. There may be several individuals here, all mixed up. Let's play it cool until we can come back and make absolutely sure that it all fits together."

We collected a couple of pieces of jaw, marked the spot exactly and got into the blistering Land-Rover for the run back to camp. On the way we picked up two expedition geologists who were loaded down with rock samples they had been gathering.

"Something big," Gray kept saying to them. "Something big. Something *big*."

"Cool it," I said.

But about a quarter of a mile from camp, Gray could not cool it. He pressed his thumb on the Land-Rover's horn, and the long blast brought a scurry of scientists who had been bathing in the river. "We've got it," he yelled. "Oh, Jesus, we've got it. We've got the Whole Thing!"

• • •

Bones

THREE WEEKS and hundreds of bone fragments later, Don Johanson's team had 40 percent of the "Whole Thing," AL 288-1 (for "Afar Locality #288-1"). The rest of the world knows AL 288-1 as Lucy. In 1974, Lucy became the strongest candidate for the world's oldest-known hominid skeleton—and oldest-known human ancestor. And despite having been buried for 3.18 million years near Hadar in the Awash River valley of Ethiopia's back-of-beyond Afar Triangle, Lucy changed Don Johanson's life. She got him a spread in *National Geographic*. She made him the Christopher Columbus of Prehistory.

Alive, Lucy had been a twenty-five-year-old female, three feet eight inches tall, who weighed a mere sixty pounds. Paleoanatomists pored over her bones. Owen Lovejoy concluded that Lucy's pelvis was so perfect that she was "even better designed for bipedality than we are." Paleoanthropologists William Jungers, Randall Susman, and Jack Stern disagreed. Instead, they said, Lucy was an imperfect biped who walked with an inefficient bent-knee, bent-hip gait. Lucy's long arms and short legs (proportioned halfway between an ape's and a human's); her primitive hands, feet, ankles, and wrists; and her long, curved fingers and toes all whispered across the millennia that she was more at home in a tree.

Either way, Johanson was on a roll. After two seasons of sieving Site 333, Afar's "First Family" emerged from the sands of time. Its thirteen Lucy-like members, including four children, seemed to have died in a flash flood that buried them in a mass grave of scrambled bones.

The "First Family" led to yet another pivotal clash of opinions. Every skull was shattered, and most of the fragments were gone. A composite skull yielded a cranial capacity that paleoanthropologist Dean Falk calculated at 400 cubic centimeters (cc), dinky compared to the 1,350 cc average in modern humans. (For comparison, male chimps average 394 cc, orangutans 411 cc, and gorillas 506 cc.) More telling, Falk saw no reorganization or expansion of the parietal-occipital areas so explosively expanded in *Homo.* In short, Lucy's kind had ape brains in skulls that also looked 99 percent ape. Even twenty years later, when Johanson's team finally turned up a good skull of her kind, Lucy's only claim to being a hominid rested on her hominid-like teeth and on having walked on two legs. Was Lucy a true missing link between apes and *Homo?* Or was her species a bipedal "evolutionary experiment" that failed?

In 1979, in a controversial but careful paper in *Science,* Johanson and colleague Tim White named Lucy and her ilk *Australopithecus afarensis* ("Southern Ape-Man from Afar"). *Afarensis,* they said, was extremely sexually dimorphic, as gorillas are, with males five feet tall but females less than four feet, half the weight of males, and verging on leprechaun size. They also suggested that *afarensis* was the ancestor of *Homo*—of all of us.

But this presents a problem. Among chimps, bonobos, and humans, sexual dimorphism is mild; males are only 120 to 130 percent the weight of females, not more than twice their weight. Moreover, this moderate difference is one trait that distinguishes humans, chimps, and bonobos from the other great apes, gorillas and orangutans, among whom males are more than twice the weight of females. Therefore, a far more moderate difference was most likely to have been present in the common ancestor of humans and chimps. This, plus Lucy's ape head, also tells us that if Johanson and White are right about *afarensis* being one extremely dimorphic species, then *afarensis* (and Lucy) is likely not a human ancestor. Having it both ways does not work.

In short, to find our own distant roots, and to understand what legacy of human nature we possess, we must look beyond Lucy. But getting a clear look based on fossils is anything but smooth sailing.

True, paleoanthropologists are gumshoes in the greatest detective story in human existence. The mystery they hope to solve entails the most fundamental and universal questions people have ever asked: Where did we come from? What manner of creature are we? What legacy of human nature do we possess?

Unfortunately, the events paleoanthropologists try to unravel and reconstruct took place millions of years ago and have almost all been lost to the winds of time. Their "time machine" consists of searching, then digging—with everything from a dental pick to a bulldozer. The difficulties in finding the "right" fossils (and in knowing that they are the right fossils) to answer the questions of our origin are hideous. Finding a needle in a haystack the size of Rhode Island might be easier. If real time machines are ever invented, paleoanthropologists will stampede to be first in line to rent them.

Since 1856, when Johann Karl Fuhlrott, a German high school science teacher, stared into the empty eye sockets of the first known Neandertal man, amateurs and professionals alike have worked under harsh conditions and with few resources, some for decades, with the single goal of uncovering key fossils that would solve the mystery of our origin. Their stories fill several books. Yet for each success, dozens of other contestants in the fossil game never managed to spell an ancestral word on paleoanthropology's "wheel of fortune." Even so, the successes have added enough fossil ape-men, protohumans, and human spin-offs to populate a bewildering lineup of suspects.

For example, standing to Lucy's left, up against a wall with vertical metric markings, is an even older fossil hominid, *Australopithecus anamensis* (this is the "newest" and, at 4 million years old, maybe the oldest-known australopithecine). Next to it, and older yet, is 4.4-million-year-old *Ardipithecus ramidus* (aka *Australopithecus ramidus,* for "root Australopithecus," but which is far too much an ape even to be an australopithecine). Standing to Lucy's right are the ever-younger *Australopithecus africanus* (aka "The Taung Child"), *A. aethiopithecus* (aka "The Black Skull"), *A. boisei* (aka "Nutcracker Man"), *A. robustus* (aka "robust *Australopithecus*"), *Homo habilis* (aka "Handy Man," "1470 Man," or "Lucy's Child"), *Homo erectus* (aka "Peking man," "Java Man," or "KNM WT-15000"), *Homo sapiens neanderthalensis* (aka "Neandertal man"), and finally *Homo sapiens sapiens* (aka anyone who can read this book). Unfortunately, exactly how this crowd of ape-men and man-apes, many of whom were contemporary, fits into our history remains as clear as mud.

Of course, paleoanthropologists do position these fossil "species" into proposed lineages. Some textbooks offer seven versions, all with the same species but arranged in different permutations of who begat whom. The critical position on these charts is the fossil identified as "The Earliest Common Ancestor" of *Homo erectus.* This is serious business, because "The Earliest Common Ancestor" is the big winner. Its discoverer enjoys the best prospects for continued grant support. Meanwhile, those fossils deemed dead ends, gone extinct with no descendants, end up as also-rans. Their discoverers get less press and receive fewer and smaller grants. Careers rise, fall, and shimmy with these lineages depicting the hominid family tree. Debates seesaw in heated exchanges over interpretations of matrix ages, morphologies, and phylogeny. Emotions flare as these lineages get continually reworked like a crossword puzzle that refuses to spell the right word. The stakes are not just knowledge but also careers.

Granted, the roots of human nature—and of human sex and violence—can be uncovered only by peering into our ancestry. But there exists more than one model of time machine we can use to travel into the past.

Before we look at these other ways, however, it is important to point out that *Homo sapiens* was no inevitable species. Instead, it was an extreme long shot. More than 99 percent of species ever on Earth are now extinct. Few species last for long. Dinosaur genera, for example, averaged only 6 million years before vanishing, and average life spans of many mammal species were less than a million years. Live-forever species such as crocodiles and cockroaches are the exceptions. Most life is fragile. Worse, even the nonfragile species have been subjected to a gauntlet of sixteen mass extinctions that caused species meltdowns in prehistory. The best known of these was the Terminal Cretaceous Event, which occurred 65 million years ago and was caused by an asteroid at least 6 miles across. This asteroid blasted out the 112-mile Chicxulub Crater on the Yucatán Peninsula to cause a "nuclear winter" effect that wiped out the last dinosaurs—along with 65 percent of all other life.

The Chicxulub Asteroid was our godfather. This is because many dinosaurs had evolved to become so advanced and efficient that they blocked the evolution of coexisting mammals for 160 million years. Highly evolved dinosaurs held our ancestors down to rat-size fuzz balls scurrying through the underbrush at night. It was only the Terminal Cretaceous Event that set the stage for the rapid adaptive radiation of mammals in the subsequent Cenozoic era. "Had the dinosaurs survived, there is no question that we would not walk the earth today," notes paleontologist Stephen M. Stanley. "Mammals would remain small and unobtrusive, not unlike rodents of the modern world."

So *Homo sapiens* owes its current role as top predator on the planet to Earth's chance impact with a big rock. Ironically, to set up an early warning system for future collisions, the National Aeronautics and Space Administration (NASA) is now trying to monitor the two thousand or so near-earth asteroids still out there. A big one could generate an explosion a million times greater than the world's entire nuclear arsenal and would wipe out most life. The asteroids giveth, and the asteroids taketh away.

And what the Chicxulub Asteroid gave was ultimately a hominid lineage leading, about 1.8 million years ago or earlier, to *Homo erectus*. *Erectus* not only blazed the evolutionary trail to nearly everything that makes us human, it also was a highly successful species for at least 1.5 million years. *Erectus* died giving birth to us.

As Darwin predicted, *erectus* emerged in Africa. In 1984 and 1985, in northern Kenya, Kamoya Kemeu, Richard Leakey's star fossil hunter, discovered the crown jewel *Homo erectus*. It was the nearly complete skeleton of a boy between 11 and 15 years old. Potassium-argon dating placed KNM WT-15000 (for "Kenya National Museum, West Turkana #15000") between 1.51 and 1.56 million years old. It is the most complete and tallest (at five feet three inches and perhaps six feet as an adult) *erectus* ever found. It is Leakey's "Lucy," and there is even more of "him" than there was of her.

KNM WT-15000's pelvis and limbs were like those of *Homo sapiens*, well designed for both walking and running. This suggests that the ancestors of *erectus* had been bipedal for a very long time. KNM WT-15000's cranial capacity was 909 cc, and the cranial vault possessed a Broca's cap, hinting at language. A ten-year-old human brain is 95 percent of its adult weight. Hence, KNM WT-15000's adult brain would have been 950 cc, more than two-thirds of a modern human's 1,350 cc.

How smart was *erectus*? The best definition of intelligence is the ability, as circumstances change, to alter one's behavior appropriately to serve one's own best interests. Stupidity, then, is the opposite trait. But intelligence defined as such is hard to measure even in living people. Jeffrey Laitman notes that the basicranium of African *erectus* is bent and flexed, like that of a modern six-year-old child, and is shaped for perfect speech. But for speech to evolve, *erectus* must have had something to say.

Verbal language was not just a major breakthrough; it was what made humans human. And it is no surprise that our propensities to learn and use language are genetic. Infants distinguish between phonetics at six months of age, and children spontaneously learn to speak entirely on their own. They can learn three or four languages, accent-free, at once. Short of raising them in isolation, this process is impossible to stop. An average child learns (without being taught) up to ten new words daily for years. By the time the child graduates from high school, he or she knows forty thousand to eighty thousand words (although one wonders why they choose to get along on just one hundred).

A high-water mark in the evolution of intelligence, verbal language is deceptive in its simplicity. It consists merely of arbitrary, "invented," discrete auditory symbols, which, when uttered by one person, can precisely communicate things, actions, and qualities—and relationships between them in the past, present, or future—to another person. Verbal language is so arbitrary that changing only syntax or inflection in identical words can change the whole message from sincere to sarcastic. But the value of human language is that it can transmit knowledge and experience between people without the receiver paying a price in sweat or pain or risk. "The development of human speech," notes entomologist Edward O. Wilson, "represents a quantum leap in evolution comparable to the assembly of the eukaryotic cell."

This quantum leap requires some serious neural hardware. To translate thought into speech, language depends on a part of the left brain called Broca's area. To comprehend and translate audible speech into meaning, it also depends on a region called Wernicke's area. And for either region to function, the two must be connected by a nerve bundle, the Arcuate Fasciculus. All three are vital to speech—and to becoming human—and all were likely well developed in *Homo erectus*.

What did *erectus* talk about? *Erectus* had the size, speed, tools, and intelligence to hunt. Wild game was abundant, and hunting was already a primate tradition. Not only do male chimps stalk and kill prey, then carry and share meat cooperatively, but thirty-eight species of nonhuman primates have been seen hunting vertebrates. Of course, no paleontologist even a century from now could reconstruct *any* of this. So with *erectus,* weak evidence of hunting is not evidence of its absence. Meat is so important in primate diets that kin groups of *erectus* likely cooperated in hunting or scavenging meat and carrying it to share with women and children, who meanwhile probably gathered plants.

But as important as hunting might have been for the protein it yielded, *erectus* could not have successfully competed against lions, tigers, hyenas, and saber-toothed cats unless *erectus* had an edge. This edge was very likely coordinated and cooperative social hunting, which would have required precise communication.

Although this may tell us several things that *erectus* talked about, does it help us to answer the even bigger question, Was *Homo erectus* human?

What does it mean to be human? Does it require self-awareness, a big brain, the ability to speak, or hands to make and use tools? Is it bipedalism? Or is it a narcissistic preoccupation with the question of what it means to be human?

Chimps are self-aware and insightful, have hands, learn American Sign Language (and use and teach it meaningfully), and make and use tools in the wild. Ostriches are bipedal; whales have a bigger brain than ours; dolphins have brains that are about the size of ours; parrots can talk. But none is human.

A single criterion defines being human. Unfortunately for paleoanthropologists, this quality is only indirectly connected with anatomy.

Imagine that an extraterrestrial visits Earth. It lands its spaceship on your lawn and oozes out to greet you. It has no hands, only tentacles. And instead of a big brain, it has three small integrated ones. It clomps along on three pseudopods. It cannot speak audibly, but communicates by writhing its six tentacles in arbitrary, symbolic patterns. Otherwise, it owes its technology and half of its lifestyle not to instinct, but to information communicated to it by its fellow aliens. It uses this information as how-to guidelines to get along in the universe. Is this alien human?

Yes. *To be human is to be a self-aware individual who relies on culture—on ideas transmitted socially—as a primary strategy to shape one's behaviors to survive, use resources, mate, and communicate with others.* Humanness is defined as cultural behavior, and culture itself evolves as each new idea multiplies as people communicate its benefits. Or culture goes extinct as people communicate its overly high costs. Most belief systems, for example, are learned in the family, and most cultural learning results merely from imitation. Although many nonhuman species also rely on culture, their reliance seems less critical than *Homo sapiens'*. Based on reliance on culture, chimps are the next most humanlike species we know of.

Clearly, humans have no monopoly on learning. Thousands of species learn by observing their parents. Even an octopus can learn by observing others. Being human is a matter of degree—of crossing some midline between reliance on instinct and reliance on culture.

Arguably the most important event in multicellular life on Earth was, as Edward O. Wilson suggests, the point at which the human brain became capable of sophisticated thought, rational insight, and cultural inventiveness, and then became able to communicate these thoughts. Linked with our ape proclivity to learn socially, our inventiveness led to an ever-increasing reliance on culture to survive and raise a family.

Culture leaped past the slow process of natural selection, which shapes adaptations only from gene mutations, with a flood of ideas and tools.

An idea can spread faster than a virus. The greater its advantage, the faster it spreads and the more addictive it becomes. Indeed, culture is exactly what genes should "invent" as their surest way to reproduce themselves and to invade new habitats.

A key insight here is that imitation is a lot cheaper than inventive genius. Any parent can tell you that the human psyche is more imitative than analytical. "Monkey see, monkey do" is more than a cliché; it is a top primate trait. Research confirms this instinct: people conform by imitating—even when in error—more often than they analyze to make correct decisions. Our bias to imitate instead of analyze is so strong that it has tricked a legion of experts into believing that humans do not act instinctively, only imitatively.

Culture can evolve like lightning. Take, for example, anthropology's favorite hunter-gatherers, the !Kung. Thirty years ago, the !Kung could carry everything they owned as they migrated across their vast territory. But now they herd cattle and goats. This animal wealth chains them to mud huts and has recast their lives. Anthropologist John E. Yellen explains:

> Once the !Kung had access to wealth, they chose to acquire objects that had never been available to them. Soon they started hoarding instead of depending on others to give them gifts, and they retreated from their past interdependence. At the same time, perhaps in part because they were ashamed of not sharing, they sought privacy. Where once social norms called for intimacy, now there was a disjunction between word and action. Huts faced away from one another and were separated, and some hearths moved inside, making the whole range of social activities that had occurred around them more private. As the old rules began to lose their relevance, boys became less interested in living as their fathers had. They no longer wished to hunt and so no longer even tried to learn the traditional skills; instead they preferred the easier task of herding.

The !Kung who once enlightened Anthropology 101 students (and charmed us in the movie *The Gods Must Be Crazy*) are no more. Instead they reveal how quickly culture can leap beyond its utility value to assume symbolic value. The !Kung now hoard beads and blankets as wealth in locked metal boxes instead of wearing them to adorn themselves or keep warm.

Likewise, culture may backfire for the rest of us. A Mercedes-Benz may be the best car, but it has also become *the* symbol of status and financial success—even if its "owner" bought it on credit that ruins him (who can tell if it is paid for?). Hairstyles, clothes, weapons, homes, cars, and watches, all have symbolic value utterly different from their original functional value. Many people strive blindly to possess symbols that raise their status or ethnic identity—even at the cost of financial ruin, divorce, children never sired, personal potential unrealized, or friends lost.

Our cultural symbols drive us this hard because our instincts command us to conform and to identify with our group. This in itself is not necessarily maladaptive. We use symbols as badges to recognize other members of our group. "Carriers" of the same symbol (a swastika, gang colors, a crucifix, an American flag, a "Kiss me, I'm Italian" button) may or may not be closely related, but they *appear* to be more closely related to one another than to someone from another group without that symbol and thus are the most likely to cooperate. Unfortunately, the instinct to bond and identify clearly with one's kin or ethnic group has functioned thousands of times for reproductive advantage via war and genocide against those carrying different symbols.

Culture backfires another way. As Darwin noted, it can reverse biological evolution. Once culture supersedes biology, as in fixing poor eyesight by wearing glasses, defective genes stop being a survival or reproductive handicap. Hence genes that cause bad eyes increase in frequency. This increase in what geneticists call "genetic load" is devolution.

Culture nonetheless is humanity's secret weapon in the conquest of Earth. Although culture can never erase instinct, and it is frequently the loser when it conflicts with instinct, culture is extremely adaptive because life is a lot less risky when individuals obey their instinct to use culture to reduce dangerous trial-and-error learning of survival technologies.

Relying on culture carried a huge price for our ancestors. The neural equipment for the verbal ability and abstract thinking required when *erectus* began relying more on culture than on instinct must have led to natural selection for a quantum leap in cognition and brain size. And the metabolic cost of this large brain must have grown in a self-feeding spiral. It no doubt forced *erectus* to increase the meat in its diet, thus increasing the importance of hunting, language, cultural learning, and then brain size again—and so on.

Archaeology reveals that *Homo erectus* was far more advanced culturally than any previous hominid. *Erectus* was an ecological jack-of-all-trades who built shelters, made weapons and other sophisticated tools, controlled fire, and probably invented language. Hence *erectus* is the only incontrovertible contender for "earliest known human." Moreover, African *erectus* seems to have been not only the mother of human culture but also the mother of all other *erectus*. This species spread across the Old World as early as 1.8 million years ago, likely in wave after wave, to pioneer regions as far apart as Georgia (in eastern Europe) and Java. The European descendants of these groups evolved into Neandertals; others into Javanese *erectus,* Peking man, and so on—all of whom were doomed when the final iteration of *Homo sapiens* emerged from Africa.

Fossil bones provide endlessly fascinating sorties in the quest to pinpoint what manner of creature we are. Yet the testimony of these bones is limited. But again, there is another way.

Molecules

IN THE 1960s, Morris Goodman pioneered an entirely new approach to viewing the vanished past. Instead of driving a beat-up Land Rover into some godforsaken desert to prowl sunbaked gulches in search of fossilized bones, he stayed in the lab and compared blood proteins of living species. Scientists since Goodman have compared the biochemistry, proteins, spermatozoa, chromosomes, and so on of primates and humans as "time machines" to calibrate a taxonomic tree and a molecular clock of evolution.

We owe the best of these clocks to James Watson and Francis Crick, who in 1953 won a frenetic race to unravel the secret of deoxyribonucleic acid (DNA). DNA is the genetic blueprint for the human body. It runs the body's metabolism and reproductive life and shapes its psyche. It even codes for sex and violence. This 6-foot-long, 50-trillionths-inch-thick molecule in each of our 10 trillion cells is a deceptively simple chain of four nucleotide bases—the amino acids adenine, guanine, thymine, and cytosine—bonded into a double helix by deoxyribose and phosphoric acid. The sequence of our 3 billion nucleotide bases contain—like a Morse code—complex information packaged into

100,000 genes to blueprint every molecule of the body. "The full information contained therein," explains Edward O. Wilson, "if translated into ordinary-size letters of printed text, would just about fill all 15 editions of the *Encyclopaedia Britannica* published since 1768."

The most elegant work comparing DNA has been done for birds. Charles Sibley and Jon Ahlquist innovated a way to compare the DNA of one bird species with that of another. Their goal was to outline the evolutionary relationships of the world's nine thousand species. They boiled DNA labeled with radioactive iodine to separate the double helix into single strands. If allowed to cool, these strands rebond into a double helix. Instead, Sibley and Ahlquist separated and placed one strand of DNA from one species with one from another species—and allowed them to cool. They measured the new hybridized double helix by reheating it to its unique melting point. Each degree lower than the pure DNA's melting point, it turned out, meant that the two original DNA samples differed by 1 percent. Greater hybridization (higher melting points) meant a closer relationship. Mismatches meant more evolutionary distance. Not only was this technique elegant, but it also worked better than any other molecular approach to reveal how closely two species are related.

In the 1980s, Sibley and Ahlquist focused on the DNA of primates. Their experiments indicated that monkeys diverged from the ape lineage 25 million to 34 million years ago. Ancestral orangutans split from the African ape lineage 12.2 million to 17 million years ago. Ancestral gorillas separated from the ancestors of chimps, bonobos, and humans 7.7 million to 11 million years ago. Finally—the date that sends paleoanthropologists into tachycardia—between 5.5 million and 7.7 million years ago, chimps and humans split from our common ancestor (bonobos later split from chimps between 2 million and 3 million years ago).

Many paleoanthropologists scoffed at these dates—and scoffed even more derisively at the idea of using molecules instead of solid, honest fossils to reconstruct lineages. But as molecular biologist Vincent Sarich points out, fossils are more questionable than molecules: "I know my molecules had ancestors; the paleontologist can only hope his fossils had descendants."

Many other scientists also greeted Sibley and Ahlquist's DNA work enthusiastically. DNA dates not only matched real fossils, but they also filled in some gaps between them. Moreover, a repeat of Sibley and Ahlquist's work by molecular biologist Jeffrey Powell duplicated their results exactly. "It is now clear," wrote paleontologist David Pilbeam, "that the molecular record can tell more about hominid branching patterns than the fossil record does."

Sibley and Ahlquist's DNA "time machine" tells us that a biped with the head of an ape started its evolutionary journey to us roughly 6.6 million years ago. The study of DNA also helps clarify what happened to *Homo erectus,* our one certain ancestor. *Erectus* appeared in Africa at least 1.8 million years ago during an arid spell and then spread to Asia. *Erectus* flourished through yet another shift to aridity a million years ago, one that saw the demise of *all* other hominid species, including two or three species of contemporary australopithecines. *Erectus* vanished in Africa 200,000 to 300,000 years ago when it evolved into an even more intelligent "chronospecies" we call archaic *sapiens.* This population evolved into us—*Homo sapiens sapiens*—which then went on to wreak havoc around the globe.

How do molecules reveal this? Human protein polymorphisms show that our ancestors descended from one very recent population. Further research on DNA from tiny micro-organelles called mitochondria, which produce energy in our cells, helps confirm this. Mitochondrial DNA mutates five to ten times faster than the nuclear DNA of chromosomes. It is a perfect research tool because eggs have mitochondria but sperm do not. This means that each of us inherited our mitochondrial DNA from our mothers. Indeed, every woman inherited her mitochondrial DNA from her mother, grandmother, great-grandmother, and so on back to her most ancient common ancestral mother. Unlike the nuclear DNA of egg and sperm that mix to become the blueprint for each of us, mitochondrial DNA never recombined with any other DNA. It offers us an uncluttered highway into the past for female ancestry.

To sort women into family trees, Rebecca Cann, Mark Stoneking, and Allan Wilson analyzed variation from mutations in mitochondrial DNA from 182 women native to Africa, Asia, Australia, New Guinea, and Europe. Their work was so intriguing that it appeared on the cover of *Newsweek,* titled "The Search for Adam and Eve." This was the best-selling issue of *Newsweek* in 1988.

On the *Newsweek* cover, two modern Africans are shown in an African Eden. The young man is beardless, the woman svelte. A green python oozes from the tree between them as "Eve" proffers an apple to "Adam." "Eve's [genes]," wrote *Newsweek* reporters John Tierney, Linda Wright, and Karen Springen, "seem to be in all humans living today: 5 billion blood relatives, she was, by one rough estimate, your 10,000th great grandmother."

Mitochondrial Eve probably looked like a blend of all women today. Cann, Stoneking, and Wilson believe that all modern humans descend from one founder population of *Homo sapiens sapiens,* itself descended, between 140,000 and 290,000 years ago, from that transitional "archaic" population of African *sapiens.* Further, none of Eve's descendants interbred with other existing primitive populations, such as Chinese *erectus* or Neandertal. And because native Africans today are the only people possessing all the variability in mitochondrial DNA across all populations, Africa is the best candidate for the location of the ancestral population. Eve's daughters (and sons) migrated out of Africa roughly 100,000 years ago.

Mitochondrial Eve was so hotly debated that several other scientists launched extensive repeats of the original work. The most careful concludes that Mitochondrial Eve lived in Africa about 143,000 years ago.

New studies of nuclear DNA, combined with strong fossil evidence, ultimately confirmed an out-of-Africa *sapiens* roughly 100,000 years ago. Archaeology reveals that African *sapiens* was then already a sophisticated big-game hunter. Archaeology and genetics also track an African *sapiens* migration into the Middle East 92,000 years ago, into Asia and Australia 40,000 to 65,000 years ago, and into Europe and Tasmania 36,000 years ago.

But modern *sapiens* did not simply waltz into uninhabited Gardens of Eden. Significant to our probe into men's violence, they likely encountered other, more primitive populations of *Homo.* What happened to these less modern children of earlier diasporas of *Homo erectus?*

One family of these children, *Homo sapiens neanderthalensis,* offers us some clues. Neandertals appeared in Europe and the Near East about 230,000 years ago, then vanished less than 30,000 years ago. Their skulls were more robust and primitive than ours, but their brains were 65 cc bigger than a modern human's. Neandertals stood erect and were athletic and well adapted to their Ice Age. Their faces were massive and powerful chewing machines. Overall, they had extremely dense bones and very heavy, massive muscles. Comparing a Neandertal skull to a Cro-Magnon (modern) skull is nothing short of shocking; they're that different.

Neandertals were hunters who ate almost nothing but meat. They crafted beautiful bifacial spear points and hand axes. They chewed hides to make clothing. They mastered fire, lived in caves and rock shelters, and hunted cave bears nearly to extinction, caching their skulls in altars. One such altar, in Drachenloch Cave ("Lair of Dragons") at eight thousand feet in the Swiss Alps, held thirteen huge skulls, seven in a stone sepulcher. Neandertals buried their dead with tools, flowers, or food. They also cared for their wounded and crippled. In a sinister vein, many Neandertal bones also show cut marks from butchering or were crushed for their marrow, both suggesting cannibalism. At least one fossil burial suggests death in combat or from murder.

That Neandertals were so well adapted to Ice Age Europe but vanished almost overnight suggests a bizarre end. Neandertals were the masters of their domain. What happened to them? Some scientists insist that they were our ancestors, but fossils and biochemistry both contradict this. Instead, it is now clear that modern humans literally stepped into their footprints.

How? Modern *sapiens'* diseases, new to Neandertals and perhaps too much for their immune systems, may have wiped them out. Even without diseases, modern humans' better hunting weapons and strategies would have elbowed Neandertals out of the game, even if only indirectly. When modern *sapiens* reached Europe's cold, grassy tundra, they brought spear-throwers (atlatls) to multiply the range, force, and accuracy of projectiles; tethered harpoons; burins for weapon making; fishhooks; and maybe even bows and arrows. Cro-Magnon also drove herds into swamps or culs-de-sac or off cliffs for easy kills. In contrast, Neandertals were still thrusting or throwing their javelin-like spears. Cro-Magnons wore well-tailored clothes made of skins, perhaps allowing them to hunt in habitats or weather conditions where even Neandertals could not. They also built elaborate shelters of mammoth bones or pitched tents, or they settled in caves in communities of fifty to seventy-five people. In short, modern *sapiens* was superior at harvesting the same resources Neandertals needed.

"A small demographic advantage [by modern *sapiens*] in the neighborhood of two percent mortality," notes paleoanthropologist Ezra Zubrow, "would have resulted in rapid extinction of the Neandertals. The time frame is approximately 30 generations, or one millennium." A millennium can look like overnight in the fossil record. Zubrow suggests that the extinction of Neandertals was an inevitable thousand-year fizzle during which they were simply outhunted by Cro-Magnons for the animals Neandertals depended upon.

As we know from current human behavior, on top of outcompeting Neandertals, Cro-Magnons also may have waged war on them. The last Neandertals on earth apparently were holed up in caves in Spain's Rock of Gibraltar as recently as 29,000 years ago. Ultimately, it seems, after coexisting for as long as 7,000 years in Europe, modern *sapiens* seized even this, the Neandertals' last piece of the rock.

A CLOSE LOOK at the molecular identity of *sapiens* shows something even more significant with regard to the dark side of man. Humanity's closest living relatives are chimpanzees and bonobos. In short, molecules reveal that people are a species of great ape. Yet even this is understating it. Fully 98.4 percent of human and chimp nuclear DNA is identical. Chimps and gorillas, meanwhile, share only 97.9 percent of their DNA. Genetically speaking, humans are not just one more ape; they are a "sibling species" so closely related to chimps that if anthropologists followed the same criteria of relatedness that mammalogists and ornithologists do when classifying genera, chimps and humans would be classified in the same genus: *Homo.*

Homo sapiens is indeed a "naked ape." He also is a very intelligent ape, an ape addicted to culture, and, most important in our quest to bare the roots of human nature and the violence of men, *an ape who carries a strong legacy of ape instincts.* But which instincts?

Behavior

THE NEXT FOUR CHAPTERS will explore in detail rape, murder, war, and genocide by human males and compare them with the violence of male great apes. Here we are ready to kick our final "time machine" into gear. As will become ever clearer, male great apes and men share some identical instincts for violence. The social behaviors of chimps, bonobos, and humans are similar but diverge from those of gorillas and orangutans by increasing degrees, which coincide with the degrees of difference in DNA. Of course, human behavior also diverges from that of chimps and bonobos. How much it diverges—or does not diverge—provides insights into the origin and functions of human male violence.

Our final "time machine" to witness the behavior of the lineage that led to *Homo* is the comparison of the behavior of human males with that of our nearest living relatives. Our assumptions in doing this are threefold:

1. The great apes and humans share a common ancestor, and thus common genes affecting our social behavior.
2. The more closely related each species of ape is to us, the more genes and behaviors we share in common.

3. Behaviors shared by chimps, bonobos, and humans are far more likely to be instinctive and to have been inherited by all three species from a common ancestor than to have evolved separately.

One trait chimps, bonobos, and humans share is the retention of males. Unlike nearly every other social species of mammal, these societies generally retain their males. Meanwhile, females marry into new groups. Anthropologist Carol Ember clarified this pattern by surveying 179 hunter-gatherer societies. She found that only 16 percent of these societies retained their young women more than their young men. Chimp and bonobo social groups also keep their males but transfer their females. Gorillas do the same, although only a few males stay with their fathers. In contrast, orangutans (and almost all other primates) disperse all their males and none of their females. This pattern is significant because this one evolutionary event in the common ancestor of chimps, bonobos, and humans—the retention of males—set the stage for cooperative ape warriors.

Retaining males not only bonds them, it also leads most women to associate with each other based primarily on the men they marry. One symptom of this, noted by physical anthropologist Sarah Blaffer Hrdy, is that women worldwide cooperate less than men do, have little solidarity, and generally fail to bond. Even since the rise of feminism, bonds among Western women remain weak. This is not to say that women do not form mutual friendships. Of course, they do. It is only to say that the strength of women's bonding does not appear to match the do-or-die bonding of men.

This instinct for kinship bonding by males likely was critical for *Homo* not just to cooperate in warfare but also to have started the risky business of scavenging or hunting together. Eskimos on dangerous hunts, for example, admit that the only men they ever trust fully are their own kin.

Another profound facet of human psychology is the opposite of bonding— independence. The availability of food is the primary factor limiting the size of primate groups. Every member must get enough to eat, or the group will disintegrate, or at least fission. The problem monkeys face in going it alone or in too small a group is a much higher risk of being eaten by a predator. Paradoxically, fissioning is exactly what groups of humans, chimps, and bonobos do to survive. These groups always come back together whenever possible, however. This pattern of fissioning and fusing, notes anthropologist Brian Hayden, is normal among people worldwide.

Normal also during fissioning is a sexual division of labor—the human predilection that ultimately made civilization possible. Men go out to hunt, scavenge, fish, or herd; women gather wild foods or farm, meanwhile keeping their dependent infants with them. The sexes travel divergent paths, often alone, to collect food that they will later share with their families. And no matter how harshly food scarcities prevent individuals from staying in their groups, they take pains to socialize in those groups anyway. Chimps even go hungry just to stay with social companions. People do, too.

Moreover, all three species are well equipped to reunite. They hug, kiss, pat, or groom each other to reestablish solidarity. The lesson (covered in detail in Chapter 6) is that *a fusion-fission social life is the only solution to the problem of needing a large social group for self-defense against enemies among one's own species but also being forced as individuals to rely on scattered foods that cannot support a large group as a unit.*

Much of this book will explore the repercussions of why the social group is so necessary and why it manifests itself in men's acts of violence. But first it is important to clarify how humans, naked apes or not, *differ* from apes.

First is our reproductive strategy. Whereas chimps and bonobos share females within their communities, men bond with, and rarely share, their women. Men, like gorillas, even kill sexual rivals. Yet men, like all apes, are polygynous and either marry or want to marry more than one wife. Anthropologist George P. Murdock cataloged 853 societies globally: 83.5 percent of them permit or prefer polygyny. In sub-Saharan Africa, 20 to 50 percent of women share their husbands. How does polygyny affect men's reproductive success? Anthropologist Monique Borgerhoff Mulder found that in eastern Kenya, a Kipsigis man's average number of wives—his degree of polygyny—was precisely what determined his reproductive success. Polygynous men sired more than twice as many children on average as monogamous men.

In contrast, Murdock found only four societies that allow multiple husbands. These societies in Tibet and Nepal allow brothers who own farmland or pasture in common to marry the same woman, one per estate, and raise her children as if each husband were the father (as chimps do). Sons inherit the family property; daughters marry out. The younger brothers marry again—each to his very own wife—as soon as they can afford to.

Sixteen percent of the societies in Murdock's sample, mostly Western, impose monogamy by law. Yet many men in these societies have multiple wives in succession, by divorcing one and marrying another. And polygyny, where possible, increases Western men's reproductive success, too. Men who belong to the Church of Jesus Christ of Latter-day Saints (Mormons) and who marry polygynously (outside the church hierarchy), for example, average 15 children, while monogamous male members average only 6.6 children. "Nothing in male sexuality—insofar as it contrasts with female sexuality," concludes anthropologist Donald Symons, "hints of an adaptation to monogamy."

Again, the difference between men and apes is marital fidelity by wives. This set the stage for a huge evolutionary advantage for men and women. They could divide labor, cooperate, and share food to raise their families. This is because female monogamy is the only relationship that allows males to feel sure of paternity. It made possible the final element of males' mating strategy: invest far more in their offspring than their ape ancestors ever did. Beyond the benefit of more food for her infant, each married female benefited by having a dedicated male to protect her children from infanticidal alien males. The price of this support was loss of freedom. Once a female married and bore infants, she could not afford the cost of trying to stop her mate from bonding with a second female. By contrast, a man could marry twice because he had the potential ability to support or protect the children of two wives. (The obverse, however, did not hold true: neither of two husbands of one woman could ever be certain that it made sense for him to support any of her infants.) All else being equal, the added parental investment by human males would have raised the reproductive success of both sexes above that in all other populations of hominoids whose males still invested little—the ape forebear. This alone would have allowed the first humans, *erectus,* to outbreed those "primitive" populations and drive them to extinction. This advantage, however, enhanced men's obsession with paternity. It led to their extreme jealousy over wives, and their proprietary emotions toward women reverted back to the level of male gorillas'.

The second huge difference between apes and humans is our adaptation to walking on two legs. The evolution of bipedalism is still only a partially solved mystery, although when it evolved is clear. Paleoanthropologists tie it to Miocene global warming 7 million years ago. This warming replaced forests with savannas and created complex mosaic habitats. Anthropologist Adriaan Kortlandt notes too that "bulldozer" herbivores—elephants, buffalo, and other megafauna—also ravaged these forests into savannas, thus opening a new habitat for opportunists. Hominids were those opportunists.

Contrary to what one might guess, however, the shift from four feet to two feet did not carry a big metabolic price. Instead, it earned hominids free travel vouchers. Physical anthropologist Peter S. Rodman and paleoanthropologist Henry McHenry explain:

> We looked at the data [on the bioenergetics of locomotion] and saw that, for a chimpanzee, walking quadrupedally was no more and no less energetically expensive than walking bipedally. So if you imagine that hominids evolved from some kind of quadrupedal ape, then you see that there's no energy barrier, no energy Rubicon in going from quadrupedalism to bipedalism. But the most important point—a new one, as far as we know—is bipedalism in humans is considerably more efficient than quadrupedalism in living apes.

But why did our ancestors bother to stand on two feet to begin with? "If you're an ape," Rodman and McHenry explain, "and you find yourself in ecological circumstances where a more efficient mode of locomotion would be advantageous, the evolution of bipedalism would be a likely outcome." What are these ecological circumstances? The fragmentation of the forest in East Africa into woodlands and ever-expanding savannas, which lengthened the distance each ape had to travel between fruit trees. The savanna, moreover, unlike the forest, was a sun-scorched thermal challenge for creatures adapted to shade. And bipedalism offered one more tremendous advantage over traveling on four legs: a 60 percent reduction in heat received, due to an upright stance presenting a much smaller solar target than does the body of an ape traveling on all fours. Again, to eat—to survive—apes *had* to travel these increasing distances between woodlands. "Bipedalism provided the possibility of improved efficiency of travel with modification only of the hind limbs while leaving the structure of the forelimbs free for arboreal feeding," Rodman and McHenry explain. The bottom line is this: *"the primary adaptation of the Hominidae [bipedalism] is an ape's way of living where an ape could not live."*

Do we know exactly *when* this happened?

In 1974, Mary Leakey began excavating at Laetoli ("The Place of

Red Lilies") in Tanzania. Nearly 4 million years earlier, a volcano named Sadiman ejected a cloud of carbonatite ash that blanketed Laetoli. Rain then transformed this one-half-inch-thick blanket into a wet cement in which a bestiary of fauna from dinky millipedes to elephantine *Deinotherium* left footprints to dry rock hard in the sun. Sadiman erupted more than a dozen times that month, depositing ash layers stacking eight inches deep.

Mary Leakey's team found bits of hominid jaws and teeth 3.6 million years old scattered in strata there. In 1976, during an elephant-dung-throwing fight among Leakey's team, Andrew Hill ducked a ball of dung and, his nose to the bedrock of carbonatite, noticed footprint-like dents. This led in 1977 to Peter Jones and Philip Leakey discovering fossil elephant prints. Then they found smudged prints that looked vaguely human. In 1978, Mary Leakey recruited Louise Robbins, a footprint expert, to oversee searches there. Unfortunately, when Paul Abell found a broken print that he thought was from a 3.6-million-year-old hominid, Robbins discounted it as a buffalo print.

Disgusted with all the furor over prints that never paid off, Leakey told her crew to excavate for bones and forget footprints. Abell, Tim White, and others pleaded with her to reconsider. Finally she relented, allowing her maintenance man, Ndibo, to peck away at the overburden of the "footprint."

It was the smartest move of her half-century career.

Hidden among the ten thousand prints left by hares, dik-diks, *Deinotherium,* and saber-toothed cats that her team would uncover in Sadiman's ash, lay an important telegram from the fourth dimension.

The next day, Ndibo described to Leakey two prints he had uncovered. She was skeptical, but when she saw them, she instantly changed her mind and told Tim White to excavate them. To avoid mistakes, White poured paint thinner on the layered rock to highlight the colors. Wet with thinner, the thin layer of calcium carbonate infill remained white, but the carbonatite layer holding the print itself turned dark—like an open-faced Oreo cookie with a footprint imprinted inside the lowest wafer. White used a dental probe to excavate the white from each footprint, inch by inch and day after day, for thirty feet.

Afterward, White cowrote that 1979 paper about Lucy with Don Johanson (mentioned at the beginning of this chapter). The two men grouped Lucy and the Site 333 hominids with Mary Leakey's two dozen hominid fossils from Laetoli and these prints, naming them all *Australopithecus afarensis.*

This was a mistake. Like her husband, Louis Leakey, who believed in an ancient age for *Homo,* Mary Leakey detested the word *Australopithecus* and hated its ape-man—not human—connotation. White's usurpation of her right to name her own finds and his insulting choice of the genus *Australopithecus* for them instantly made him persona non grata at Laetoli.

Ron Clarke replaced White on Mary Leakey's footprint excavation. The fruits of all the scientists' work were eighty feet of trail and seventy footprints. At least two hominids, an adult and a child walking side by side, strode across Sadiman's ash—and across the pages of *National Geographic*—into the sunset.

"The Laetoli footprints," says White, "are probably the most precious discovery that has ever been made in this science, or ever will be made."

Mary Leakey agrees. After casting and photographing the prints, she protected them by backfilling them with soil. Ironically, the backfill was so fertile that acacias and other plants sprouted as if potted in a greenhouse. White now ruefully calls the prints "Laetoli National Forest."

The upshot? The Laetoli footprints reveal that hominids 3.6 million years ago were fully bipedal. Richard Hay and Mary Leakey also reckon from the stride lengths between the prints that the midsize hominid was four feet seven inches tall and the smallest one was three feet ten inches tall. Prints for the suspected third and largest individual were too obliterated to measure its stride and height. The Laetoli prints also tell us that for at least a million years before Sadiman erupted, hominids had been bipedal. Were these Laetoli walkers our ancestors?

Not if they were the same Lucy-type *afarensis* as Johanson discovered in Ethiopia. Paleoanthropologist Henry McHenry analyzed the limbs of *afarensis.* "Except in relatively minor details," McHenry concludes, "[they] are very similar to one another and unlike any living hominoid [apes and humans]." *Afarensis* used "different firing patterns of the muscles, different movement of the hip joint, and so on." Although *afarensis* was likely an ancestor of all later australopithecines, its connection with *Homo erectus* still seems tenuous.

If not *afarensis,* whoever did make those Laetoli footprints 3.6 million years ago may indeed be the forebears of *erectus* and ourselves.

Either way, bipedalism was *the* pivotal adaptation leading to hominid success. It freed early hominids' hands to carry food, and, when combined with the braininess of the apes, it allowed them to make, use, and carry tools and to use sign language. It was the second critical adaptation—along with female fidelity—allowing hominids to divide labor and share food.

It is important to remember, however, that bipedalism neither *caused* the evolution of humans nor *made it inevitable.* Being a two-legged hominid does not equal being a human, or even a human ancestor. At least three, and maybe four, bipedal hominid species lived in Africa 1.5 million years ago, as do three species of great ape today. Two of these hominids, *Australopithecus africanus* and *A. robustus,* were tiny-brained creatures whose cultures could not have been much more sophisticated than a chimp's. In contrast, *Homo erectus* was apparently a culture addict.

This leads us to the third difference between apes and humans: brain size and intelligence. To be human, to rely more on culture than on instinct, takes a brain twice the size of an ape's. From whence came this big brain? One reason cited for mammals having evolved larger brains is to cope with the demands of making a living in a complicated ecology. But research reveals this not to be true for most primates. Instead, primatologists Dorothy Cheney, Robert Seyfarth, and Barbara Smuts examined primate intelligence and concluded that it evolved under the evolutionary pressure to outwit and outcompete one's own kind. Among apes, to be stupid is to die at the hands of other apes. As primatologist Meredith Small explains, "It is incontestable that we are social animals and that knowing who we are, where we stand and how to get what we want—even at the expense of others—is the key to survival for all primates, monkeys and humans alike."

Did the doubling of *erectus*'s brain power occur merely because its larger version of an ape brain gave *erectus* just enough culture to become a culture addict, which then drove natural selection for a still larger brain in a spiraling cultural arms race to feed that protein-demanding brain with meat? Or was brain expansion part of a primate arms race against all other hominid populations, among whom *erectus* itself was its own worst enemy?

The answer is likely yes on both counts. *Homo,* notes biologist Richard Alexander, eventually whittled down the major players on the field until *Homo* itself was left as *Homo*'s "only significant hostile force of nature." The natural arms race that this fostered, suggests anthropologist Robin Fox, "may well have benefitted *[Homo]* from the selection pressures for intelligence, foresight, strength, courage, cooperation, altruism, comradeship, and sociality that this intercommunal violent activity involved."

With this superintelligence came the fourth main difference between humans and their chimp and bonobo relatives—and the final ingredient in being human: symbolic language. This was likely invented to coordinate hunting, foraging, and war—and eventually was vital to courtship. Language gave the biggest boost to the cooperation, coordination, and division of labor necessary to human lifestyles.

In short, combining big brains and this violent but cooperative male social system and reproductive strategy with DNA made human nature inevitable. Human nature was forged when males of the genus *Homo* stayed together, when females were faithful to one male, when bipedalism freed hands during travel, and when *Homo*'s ape brain expanded to be capable of cultural inventiveness and symbolic language. Apes became humans only when culture became their top competitive advantage. Instantly, the less brainy were at a severe disadvantage. With culture, humans became the most dangerous creatures in the universe.

Given this rare combination of ingredients, natural selection, and especially sexual selection, would have driven the behavioral evolution of *Homo* down this fairly narrow road to sophisticated sexism and cooperative violence no matter what. It would have happened the same way in another galaxy. To see why, consider the alternative strategies. No lone male, no matter what strategy he used, could compete for long against a male kin group that cooperated in lethal territorial combat. The evolutionary "invention" of primate "armies" was a one-way street with no turning back.

All of this paints an interesting picture of human nature and its dark side in males. But do rape, murder, warfare, and genocide truly emerge from men's nature?

The next part of this book, "Violence," explores this question and offers some surprising answers.

• • •

PART TWO

VIOLENCE

For my own part I would as soon be descended from that heroic little monkey, who braved his dreaded enemy in order to save the life of his keeper, or from that old baboon, who descending from the mountains, carried away in triumph his young comrade from a crowd of astonished dogs—as from a savage who delights to torture his enemies, offers up bloody sacrifices, practices infanticide without remorse, treats his wives like slaves, knows no decency, and is haunted by the grossest superstitions.

— Charles Darwin, 1871

It is a law of nature, common to all mankind, which time shall neither annul nor destroy, that those who have greater strength and power shall bear rule over those who have less.

— Dionysius

CHAPTER FOUR

Rape

I'm ready to call it a night," Kay admitted to some other volunteers.

It was midnight. Kay and fifteen others were celebrating the end of their three months of training on the coast of Ecuador. The next day, they would be sworn into the Peace Corps. But being sworn in with a hangover was not in Kay's plans. She had dreamed of joining the Peace Corps since she was five years old, and nothing was going to spoil it.

Two male volunteers told Kay that they also were ready to leave. They offered to walk her as far as their hotel; hers was only three blocks beyond.

The streets of Quito were almost silent. Nearly a million people lived in the capital of Ecuador, but at half past midnight on Thursday night, most of them were asleep, oblivious to the glittering display of stars above their equatorial, two-mile-high city.

Kay thought about her swearing in. Her degree in geography would qualify her to work as an animal science volunteer to help alleviate severe overgrazing among the sheepherding Saraguro people. Tomorrow she would travel farther into the Andes and meet the Saraguro. It could not happen soon enough.

Her two friends said good night. For a moment, she wanted to ask them to walk her the last three blocks to her hotel. It would take them only a few minutes. No, she decided, she was a big girl. She would go alone.

She found the gate to her hotel locked, and she had no key. As she wondered what to do, two men approached. Both swayed drunkenly; their rapid Spanish slurred menacingly. To avoid these two, Kay circled around to the back of the hotel. It also was locked up tight. Suddenly, Kay felt helpless.

A car pulled up. The man in the passenger's seat demanded her papers. "Who are you that you need to see my passport?" she asked in Spanish.

"Don't worry about that," he answered in Spanish, pulling his leather coat open to reveal a pistol. "Get in."

This man with the leather coat had "menace" written all over him. No way would she get in his car, gun or not. Staying a few feet from the car, Kay showed him her passport by holding it up. Then she hurried back to the front of the hotel.

Both drunks were still loitering near the gate. Now she wished more than ever that she had asked her friends to walk her here. This was beginning to seem like a grade B horror movie—one in which she would be the victim.

Suddenly, the car reappeared alongside her. The back door swung open. Before she realized what was happening, the man in the leather coat reached out, grabbed her, and dragged her inside with him. She crashed on her back on the floor. That he could grab her like this seemed impossible, but the reality was that he was kidnapping her—and fast. He slammed the car door behind her.

Kay tried to pull herself up to the door. He punched her, then he shoved her back down on the floor. She struggled fiercely to claw her way back up. He punched her again. Kay felt the adrenaline of fear surge through her. She had to escape, but his thumbs dug into her face to gouge her eyes. The engine roared and the gears meshed as she struggled, screaming. But he persisted, gouging.

Finally, thwarted by the sheer violence of her resistance, he stopped gouging her eyes. Kay felt a glimmer of hope. But immediately he thrust his fingers into her ears. Then he pounded her with blow after blow as if trying to beat her to death. She knew he was far stronger than she was—he seemed nearly twice her weight—and as she kept trying to claw her way back up, she felt her strength ebbing. Against her fiercest efforts, he shoved her right back down. Then he sat on her as the driver drove on. She could hardly breathe.

This is unbelievable, Kay thought. He's going to rob me and maybe even kill me, assuming I don't suffocate first.

After what seemed like hours, the driver stopped the car. They were far outside town. Kay was now convinced that "Leather Coat" would kill her. She was just beginning to catch her breath and marshal her strength when he yanked her pants off.

She fought to remain conscious, to resist to her utmost, fearing that if she lost consciousness, he would kill her. She grabbed the door handle. He mocked her in broken English and pried her fingers back, one by one. Kay felt, then heard, the cartilage tearing in calculated torture.

Kay pleaded in Spanish, "I'm too young. I'm a virgin. What will my family think? What will God think?" Her captor mimicked her in broken English, savagely assaulting her through her rectum with his fist for nearly an hour. Her blood and feces spilled into the car.

Kay begged him to stop. She cried that the pain was too great and that she could not go on. He mocked her again. Even so, she never stopped begging him to stop, to have pity on her. She prayed to God.

Kay smelled cigarette smoke and saw a shadow outside the window. She realized it was the driver, standing and smoking impassively while she was being brutally destroyed only three feet away. Meanwhile, as "Leather Coat" continued assaulting her, she thought the pain alone might kill her before she bled to death.

Still she fought, pleaded, and cried. But nothing could stop the nightmare.

The driver returned to his seat and leaned his forehead against the steering wheel.

"Leather Coat" never entered her with his penis, but he did ejaculate on her. And as he rammed his fist again into her shredded insides, he hissed to the driver, *"Cuerda"* (rope).

"Leather Coat" was a strangler.

Now Kay knew she was about to die.

• • •

Rapists and Their Victims

DESPITE HEADLINES we see splashed across newspapers—and in contrast to Kay's sadistic nightmare in Ecuador—most rapes are not grisly torture rapes or rape-murders. Partly because television news producers tend to report the most heinous and bizarre violence, and do so ever more often, and partly because such crimes form the most lasting images, many of us equate "rape" with this sickening sexual brutality, like Kay's horrible experience related above. But the crime of rape is normally less brutal than this example—yet, at the same time, it is far worse. How could it be worse? Sheer numbers.

This chapter explores the nature of the majority of rapes in the United States and elsewhere. It also identifies the types of men who commonly rape, who their most common victims are, and why these men rape these women.

First, let's distinguish between rape and rape-murder. The Federal

Bureau of Investigation's (FBI's) *Crime in the United States* (an annual publication that, along with the U.S. Department of Justice's *Sourcebook of Criminal Justice Statistics,* provides the most complete, albeit tardy, data on crime in the United States) notes that rapists kill only 1 in 1,596 reported victims. Sixty women were killed by rapists in 1996. Taking unreported rapes into account, rapists likely kill fewer than 1 in 10,000 rape victims in America. As we will see, this high survival rate helps explain why men rape to begin with.

Lethal or not, rape has become an epidemic of terror for all women. In 1996, the most recent year for which data are available, the FBI received 95,769 reports of forcible rape. This number is down 1.7 percent from 1995 and 12 percent from 1992, when the number of reported rapes peaked. Yet despite this being the lowest rate since 1987, reported rapes have leaped 400 percent in the past 40 years, from 9.3 per 100,000 population in 1958 to 36.1 per 100,000. This translates to 71 rape victims per 100,000 females in 1996. Overall, rape now accounts for 1 in every 19 reported violent crimes. In 1996, the National Academy of Sciences' Panel on Research on Violence Against Women concluded on page 1 of its report, *Understanding Violence Against Women,* that "between 13 and 25 percent of all U.S. women will experience rape. These figures are believed to be underestimates."

Due to incomplete reporting, the true number of rapes remains unknown. Studies of rape in the United States report rates varying from only 2 percent of all women to fully 50 percent. Either way, no sociological class is immune. In a survey of middle-class women under age forty in Los Angeles, for example, 22 percent reported having been sexually molested or raped.

In America today, a new victim reports a rape every five to six minutes. But this is only the tip of the iceberg. The FBI also estimates that between five and twenty unreported rapes occur for every reported one. *On average, at least one woman is raped in America every minute.*

Rapists stalk and prey on women for a reason. Unfortunately, explanations by many rape experts are so off the mark that they may even enhance a gullible woman's chances of becoming a victim.

Compounding this misinformation (see below) is the following disturbing reality: only a small fraction of men who have forced a woman to submit to a sexual act against her will are punished. In part, this is because too few victims report being raped. But it is compounded by minimal sentencing in the U.S. justice system. By the mid-1990s, for example, although 51.3 percent of all reported rapes were cleared by arrest, only about half of those arrested were convicted, and only 88 percent of those convicted were sentenced to prison. In short, fewer than one in four accused rapists can expect to serve a prison sentence, which will average only 7.25 years. The rest go free. If unreported rapes are added into the equation, a mere one in twenty to eighty rapes leads to a rapist being sentenced. Clearly, for most sexual predators in America, rape is a free ride—so free, the Senate Judiciary Committee found, that a woman is eight times more likely to be raped in America than in Europe and twenty-six times more likely to be raped in America than in Japan.

Although weak punishment undoubtedly contributes to America's high rape rate, it does not tell us why men rape to begin with. Is rape a product of socialization? Or, disturbing though the thought may be, is rape an instinct in the human male psyche? Or does rape result from both socialization and instinct? And if rape is instinctive, why is it so?

To answer these questions, we must acknowledge that we each possess a complex brain ruled by neurotransmitters that spur instinctive emotions and influence our behavior. As uncomfortable as this perspective may be to some people, no understanding of rape is possible otherwise. From this perspective, the two essential keys to understanding why men rape are identifying which women become victims and which men rape them.

First, the victims. Significantly, not all women stand an equal chance of being chosen as a target. The U.S. Bureau of Justice Statistics reports that nearly all rape victims are young: 88 percent of the 1,634,000 female victims reported in a huge sample from the 1980s were between the ages of twelve and thirty-five years old. By the mid-1990s, 90.3 percent of all known rapes occurred in this age-group, which encompassed fewer than one-third of all females in America, but included nearly all fertile women. Even more revealing, although 77 percent of all rape victims were between sixteen and twenty-four years old, these women accounted for only about one-tenth—again, the most fertile and sexually attractive tenth—of the entire female population. As we have seen, these young women are precisely the ones whom men worldwide prefer most as sexual partners—and, incidentally, whom the U.S. pornography industry prefers as actresses.

If rape victims tend to be young and sexually desirable, what about the sexual predators? Two-thirds of the men who raped the 1.6 million victims above were strangers. By 1990, 88 percent of reported rapists were solitary sexual predators who shared their victims with no one.

And, like their victims, sexual predators were typically young. Forty-four percent of men arrested for rape were under twenty-five years old; most of the remaining rapists were men in their late twenties or thirties. Only a third of these men were under the influence of alcohol, drugs, or controlled substances during their offenses.

A clue to the factors that lead to most men's decisions to rape is their socioeconomic status. Although sexual predators are at their physical apex, they typically are at their financial nadir. Like most felons, sexual predators are uneducated, are unemployed or underemployed, and have low incomes. Indeed, the most common trait of men arrested for rape is that they are early losers—or at least not yet achievers—in the socioeconomic arena. (This may help to explain why blacks account for a disproportionate, but stable, 42 percent of arrests for rape.)

Rapists fit into what sociologist Marvin Wolfgang calls the "subculture of violence." Simply put, they belong to the cohort of men who decide to use force once they learn they cannot count on a job to get them what they want. "Rapists are usually all-round offenders with a long list of convictions," explains Clinton Duffy, warden of California's San Quentin prison for thirty-four years. "The opportunity for rape often crops up while they're in the course of committing another felony, and they take advantage of it. Actually, practically every offender who is not an overt homosexual is a potential rapist."

Sexual predators are the "bad men" our parents warned us about. Two-thirds of rapists, for example, have prior arrest records, and 85 percent will be arrested again. Moreover, they committed their first crimes at a young age: 94 percent of convicted sexual predators were convicted for their first crimes at age fifteen; they first raped, on average, at age eighteen. After prison, half of all convicted rapists rape again.

Close though Duffy was to profiling the archetypal career rapist, he missed one key nuance. Although nearly all sexual predators threaten to use force, only 14.8 percent of U.S. rapists use a weapon. In the mid-1990s, these weapons were mostly knives. Only 5.9 percent of known rapists use a gun. One-quarter of rapists use threats but commit no violence. Surprisingly, the use of a weapon raises the success of rape attempts by only 9 percent. Yet despite their limited use in rapes, weapons do reveal an important facet of the mind and motive of the rapist. This is where Duffy was blindsided.

In their excellent nationwide study of convicted felons, *Armed and Considered Dangerous: A Survey of Felons and Their Firearms,* James D.

Wright and Peter H. Rossi classified criminals by their habitual weapon of choice: Unarmed, Improvisor, Knife, Sporadic Handgun, Handgun Predator, Shotgun Predator, and Shoulder Weapon Predator. Weapons did make a significant difference in these felons' incomes. Robberies by Predators netted $164 each, for example, while those by Knife felons collected only $60 each. More unexpected, Wright and Rossi found that Knife felons raped women twice as often as any other felon. And a full 90 percent of these men actually wielded their knives during the rape. Knife felons also were the most common repeat rapists; 48 percent of them were habitual sexual predators.

The average rapist with a handgun wielded it only a third of the time that he raped. More intriguing yet, Handgun and Shotgun Predators, who made substantially more money from their crimes, rarely raped at all. Instead, Wright and Rossi explain, Predators were "omnibus felons . . . men prone to commit virtually any kind of crime available . . . excepting rape." Why these criminals who choose guns might "court" women traditionally, while those with knives rape them, will soon be clear.

The upshot is this: *men who rape women are typically the least successful economically in their society—even the least successful among criminals—but the age cohort of women whom rapists choose as victims is the most desirable to the most affluent men in societies worldwide.*

Myths of Rape

KNOWING THAT predominantly young men rape and that they seek young women as victims does not answer the critical question, Why do men rape? Of all questions concerning human violence, rape seems the most deeply mired in myths. Indeed, no discussion of rape seems possible without these fallacies emerging as unspoken truths, which then cloud all objectivity. Since the goal of this chapter is to uncover the true nature of rape, let's look at what is wrong with these myths.

The three most pervasive myths are as follows: (1) the woman is to blame, even if she truly did not want to be raped; (2) the victim is willing (she wanted to be raped); and (3) rape is a crime motivated by power and control, not by sex. Each myth can be seductive, yet each is wrong.

Surveys reveal that significantly more American men than women believe that the victim must share the blame for her rape. But Americans hold no monopoly on this delusion. Men in India and in other cultures also feel that female victims contribute to their own rapes. These men generally feel that women, simply by being women—especially young, attractive ones—automatically qualify as potential victims. This may seem to be trivial or shallow logic, but it is definitely not trivial to rapists. This myth becomes more plausible when tied to specific cases: a woman took a wrong shortcut home, wrongly forgot to lock her door, wore the wrong clothes, or trusted the wrong man. In a climate of rape, a woman who does not protect herself by avoiding all situations in which rape is even a remote possibility (even though by doing so she forfeits her freedom and uses up most of her money) is just as guilty as one who forgets her umbrella and gets drenched by a sudden rain.

Whereas men only believe this logic, women are forced to live it. In one study, 41 percent of urban women admitted that to avoid being raped, they never went out after dark alone. Many people believe that it is unfair and very inconvenient for women to have to plan around rape—by keeping their windows closed even on hot nights, by not being able to park their cars in terraced parking, and so on. Regardless of the fact that women live in a society in which some men rape (in all societies, by the way, some men *do* rape), it is impossible to logically conclude that women are to blame for men's decisions to rape.

Another popular myth is that of the willing victim. Surveys reveal that more men than women also believe that at least some women (if not all) want to be raped, despite their resistance during the actual crime. That forced sex is a woman's fantasy is a common claim.

To find out whether women really fantasize about being raped, Lisa Pelletier and Edward Herold interviewed 136 single women, all unmarried college seniors. The most common sexual fantasy of these women (90 percent) involved intercourse with a boyfriend or husband. Other popular fantasies included being undressed by a man (79 percent), reliving a previous sexual experience (78 percent), intercourse in an exotic place (72 percent), undressing a man (71 percent), cunnilingus (66 percent), and intercourse with a male friend (60 percent). Number 8 of the twenty-four most common sexual fantasies, however, was forced sex with a man (51 percent). Number 19 was forced sex with more than one man (18 percent).

An important question here is, Were these 51 percent of women who fantasized about being raped fantasizing about forced sex in which they were an unwilling victim, or were they merely imagining rough sex as a willing partner? Susan Bond and Donald Mosher sought an answer to this question by presenting 104 undergraduate women who had never been raped with different versions of the same rape story. Version *a* was an erotic fantasy of "rape." Versions *b* and *c* were realistic stories of forcible rape. In version *a,* the "rapist" was desirable and desired his victim; the woman was braless and feeling more daring than usual and also could influence the "rapist's" actions, because "she enticed and permitted the man to 'rape' her for her own purposes." The critical difference between version *a* and versions *b* and *c* was in the woman having control over her own sexual behavior.

For this study, significantly more women who imagined erotic "rape," version *a,* were sexually aroused and experienced pleasure than those imagining other versions. Half of them, however, also felt guilty and angry over being aroused. Meanwhile, no woman imagining version *b* or *c*'s realistic rape reported sexual arousal. Rather, they felt "intense pain, disgust, anger, fear, and shame with mild depression and some guilt." Bond and Mosher conclude, "Not only do women not enjoy the experience of being raped, they do not even enjoy the experience of imagining being raped."

This may or may not seem consistent with the research mentioned above revealing that at least some women *do* enjoy fantasizing about forced sex. An important distinction here is that, for most of us, the realms of personal fantasy and the real world are very different. Either way, the acid test of the "willing victim" myth is how women feel about men who have forced sex on them in spite of their protests and struggles. F. Scott Christopher surveyed the occurrence of date rape among 275 single college women. More than half admitted having been pressured into oral-genital sex against their will; 43.4 percent had been pressured into intercourse.

More than half of the men pressuring these women were serious dating partners. The rest were casual dates, friends, or strangers. The pressures men used escalated. These rapists typically started with persistent physical attempts. For forced fellatio and sexual intercourse, this worked roughly half the time in these date rapes. When it did not, they promised the women a secure future relationship (one that never materialized). This yielded forced fellatio 44.9 percent of the time and sexual intercourse 35 percent of the time. When both these strategies failed, rapists threatened force. Threats worked 3.8 percent of the time for fellatio and 3.4 percent of the the time for intercourse. And when all of the above failed, many men used force. For fellatio, 4.5 percent used force. For sexual intercourse, 6.8 percent used force.

How did these victims of date and acquaintance rape feel? They felt "really upset and used," "guilty, sleazy, and violated," and they said that they "hated it." "Women's responses to being pressured into sexual behavior," Christopher concludes, "were overwhelmingly negative."

Whereas most women loathe rape, most men are turned on by it—so much so that psychologists Alfred Heilbrun and David Seif wondered whether women under duress sexually excite men even more than do willing women. They tested this idea by showing 54 adult male students sexually explicit photographs. Heilbrun and Seif were surprised to find "an overall sadism effect" in most of the men. They defined this as an "enhanced sexual attractiveness of women who are experiencing distressed emotions."

This sadism made a difference in men's behavior. In another study by Heilbrun, this time with M. P. Loftus, twice as many college men (60 percent versus 29 percent) were sexually sadistic than were not. Moreover, *60 percent of men admitted having used force, as early as age sixteen, to achieve sexual intimacy with a girl despite her negative response.* Surprisingly, these sadists also reported feeling the least peer pressure to rape women. Peer pressure and "macho" masculinity, Heilbrun and Loftus found, were irrelevant to them; only sex was important. These sexually sadistic and aggressive men were both more numerous than nonsadistic men, and, by their own admission, the most likely to rape.

The newest and currently most pervasive rape myth—the one that explains rape as motivated by violent power and control, not sex—may also be the most important myth. This myth is thought by many sociologists and social workers to be the "Ultimate Truth"—which, tragically, may have led to many women being raped who otherwise probably would not have been.

In 1975, feminist Susan Brownmiller expressed this conclusion in her book *Against Our Will:*

> Indeed, one of the earliest forms of male bonding must have been the gang rape of one woman by a band of marauding men. This accomplished, rape became not only a male prerogative, but man's basic weapon of force against women, the principal agent of his will and her fear. His forcible entry into her body, despite her physical protestations and struggle, became the vehicle of his victorious conquest over her being, the ultimate test of his superior strength, the tri-

> umph of his manhood. . . . From prehistoric times until present, I believe, rape has played a critical function. It is nothing more or less than a conscious process of intimidation by which all men keep all women in a state of fear.

Brownmiller also claims that the real reason women are monogamous is their partners' forcible abduction of them, the men's sexual entry into their bodies against their will, and the "female fear of an open season of rape." Monogamy, Brownmiller says, is the only thing that protects a woman from other men who would otherwise rape her, too. Men are so bad, Brownmiller adds, that even these desperately monogamous women are raped.

The fact that rape does erupt and flourish during war—Brownmiller's "open season of rape"—far more than it does under any other conditions seems to support her idea that rape is merely a tool of power and control. Men everywhere seem more likely to rape women who have lost their defenders. "The opportunity for free play to sexual passion has been considered a perquisite of soldiers, especially in the taking of a besieged city," notes Quincy Wright in his classic *A Study of War.*

Rome, for example, stood for eight hundred years before Alaric's two-year siege finally broached the city's defenses on August 24, 410. Alaric's Visigoths looted and killed the starving Roman citizens and raped the Roman women. Likewise, Nazi invaders mass-raped Russian women in village after village in 1941. Russian soldiers who later occupied Germany reciprocated.

Internal wars (pogroms)—such as those waged against the Jews by the Nazis, against Armenians by Turks, against American blacks by the Ku Klux Klan, and against whites by native Congolese during the shift to Independence—also include the sorts of rapes perpetrated during these invasions. Invading West Pakistani soldiers, for example, raped 200,000 to 400,000 Bengali women over nine months in 1971. Saddam Hussein's Iraqi soldiers mass-raped Kuwaiti women in 1990. Hutu troops raped Tutsi women en masse in Rwanda in 1994. In Bosnia, ethnic Serbian paramilitary troops systematically raped Muslim women from 1991 to 1995, imprisoning many of their victims in sex camps for repeated raping. In 1997, Algerian women were mass-raped as sex slaves by Muslim revolutionaries. In 1998, Indonesian Muslim security troops were accused of participating in gang rapes of hundreds of ethnic Chinese Indonesian women.

Whenever women lose their protectors, mass rape occurs. It is probably happening somewhere now as you read this.

The "Rape of Nanking" shows how bad this can get. Japan invaded China's capital in the wake of General Chiang Kai-shek's retreat in 1937. American missionary James McCallum estimated that Japanese soldiers raped at least one thousand Chinese women (65 percent of them between sixteen and twenty-nine years old) every night and raped even more women during the day. When these rapists were satiated, they sometimes bayoneted their women victims to death (they had been ordered by their high command to silence their victims by either paying or murdering them). During the first month of occupation, the Japanese raped at least twenty thousand women. (If this is shocking, bear in mind that during each month of this year, it is likely that even more American women will be raped.)

Why do soldiers rape? Is it truly because, as Brownmiller claims, men are designed to hatefully grind women into the mud and to control them in a state of perpetual fear, and war offers men a good opportunity to do this?

A closer look shows the issue to be far more complicated. During war, most combat troops are uncertain about surviving. Furthermore, most are young and have not yet sired children. Their prospects for fatherhood, therefore, are doubtful. Yet under these uncertain conditions, they are constantly meeting young, fertile, pretty, and *unprotected* women. Moreover, rape in war is rarely punished. And if a rape does produce children, the rapist invests neither energy nor resources in child support, because he is never sure the child is his. On top of all this, soldiers who are able to rape their enemies' wives and daughters have obviously won. Such rape is both a vindication of victory and a prize of that victory. In short, mass rape is a massive reproductive victory.

Could both of these be the real reasons for rape during war? In *About Face: The Odyssey of an American Warrior,* Colonel David H. Hackworth explains:

> There's a thing about combat soldiers and sex. On the one hand, it's the most important activity in the world. On the other, it means nothing at all. . . . You were always horny and never discriminating; you weren't looking for love, you were looking for pussy. . . . Getting as much pussy as you could was part of playing the role: proving yourself in the

> cot was as important as proving yourself on the battlefield. Tribal behavior, I guess—the great warrior, the great conqueror of all lands and all broads—or some deep psychological thing: knowing you might get killed and wanting to plant the old seed before you went.

Although Hackworth is talking about consensual sex (even with prostitutes), not rape, his message is clear: soldiers want to "plant the old seed" before they go. This concept is difficult to accept for people educated in the politically correct late twentieth century. But it has been driving male behavior for millennia.

Journalist Barbara Crossette looked at the epidemic of military mass rape in 1998 and concluded:

> It is becoming increasingly apparent that the new style of warfare is often aimed specifically at women and is defined by a view of premeditated, organized sexual assault as a tactic in terrorizing and humiliating a civilian population. In some cases the violators express a motive that seems to have more in common with the tactics of ancient marauding hordes than with the 20th century—achieving forced pregnancy and thus poisoning the womb of the enemy.

Rape in war seems driven by men's sex drive. It seeks not to "poison" the womb of the enemy but to plant that seed—while simultaneously demonstrating victory over those males no longer able to protect the victims. During war, rape often offers young men their very best opportunity for sex and for siring offspring. And, after all, fertile women are the rarest of all reproductive resources for men. *Indeed, rape during war may be an instinctive male reproductive strategy.*

I mention this because Brownmiller's explanation of rape as a means of power, not sex, is now accepted and taught—in whole or in part—as the central dogma of rape in sociology, psychology, and other social sciences. In 1980, for example, C. G. Warner explained the social sciences' majority view: "It is now generally accepted by criminologists, psychologists, and other professionals working with rapists and rape victims that rape is not primarily a sexual crime, it is a crime of violence."

Many people cling to this dogma so blindly, notes psychologist Craig Palmer, that even when rapists during the 1980s admitted that they had raped "for sex and that's all," the politically correct psychologists who interviewed them often discounted these frank admissions. Professor Nicholas Groth claims, for example, that because rapists' "efforts to negotiate the sexual encounter or to determine the woman's receptiveness to a sexual approach are noticeably absent, as are any attempts at lovemaking or foreplay," they were not, despite their statements to the contrary, interested in sex. Groth justifies this bizarre thinking with his special definition of "sexual motivation" (as many social scientists define it) as *only* honest courtship and pair-bonding in which men feel tenderness, affection, and joy. (Remember, however, that the research by Heilbrun and Loftus revealed that most men are turned on instead, in a purely sexual way, by damsels in real distress.) Because rapists feel none of those tender emotions, insists Groth, by definition rape must be an aggressive quest for power and control, not sex. Rape, he contends, cannot be aimed at something so simple as sex.

Brownmiller insists that rape is nothing less than a political tool used by men to slam all women facedown in the mud, then hold them there. "Men who commit rape have served in effect as front-line masculine shock troops," she explains, "terrorist guerrillas in the longest sustained battle the world has ever known." Brownmiller's conclusion that men evolved somehow to hate women emerged from her intuition, not from scientific study. Revealingly, no evolutionary biologist, man or woman, thinks that Brownmiller's idea that men evolved to *hate* women is right—although most would agree that both sexes have evolved to *exploit,* in many reproductive ways, the other.

Moreover, Brownmiller's "men hate women" model does not jibe with the fact that for every three women violently assaulted in America in the 1990s, four men also were violently assaulted.

Even formal gang rapes and rapes admittedly committed as a form of punishment may be primarily for sex. As Donald Symons points out, rape as punishment "does not prove that sexual feelings are not also involved, any more than deprivation of property as punishment proves that property is not valuable to the punisher." Symons's point here is that sex is the only way that men procreate, thus sex is very valuable, and rape merely is *coercive* sex—a "stolen" copulation which may enhance the rapist's reproductive success.

Two other facts, however, may still make Brownmiller's idea appear reasonable to some women. First, just as there are sadistic and murderous rapists, a few men do rape women to dominate them. Second, to normal women, this small fraction of power rapists embodies the essence of rape. This is because the average woman—who is not driven by the same incessant sexual dictates as a man—cannot understand how a man could put a knife to a strange woman's throat, threaten to kill her, violate her sexually, and torture her emotionally, without his also being full of hate for her or for all women. Although women can well understand hate, most women are not equipped with a psyche that can truly relate to men's insistent sexual psychology and the psychology of rape that can spring from it.

Real episodes of nightmare rape—the sort of sexual assaults that the media thrive on—also increase the appeal of Brownmiller's hate model. Patricia (her last name is protected) was a white investment banker. Her job in New York's financial district left her too few daylight hours to exercise, so she jogged in Central Park in the evenings. On April 19, 1989, she was running alone past a grove of sycamore trees when a pack of six teenage boys attacked her. The six boys chased her into a gully and beat her for half an hour with a rock and a metal pipe. They raped her repeatedly and left her for dead. Three hours later, a passerby found her in a coma. So much of her blood had drenched the soil that rescuers could not believe that her heart was still beating. (She is now recovering from a long coma and, as of this writing, has no memory of being raped.)

New Yorkers were appalled, despite Patricia's being just one rape case of 1989's 3,400 in their city. Mayor Ed Koch asked the world of human behavior experts to "name one societal reason that would cause people to engage in a wolf-pack operation, looking for victims." Sociologists told Koch that Patricia was raped and left for dead because of broken families, crumbling communities, poverty, easy access to drugs, the commercialization of sex and violence in our culture, racial tension, peer pressure, and lack of equal opportunity.

But the six black youths who raped Patricia were only fourteen to sixteen years old, and none of these stereotypical reasons fit them. No drugs were involved. Neither was poverty or alienation: most of the rapists came from working families with a community identity. Only one boy had been in trouble with the police before. One was a born-again Christian. Another was a teenage social paragon. The six included Little League baseball players and private school students. Racial tension also didn't seem to apply, as on that same night the group also had terrorized a black man.

So why did these boys rape? Consider their attitudes. The six not only felt no remorse, but they even were smug. They explained that they had been on a "wilding" spree and had beaten and raped Patricia out of boredom. "It was something to do," one boy said. "It was fun," shrugged another.

Were they smug due to the slim odds of their being severely punished? Or was their smugness due to a deeper attitude—likely incomprehensible to many women—that rape is reasonable under certain conditions? A late 1980s survey of boys at a Rhode Island junior high school may help answer this question. Half of the boys interviewed said that forced sex was OK if a man had spent at least $15 on a woman.

The rapists in Patricia's case displayed a similar attitude. These boys revealed that their decision to rape was motivated neither by hate nor by control, but instead by powerful sexual urges and by an attitude that violence is reasonable in a milieu where rape's consequences include, at worst, merely trivial punishment to the rapist.

What appears to be a crime of hate or control is actually a male strategy to steal a copulation no matter what the cost to the woman. That rape is not unique to *Homo sapiens* supports the argument that men rape women for sex and not because they hate or wish to dominate women.

Natural-Born Rapists: Orangutans

FOR THE FIRST DECADE of field research on orangutans, primatologists focused on the apes' natural history, dutifully recording every detail of their solitary lives in the rain forest. But right from the start, orangs, who share 96.3 percent of identical DNA with humans, were doing things that were politically incorrect, to put it mildly.

Half an orangutan's diet, fieldwork showed, is fruit. Evolutionarily, the sole reason an orangutan eats anything is to create more orangutans. But the conversion of fruit, bark, leaves, and blossoms into more orangutans is not easy. This is because an orangutan's feet are a second pair of curved hands. This adaptation to foraging for fruit in tall rain forest trees makes walking on the ground so slow and clumsy that on an average day, an orangutan sticks to the tree tops and rarely travels half a mile.

Thus the social and mating problems orangs face are bioenergetic: any companion would dictate that the pair would have to travel farther than a solitary ape would to visit enough trees to feed them. This is so expensive in time and calories that companions are simply not worth the company. Mothers and grown daughters often forage in the same neck of the woods, but not together. Females and males are like trains running on different tracks. Only once in a while do they stop at the same time at the same station (a large, fruiting tree). Every orangutan spends its life mostly in solitude.

Solitary male orangutans in Sumatra and Borneo use three strategies to mate, all of which hinge on violence (described below). Mating also hinges on disparities in parental investment (as we saw in Chapter 1). All wild female apes face a daunting investment in each infant. Because the average birth interval in wild orangutans is an amazingly long six to seven years, a female can produce only three to five young in her lifetime. Most females produce fewer. By contrast, males are free to impregnate a new female every day. The limiting factor is that "extra" females are difficult to come by. This predicament leads to those three violent mating strategies.

The first strategy is based on territory. Territorial, resident males—seemingly the most successful in mating—patrol a region overlapping the home ranges of two or more adult females. These males scream long calls that carry nearly a mile to warn off male interlopers and to attract females. Resident males also challenge interlopers to no-holds-barred combat, especially if a female is near.

The second male strategy is to stay on the move. Because females are scarce and also rarely sexually receptive, some wandering males may meet more receptive females than territorial males do. But to mate with any of these females, a wanderer must beat a resident male in very risky combat—or else be a fast suitor.

Adolescent male orangutans are also wanderers, but they use a third strategy to mate. While too small even to consider trying to vanquish a 180-pound adult male, subadult males are bigger and stronger than an 80-pound adult female. Propertyless subadult males attach themselves to unwilling females, sometimes for days, and rape them even more often than adult males copulate with willing mates. One-third to one-half of the hundreds of copulations seen by researchers among wild orangutans have been rapes. Typically, these unlucky females cannot shake an adolescent male suitor/rapist. The male lurks within a few meters for days on end, and as the mood strikes him, he roughly forces the unwilling female to copulate. Anthropologist Biruté Galdikas even reports that an ex-captive male named Gundul, a nearly 100-pound subadult, raped her very unwilling cook.

> I attacked Gundul with all my strength, trying to jam my fist down his throat. I shouted to the visitor to take the

> dugout back to Camp Leakey for help. My repeated blows had no effect on Gundul; but neither did he fight back very aggressively. I began to realize that Gundul did not intend to harm the cook, but had something else in mind. The cook stopped struggling. "It's all right," she murmured. She lay back in my arms, with Gundul on top of her. Gundul was very calm and deliberate. He raped the cook. As he moved rhythmically back and forth, his eyes rolled upward to the heavens. I was in shock. I felt as though this were happening to other people somewhere else, and I was watching from a distance. I have no idea how much time passed.

In short, orangutans reveal to us a natural context in which male aggression and violence occur, and they especially reveal rape as a *primary* reproductive strategy for males too young to have established themselves in such a way as to be attractive to females.

Moreover, orangutans are not unique. Rape is rampant in the wild. Male scorpion flies, mallard ducks, fish, snow geese, mountain bluebirds, and, as we will see, even chimps and captive gorillas rape females. A big clue as to why they rape is that such males, who are incapable of "hate" or gratuitous violence, rape only females of their own species (Gundul had been captured as an infant and raised by humans and very likely had imprinted on women as appropriate sexual partners), females who are of fertile age, and females who refuse to mate with them.

Why Men Rape

THE KEY TO UNDERSTANDING why men rape, explains psychologist Craig Palmer, is the rapists' motives. Unfortunately, whereas the *behavior* of a rapist is visible, his motive is locked in his mind and must be surmised. Therefore, to answer the question of why men rape, we must first rephrase it to make it answerable in the real world. Instead, we must ask, Is rape a means or an end? In other words, which is more important to the man who rapes, dominating women or copulating with them?

The facts of rape, Palmer says, show that the driving motive is copulation. At least two-thirds of rapes involve full sexual function by the rapist. In addition, whereas sexually functional rapists have a repeat rate of 35 percent, castrated rapists have a repeat rate of only 1 to 2 percent. Palmer critiques every argument used to bolster the myth that rape is motivated by power and control, not by sex, and he solidly proves them all false in logic and/or fact. Instead, he concludes, the facts tell us that nearly all rapists rape for sex.

But because a few rapists do injure women physically, the power myth has staying power. Palmer explains that to understand rape, we must discern between *instrumental* force used to control the victim so as to perform the rape (and perhaps to deter the victim from resisting or reporting it) and *excessive* violence that seems an end in itself.

One clue as to which is more important is that men who rape women rarely intend to injure them seriously. One study found that whereas only 11 percent of 100 women who had been raped were injured badly, 53 percent of heterosexual men raped by bisexual men received serious injuries. And nationwide, the U.S. Department of Justice bulletin *The Crime of Rape* reports, although rapists did injure 58 percent of 479,000 rape victims in some way, "an extremely small proportion of the victims suffered gunshot or weapon wounds or broken bones . . . too few cases in the survey sample for statistical analysis." The medical expenses of these 479,000 victims averaged only $115. In another study, 88 percent of rapists used only instrumental force to rape. And in yet another study, 61.7 percent of rapists said that they had not intended to use violence at all. This is not meant to minimize the victim's emotional trauma, which may be severe, or her injuries. Rather it simply shows that rapists normally use only instrumental force and that copulation, not control, is their goal. Control is only a tool of rape.

Particularly revealing is a typical multiple rapist's response when asked what the difference was for him between sex with a willing woman and sex with an unwilling one. "There was no difference at all," he said, and he went on to explain:

> All I wanted was a place to get rid of my thing. . . . Some [women victims] had to be coerced, but I didn't enjoy doing it. It wasn't a turn-on. I wanted things as easy as I could get them. And if they didn't give in, I would threaten, and if I had to go through a big hassle, or exert any kind of violence, well, that was nothing for me then, but I didn't like it.

Even more telling is the fact that although most rapists rely on force to rape, their willingness to use violence is so limited that women who fight back often win. "Most individuals using self-protection were victims of attempted rape [only 27 percent were raped]," notes *The Crime of Rape,* "while most not using self-protection were victims of completed rapes [56 percent were raped]." This huge survey of rape victims reveals that a *woman who protects herself, whether by arguing, screaming, running away, and/or fighting back, doubles her chances of escaping rape, but she raises her odds of being physically harmed by only 10 percent.*

Clearly, rapists themselves want to avoid being injured. Lawyer Don Kates noted, for example, that when Orlando, Florida, police trained 2,500 women with handguns after a nine-month period when thirty-three rapes were committed, only three rapes occurred in the next nine months, a decline of 88 percent.

Beyond rapists' desire to avoid self-injury lies the larger insight into what they actually seek. In their study *Stopping Rape: Successful Survival Strategies,* Pauline Bart and Patricia O'Brien examined this question from the victims' side by comparing victims of rape with victims of attempted rape. One big difference emerged. Women faced with a rapist who succeeded in raping them were primarily concerned with *not being killed or injured.* In contrast, the main concern of women who were confronted with a rapist but avoided being raped was *not being raped.* These latter women screamed, fought, and used every defensive tactic that occurred to them. Meanwhile, women who focused most on not being killed or hurt used substantially fewer and weaker tactics—they remained "nice girls" caught in an ugly situation—to try to dissuade the rapist, who went ahead and raped them anyway.

That so many women can, through unbridled resistance, prevent a more powerful and often armed male from raping them suggests that most such assaults constitute a quest for sex, not an act of hatred or domination, which could be enacted simply by beating a woman senseless or by wounding or murdering her. Again, the facts show that *rapists normally use only as much weaponry or force as is needed to coerce or control women into sex, and such rapists seek sex well above inflicting any sort of injury.*

Even in cases where rapists *do* use excessive violence, the very nature of the assaults is always sexual rather than violent. A skeptic might argue that a rapist uses violence to control his victim. Granted, but a more immediate and self-serving reason for a rapist's use of violence is to terrorize his victim so that she will not bear witness against him. Even the schizophrenic and psychotic "Hillside Strangler," Ken Bianchi, admitted that he murdered his seventeen rape victims to silence them. Hence not even murder proves that a rapist hated or wanted to dominate or control his victim. Rather, it tells us that the rapist wanted sex and that he did not mind getting it sociopathically.

The final argument used to bolster the "rape as control" myth is the fact that some men rape older women or little girls instead of fertile women. Two points must be remembered here. First, rapists choose their victims based not only on desirability but also on vulnerability. Although nearly all rapists prefer attractive young women (again, about 90 percent of known American rape victims are between twelve and thirty-five years old), in spite of most such women being best able to resist rape, older women (less than 5 percent of reported victims are over fifty) and little girls are far more vulnerable to rapists and can be raped more safely and easily. The second point is that men who rape very young girls or very old women do not fit the usual profile for rapists. Indeed, rape psychologist Nancy Thornhill and biologist Randy Thornhill insist that such rapists are crazy: psychotic or psychopathic. For this reason, rapists who attack little girls or older women also are far more horrifying.

Examples of archetypal full-spectrum rapists include the "Hillside Strangler" and the "Night Stalker" (Richard Ramirez, a twenty-five-year-old drifter who terrorized Los Angeles in 1985), who murder their victims, give millions of women nightmares and hold them prisoner behind their own locked doors. The "Night Stalker" invaded homes after dark, shot male victims in the head with a .22 pistol, thumb-cuffed women victims, and then raped them beside their dead husbands. He also stabbed, slashed, and mutilated his male victims, nearly decapitating some. After demanding loot from the survivors, he drew pentagrams on the walls—and on the victims. Ramirez later bragged of murdering twenty victims, some in northern California. "More," he claimed proudly, "than the Hillside Strangler." Such men are clearly insane. Moreover, they are a tiny, very nonrepresentative fraction of the rapist population. As with men who rape children or old women, those who choose anal or oral penetration or other forms of rape with no reproductive potential are mentally ill.

A big clue as to what the average rapist's true motives are lies in his deceit—as opposed to the "Night Stalker's" boastfulness or the Central Park teenage rapists' smugness. The typical rapist rarely brags about or confesses to raping on his own turf. Instead, he usually denies it. Up to 80 percent of convicted rapists in prison have been reported to deny their crimes, whereas only about 25 percent of other violent criminals deny theirs. Rapists even deceive themselves. Most men in college who admit to date rape claim in the next breath that they are not rapists.

Rapists are canny in their deceit. Donald Symons watched interviews of convicted habitual rapists under treatment at Atascadero State Hospital in California. At first, these men claimed that they had raped for power and control. They complained that they themselves were victims of a society that glorifies machismo but denies them desirable women. (These sorts of responses allowed them to enter the ranks of the "rehabilitated" slated to be released.) But as each interview progressed, the rapists focused more and more on their sexual desire for the victims, until it became clear to Symons (but unfortunately not to California state social psychologists) that it was their desire for sex alone that drove them to rape. "Control" per se had never been their goal. This clinical error in diagnosis can yield tragic results. "Given the extremely high rate of recidivism for men released from the Atascadero State Hospital," Symons warns, "fashionable efforts to minimize sexual motives for rape—if taken seriously by Hospital authorities—might actually promote rape."

Unfortunately, Symons's advice has gone largely unheeded by sociologists. In their 1990 book *Violence: Patterns, Causes, Public Policy,* Neil Weiner, Margaret Zahn, and Rita Sagi write, "Rape must be understood as nonsexually motivated so that effective responses can be organized." The "effective responses" to "nonsexual" rapes that these experts recommend to legal systems is rehabilitation of rapists, and the responses they recommend to women targets is trying to talk the rapist out of rape rather than physically resisting him.

Because rape is such a hideous crime, it is a hard one for many of us to understand, even when the facts are clear. Many of the ideas, ideals, and paradigms we hold about human nature become stumbling blocks to understanding rape when the very idea of sexually motivated rape clashes with those ideas. The problem is that many of us still prefer our own ideas over the facts. Some feminists, for example, still argue that men rape to dominate and control women because society trains men to be superior to women and to dominate them. If this were true, we would see three trends: First, men would rape older, more powerful women more often. (They do not.) Second, rapists would come in all ages and from all walks of life. (Again, this is not the case.) Third, when socialization changes, rape should change. (It does not.) For example, some feminists' ultimate solution to rape is to create sexual equality in earnings, education, employment, and prestige so as to minimize male domination of politics and economics and to equalize power. But in twenty-six large American cities that have made progress toward sexual equality in their police departments, researchers found the highest, not the lowest, rates of rape.

This much is clear: the only factor common to all rapes is sexual assault. Indeed, in *Exploring Human Sexuality,* psychologists Kathryn Kelley and Donn Byrne define rape as "a violent sexual crime in which threat, force, and intimidation are used to coerce an unwilling victim to engage in sex-related acts." Moreover, to blame "aggression" instead of sexual impulses for rape, notes psychologist Herant Katchadourian in his *Fundamentals of Human Sexuality,* presents three problems.

> First, to claim that coitus can be a nonsexual act undermines the concept of sex as objectively definable. Second, it has us pretend that everything about sex is wonderful, when in reality sex can be awful, and rape is a good example of it. Third, it does not allow a context for experiences like date rape or acquaintance rape, which may or may not be violent, but certainly center around the issue of sexuality.

Even more enlightening is how women victims themselves define rape. Pauline Bart and K. L. Scheppe interviewed victims to determine their opinions. What they found should come as no surprise: "Although all the women studied were victims of acts legally defined as rape, those who were subjected to non-phallic sexual acts were likely to have labelled themselves as having *escaped* rape." In short, no penis, no rape. According to this study, power and control are irrelevant to women victims in *defining* rape.

Revealingly, even such penetration itself may not constitute rape in the opinion of the rape victim *and* that of her husband. Remember the rape of Biruté Galdikas's cook by the young male orangutan Gundul? Immediately afterward, the woman expressed relief that it was over and that she was unscathed (other than just having been inseminated by an orangutan). She reassured Biruté that she was "all right." Her husband later told Galdikas, "It was just an ape. Why should my wife or I be concerned? It wasn't a man."

So, it seems, rape is defined by most people simply as *an act of sex in which a man forcibly enters a woman with his penis.* Clearly, then, most men who rape women rape for sex. *But why do they do so?*

Imagine that we are members of an extraterrestrial team of biologists whose mission it is to understand and explain human violent behavior. We are free of preconceptions. We do know that natural selection operates in the same way on every planet with life in the universe, and we have no doubt that human behavior emerged from the same processes of natural selection that produced human legs and arms and the human brain. Further, because rape is so widespread and rampant around the planet, by males both human and nonhuman, it is clearly a male biological adaptation. But how on Earth could rape be an adaptation?

First, rape is universal. Men rape in all societies—whether in New York's concrete jungle or in real rain forests in Irian Jaya. Among Yanomamo Indians who raid one another in the Venezuelan rain forest, rape is simply one part of their overall reproductive strategy. Yanomamo villages, averaging fewer than one hundred residents, are structured much like chimpanzee communities. Males born in a village remain together during adulthood, while females expect to spend their adulthood in a different social group. Also like chimps, Yanomamo males share newly captured females, at least at first. "[A] captured woman is raped by all the men in the raiding party and, later, by the men in the village who wish to do so but did not participate in the raid," explains anthropologist Napoleon Chagnon. "Afterward she is then given to one of the men as a wife."

Why does such brutal sexual behavior occur? Randy and Nancy Thornhill propose that rape evolved as a "condition-dependent" sexual strategy of males in a social system in which men win wives by controlling resources. The "condition" that leads to rape is the failure by a male to win the resources and status needed to attract a female. Two facts support this hypothesis. First, men almost exclusively rape women who are the most fertile and desirable as wives. Second, most rapists are socioeconomic losers, or at least not yet winners, with an inferior ability to attract desirable women through honest courtship.

What rapists themselves say about why they rape illustrates this. Donald Symons notes that interviews with rapists routinely show that they are frustrated "because women incite ungratifiable sexual desire . . . simply by existing." Further, interviews also reveal that "the rapist wanted to copulate with the most physically attractive women; his victims were attractive not just to him but to most men . . . [and] the rapist's belief that he could not have sexual intercourse with his victims in any other way probably was correct."

One source of rape is that women are significantly pickier when it comes to choosing a sexual partner than are men. Men, explain the Thornhills, are both more indiscriminate and more easily sexually aroused than women. Women tend to select a mate more carefully, because the choice of a poor mate carries a far higher price in reproductive success for women than it does for men. The inevitable result is a world in which most women are not interested in some men who are sexually interested in them.

Men employ three courtship strategies to combat women's reluctance: honesty, deception, and coercion. Rape boils down to where each man draws the line between a woman being reluctant and her being unwilling. For many men, no line exists.

"Most men use both noncoercive and coercive tactics in combinations," the Thornhills note, "and these tactics grade into one another to the point where there is no distinct boundary between them." All of this behavior by men is so deeply rooted in our past that "direct selection for rape by men" must have occurred, "allowing penile erection, copulation, and ejaculation with a woman not sexually consenting or enthusiastic, or even willing. . . . And what has not evolved in men is sexual arousal only by women who are sexually receptive."

Just how interested men are in unwilling women is scary. The Thornhills reviewed experiments, such as those by Heilbrun and Seif mentioned earlier, in which young men watched depictions of consensual sex and of rape. All of the men were sexually aroused by the depictions of consensual sex, and many, though not all, also were aroused by the depictions of rape. *But all of the men not aroused by rape became aroused by it after either drinking alcohol or believing they had drunk alcohol, listening to a woman narrate the rape depiction instead of a man, and/or being told that sexual arousal during the depiction of rape is normal.*

Do these findings reveal that men are depraved, that male primates are rape-oriented, or both? Physical anthropologist Barbara Smuts (Chapter 6) notes that male chimpanzees commonly coerce females into mating. This coercion goes well beyond threats (the second most common tactic men use to rape women) to include attacks (the most common tactic men use). Smuts concludes that mating among apes and other primates is commonly conducted in an arena of sexual coercion laced with violence. Even when a female chimp in estrus goes "on safari" with a dominant male apparently voluntarily, she may well be responding to previous attacks on her by this male. In short, men did not invent rape. Instead, they very likely inherited rape behavior from our ape ancestral lineage. Rape is a *standard* male reproductive strategy and likely has been one for millions of years. Male humans, chimpanzees, and orangutans *routinely* rape females. Wild gorillas violently abduct females to mate with them. Captive gorillas also rape females.

So it is no surprise that, in the natural history of women, counterstrategies to escape or cope with rape have evolved. Intriguingly, married victims employ one negative counterstrategy: they are the least likely of all victims to accuse a rapist. Their silence is so common that biologists Richard Alexander and Katherine Noonan suggest that, by "concealing" rapists from being detected and killed by other men, such women have enhanced the evolution of rape as a male strategy.

Why are married women so reticent to report rape? The Thornhills found married women to be the most traumatized of all rape victims. They suggest that this is because rape causes married women to suffer a greater loss. Of course, all rape victims suffer, via shock, fear, humiliation, guilt, a sense of helplessness, and in many cases sexual dysfunction. Some victims also are robbed of their options to choose the timing and father of their children and to use their willingness to mate with just one man to secure his full investment in their children. But a married rape victim also may find that her husband now questions the paternity of her children to the point where she loses his commitment. So, although rape is the ultimate robbery for all women, it can be a total disaster for married women, hence their greater reluctance to admit it ever happened.

This tremendous cost of rape for married women, the Thornhills note, leads to four consequences for both sexes. First, married women are more suspicious of male strangers than are single women. Second, married women resist rape more aggressively than other women, thus requiring rapists to use more force—and often to abandon the attempt. Significantly, of all rape victims, married women are the most likely to be raped at gunpoint. Third, married women raped by force are *less* traumatized psychologically than those raped by more subtle and nonviolent coercion. The Thornhills suggest that these latter women are far more worried that their husbands will doubt that they were raped and instead suspect infidelity. Fourth, men abhor and punish rapists of their wives or female kin with the greatest retribution they can muster. The most common reaction by men to the rape of their wives is an overwhelming rage and desire for retribution. Often no less than the killing and/or castration of the rapist can satisfy them.

Whether we are extraterrestrial biologists or just people on Earth who want a clear answer to the question of why men rape, the Thornhills' hypothesis of rape as a male condition-dependent strategy for sex holds up. Men do rape women when sex with them is otherwise impossible, and these rapes do function very clearly in many cases to increase the reproductive success of the rapists. After the nine-month mass-rape of Bengali women by West Pakistanis in 1971, for example, Bengali health officials estimated that the Pakistanis had fathered at least twenty-five thousand children.

Grim and accurate as the Thornhills' explanation is, it may in fact be too optimistic. This is because not only economically "unacceptable" men rape. Even some successful, married men do so. But this too is easily explained. Such men are normally limited from copulating with *other* women, whether by cultural laws or customs, or by their wives' jealousy, or by the extra women's refusal—or by all of these situations. But whether they are professionals with all the trappings of success or sleazy petty thieves with none, men rape only when they expect to escape punishment for it.

Rape proves yet again that the psyches of men and women differ fundamentally. But it also reveals far more. Men, the Thornhills conclude, have a "psychological adaptation to rape," to steal copulations from unwilling women and thus increase their odds of siring offspring. This becomes even more obvious if one turns the question around: does is make sense to postulate that men rape in every culture on Earth—and do so frequently—because they *lack* a natural predisposition to rape? Instead, the facts on rape tell us that rape is one more natural product of macho sexual selection, an extra "tool" or adaptation in many men to help them "win" natural selection's reproductive contest.

• • •

The driver never brought "Leather Coat" a rope.

After an eternity, "Leather Coat" commanded Kay to put on her pants. When she groped for her underpants, he looked into her eyes, picked up her underpants, and put them in his pocket.

She could not believe it was over. Would he now shoot her? She was maimed and torn: her eyes, ears, face, anus, vagina. Her hands, swollen from torn cartilage, were difficult to move. But her will to survive still surged.

Somehow she pulled on her jeans.

In a gesture as bizarre and surreal as everything else that night, the driver stepped out and opened the door for her.

Kay ran across the road to flee the lights of the car and escape into the darkness. She felt as if her insides were spilling onto the ground.

"I know that he expected me to die out there," Kay said later. "He was leaving me to die."

She squinted her swelling eyes to peer into the valley below. The lights of Quito twinkled in the distance. The car's taillights became tiny, then vanished. Kay followed, bleeding so heavily that she knew she would die unless she found help.

After two miles, she reached some apartments. It was now 2:00 A.M. She rang the bell. The janitor was appalled at her bloody face. She begged him for help, but he and his wife turned away, offering only directions to her hotel nine miles away.

Again she walked. She found a nursery school. Two drunk men sat in the doorway. She avoided them as best she could.

Still bleeding heavily, she approached a house. A man and his wife answered. They drove her to her hotel immediately.

Along the way, the man told her, "You must remember that rape is something that happens to your body, not to your mind."

His statement gave her strength—and sanity.

Two other Peace Corps volunteers at her hotel called the hospital. Her time there was a blur. Though small, Kay was a strong distance runner. Now, despite her struggles, she could not even sit up. She knew she was dying. She could not stop focusing on being filthy with her own blood and feces.

The police questioned her. She begged to be put to sleep, but they forced her to tell her story in minute detail.

Kay spent five hours in surgery as Ecuadoran doctors trained in America reconstructed her vagina and anus, repaired her perineum, and rebuilt her sphincter muscle. They equipped her with a colostomy bag to be used for the next three months. One surgeon suspected that "Leather Coat" had used the gun inside her.

The police posted two armed guards outside her room and stationed a twenty-four-hour nurse inside for three days. At one point, two strange men managed to bypass the guards and enter her room, but the nurse scared them off. The police decided to fly Kay to the United States as soon as possible to prevent "Leather Coat" from killing her so that she wouldn't be able to testify against him.

Kay arrived in Washington, D.C., the next day and was whisked off to a hospital. Three days passed before she could open her gouged eyes. Three weeks passed before she could walk. Her torn fingers did not heal for a year. She spent weeks in physical therapy and was eventually cleared by an ophthalmologist. Kay had survived.

The FBI, the American ambassador, and the Ecuadoran police joined in the hunt for "Leather Coat." Kay's rape matched a crime three years earlier perpetrated against a male victim. The rapist, a bisexual hairdresser, had been caught, convicted, and sent to prison.

The police checked the prison. The rapist was no longer there. He had bribed his way out.

Determined to put "Leather Coat" behind bars again, a policewoman walked from the rape site back to Quito, retracing Kay's steps in a replay of her struggle to survive. She even risked going undercover as a hairdressing client of "Leather Coat." The game finally ended when the police found Kay's underpants mixed in with some other women's clothes in "Leather Coat's" house.

They arrested him.

Two months after the vicious assault on Kay, the police called her in the United States. "Can you be back in Ecuador tomorrow?" they asked.

Kay flew to Ecuador. The police escorted her to the American embassy in a limousine protected by bullet-proof windows. She sat between two guards armed with Uzis. She felt panic well up inside her.

At the police station, behind one-way glass, she identified "Leather Coat" in a lineup of fifteen suspects. Under Ecuadoran law, she had to identify him again, this time face to face in front of the judge. Even though she was surrounded by men armed with guns ready to kill "Leather Coat" at his slightest move, Kay was terrified. She could not stop trembling. She became hysterical as she faced him a few feet away.

He looked her in the eye and said, "I'm not the one who did it."

Two days later, "Leather Coat" agreed to lead the police to the driver. But after he stepped out of the car in what police suspected was a wild goose chase, they simply shot "Leather Coat" five times, breaking both legs. This was to "soften" him up. Then they asked him, "Who was the driver?" He would not talk. The police put him in the hospital for a month without painkillers. Still he was silent.

"I don't think he should be tortured," Kay told me pointedly. "I think he should be dead. For him to live is unjust."

The court found "Leather Coat" guilty of the most serious crime in Ecuador. The maximum sentence is fourteen years in prison.

The driver remains free.

Kay, a dedicated survivor, decided that the Peace Corps does not prepare volunteers for what they might face overseas. She joined with two other Peace Corps rape victims to produce a video to teach volunteers how to avoid sexual assault.

Kay is now fully recovered, works with the National Park Service, is happily married, and has a young, healthy child.

CHAPTER FIVE

Murder

It was one of those quiet nights when you don't expect anything," Sergeant Ray Martinez explained softly. Midnight had come and gone without incident. Nights like this were beautiful but deceptive. Stars winked in the frosty air above an empty Interstate 40. Ray steered his squad car off I-40 at the north end of Flagstaff, Arizona, to circle a minimart—a "stop-and-rob" as he thought of them.

It was an hour past midnight on Sunday. Even so, the store was hopping. Four vehicles were parked in front.

Ray thought about the elderly clerk inside. These people were sitting ducks. Ray was still haunted by an event two years earlier, when he had seen a 1977 Nova parked beside a minimart. Because the car was parked in a suspicious way, Ray had run a check on it. Before the check had cleared, however, Ray had received a radio call requesting him to respond at an injury-accident. It was a false alarm, but before Ray could return to the minimart, two teenagers from the Nova had robbed it and fired shots at the clerk. Luckily, the clerk had survived the wounds. Both robbers had been caught because Ray had identified the Nova by its plates. But they had still shot the clerk, a thing he might have prevented.

Ray now spotted a 1978 Buick parked on the unlighted side of this minimart. After circling the store once, Ray ran a check on the Buick's California plates.

Alone and without backup, Ray decided to check the car, then the store. A glance revealed a teenage male in the driver's seat.

Ray walked past the Buick to peer into the minimart. The elderly female clerk was behind the counter as usual. The foot traffic inside looked normal. Satisfied for the moment, Ray returned to the Buick.

The teenager had shifted to the passenger's seat. Ray radioed dispatch. The Buick was clear, but the situation still felt wrong.

Deliberately keeping his right hand free (Ray was a nineteen-year police veteran who taught officer safety), he knocked on the window. He instructed the teenager to keep his hands in front of him and slide across to the driver's seat. Ray noticed a middle-aged woman asleep in the backseat.

The teenager slid across the seat. He dislodged a .22-caliber semi-automatic concealed between the seats. Ray also saw a survival knife on the seat and a .22 Magnum revolver on the floor.

"This kid," Ray told me, wryly, "was not your average shopper."

Ray thumbed his radio to tell dispatch, "I think I have more than an FI [field interview]." He then told the teenager to step out of the car.

The boy was nervous and fidgety as Ray felt his belt for more weapons. "Where's your partner?" Ray asked.

"Inside," the teenager said.

Ray handcuffed the teenager to himself. He knew he had no time to question the woman in the backseat. He had to check the store again.

A private security guard passed by and noticed Ray's police car parked beside the store. He parked in front of the store, just in case.

Inside the store, the elderly clerk was befuddled. One of her customers, Donald T. Hawley, age twenty-six, had been piling unusual items on her counter for twenty minutes: toys, gum, magazines, junk food—dozens of things.

Ray hurried to his car to request backup. He hoped support would arrive before he was forced to act.

Frustrated because the store was still too crowded to rob, Hawley looked out the window and saw the security guard's car. He exited the store and walked around the corner toward the Buick. Hawley saw Ray's backup unit as it entered the parking lot behind Ray. Then, against the glare of the approaching cruiser's headlights, Hawley saw Ray coming toward him.

"He was startled to see me," Ray told me. "He froze. But he had the kind of look on his face that told me he was not going to do anything I was telling him to do."

"Put your hands over your head," Ray commanded, radio in hand.

Hawley deliberately pulled his left glove off and stepped forward.

"Stop!"

Hawley now rolled his right glove off. Then, still stepping forward, he unzipped his black leather jacket.

Ray saw the gun butt protruding from Hawley's belt as the man reached to draw. Ray's backup saw it, too, and drew his own gun. But he was afraid to shoot because of bystanders in the store behind Ray and Hawley.

Realizing that any further words were now useless, Ray yanked his Glock 9 mm from its holster and simultaneously stepped to the left to present a moving target. With barely a dozen feet between them, Ray was facing a point-blank gunfight.

Ray watched the muzzle of Hawley's gun flash as the man fired at him.

• • •

Murderers and Their Victims

HAWLEY WAS WORKING HARD to become a murderer. Unfortunately, he was far from unique. Worldwide, tens of thousands of men murder each year. In America, a man murders someone every half hour. We ought to hate it, but strangely we don't.

For some reason, buried deep in the human psyche, murder—especially creative, bizarre murder—fascinates us like a cobra charms a bird. Murder mysteries sell well in bookstores, at the box office, and on TV. The drama *Murder, She Wrote,* for example, was as lighthearted as a situation comedy and was profitably popular for years beyond the wildest hopes of its producers. But most producers know that real murder is better than fiction. They trample each other to purchase the production rights to bizarre real-life homicides because these stories are guaranteed to rivet viewers to their sets. By one count, each evening sixty-two victims are murdered as entertainment on prime-time TV, and an increasing number of these murders are real-life reenactments. Details of the 1994 murders of football and media star O. J. Simpson's ex-wife Nicole and her friend Ronald Goldman were televised daily for years. They received more airtime than all coverage of the Vietnam War over the past thirty years!

Why do flashy murders kick environmental destruction and global political mayhem off the front page? Is writer Robin Winks right when he asserts, "Mystery and detective fiction [are so successful because they] deal with society's deepest fears"? Maybe.

More in our fears, but furthest beyond our comprehension, are the motives—if not the very existence—of mass and serial killers. Oddly, these two types of murderers have nothing in common. Typically, mass killers are white men in their prime (and all too frequently now, white juveniles in schools) who lose control in one murderous rampage that ends in their own death from a police officer's bullet or suicide. Charles

Whitman is a classic example. A straight-A college student, psychiatric patient, and wife beater, Whitman in 1966 killed his wife and mother, then climbed a 307-foot tower at the University of Texas at Austin and shot forty-four people, fourteen fatally. His autopsy revealed a brain tumor. Other famous mass killers include Patrick Purdy, Joseph T. Wesbacker (a Purdy copycat), and America's worst, George Hennard, who killed twenty-two people in a Luby's cafeteria. The world's worst solo killer is an Australian, Martin Bryant, who killed thirty-five victims in Tasmania. Mass killers normally share two traits: first, most of them see psychiatrists or otherwise send up red flags before they commit wanton slaughter; second, their modus operandi (MO) is to commit suicide in a nihilistic, "I'll take everyone with me" way.

In contrast, serial killers, though also usually white males, are normally sexually motivated, but in the sickest way. Their typical MO is a gruesome rape and strangling of a series of vulnerable female victims, whether prostitutes, hitchhikers, elderly women, or children. Homosexual serial killers rape and strangle men or boys. Some of America's worst serial killers include Donald Leroy Evans, Jeffrey Dahmer, Richard Ramirez (the "Night Stalker"), and Ken Bianchi (the "Hillside Strangler"). Ed Gein, an insane resident of Waushara County, Wisconsin, is America's archetypal serial murderer. Gein was the model for Robert Bloch's book *Psycho,* which Alfred Hitchcock turned into one of Hollywood's scariest movies, and also for the movie *The Silence of the Lambs.* I guess I should not be surprised that this latter tale of sexual murder, mutilation, and cannibalism won an Oscar, but I am.

As evil and sick as mass murderers and serial murderers are, they commit only a fraction of 1 percent of all American murders. They are *not* "normal" killers. And because the goal of this chapter is to identify what drives "normal" murderers to kill, we'll move on to them now.

Whether or not "mystery and detective fiction deal with society's deepest fears," the 54 real murders per day in America (19,645 in 1996, down 20 percent from 1991's record 24,700 murders) are not entertainment. They are a tragedy that forces us to ask, Is murder a psychopathic epidemic spawned in the crowds of "civilization," or is it instead simply human nature? If we are natural-born murderers, *why* are we? And why are most murderers men?

Despite agreement by people worldwide that the taking of an innocent life is the ultimate crime, no nation or tribe exists murder-free. Understanding the dynamics of murder first demands a look at its basic statistics.

Murderers and Victims: The Numbers

TABLE 1 SHOWS how rates of murder differ between societies and also cycle within them. The world's highest murder rates are in primitive tribes and in Colombia, El Salvador, and Mexico. But certain cities in the United States outstrip even these. FBI records reveal that in 1997 an estimated 6.7 homicides per 100,000 people occurred in the United States, down from 7.4 per 100,000 in 1996. These are record lows for the past twenty years, down from a record high of 10.2 per 100,000 in 1980. Even with this nearly 30 percent drop in U.S. homicides in the 1990s, the United States still consistently produces a world record tally of murders.

In 1996, murder rates were six times higher in U.S. cities with populations of 250,000 and up than in towns with populations of 10,000 and under. And although the United States is currently on a down cycle in murder, only a few years ago, in 1991, America saw its highest rate of violent crimes ever—758 per 100,000 people. This equates to 1 victim per 132 people, a rate double that of 1970 but nearly identical to those of 1992 and 1993. Again, violent crime rates in cities were many times worse than those in rural areas.

All of this violence remains unfathomable, however, until we identify the victims, the killers, and their motives. First, the victims. Young black males face the greatest risk of murder. Thirteen- to twenty-four-year-old black males are eight times more likely to be murdered than white females over fifty-five—and eighty times more likely to *be* murderers. Murderers may kill anyone of any age or sex, but significant patterns among victims exist. And even though not everyone in America is at equal risk of being murdered, the risk that any person faces is scary.

Overall, roughly 1 in every 15,000 Americans is murdered each year. If these odds seem remote, bear in mind that they compute to 1 in every 200 Americans being murdered during an average seventy-five-year life span.

The specific risks of being murdered vary wildly by race and sex. Most victims are males, who account for a fairly stable 75 to 80 percent of all U.S. murder victims over time. On average, as of the 1980s, 1 white American male in 133 was murdered in his lifetime. Far worse, 1 of every 21 black men faced death by murder (see Table 1). Black women were murdered one-fifth as often as black men (1 in 104), and white women were murdered one-third as often as white men (1 in 369).

Today murder is still the fifth leading cause of death among all blacks, the second leading cause of death among all Americans between fifteen and thirty-four years old, and the leading cause of death among young black men.

What are the relationships between victims and killers? The most careful study of murder in North America is *Homicide* by psychologist/biologist Martin Daly and psychologist Margo Wilson. They focused on data from Canada, Detroit, Miami, and Philadelphia. Overall, three-quarters of the 1,552 victims were known to the killer; only one-quarter of the victims were strangers. However, only 1.8 to 6.5 percent of the victims were blood relatives. The FBI's *Uniform Crime Reports* show much the same trend in the United States for 1996. Of 10,350 victims for whom the relationship to the murderer was known, 78 percent were known to, married to, or (a very small fraction) related to the murderer. Nearly half of victims were only "vague" acquaintances. Strangers made up 22 percent of the victims. Hence, contrary to what is commonly assumed, blood kin accounted for far less than 10 percent of victims (unfortunately for our full understanding, FBI records do not discriminate between stepchildren and natural children).

None of this tells us why killers choose these victims. All it does say is that murderers normally kill nonrelatives and most often kill males who are mere acquaintances, sometimes strangers. We will probe more deeply into the victims of murder in the following sections of this chapter, after a look at their killers.

The FBI reports that a stable nine out of ten known murderers are males. Young men from ethnic, racial, or religious minorities top the list. The FBI *Uniform Crime Reports* note that a relatively stable 52 to 56 percent of murderers are black, a representation five times higher than chance would dictate. This bias for crime is so severe that, "on a typical day in 1994," writes Glenn Loury, "nearly one-third of black men aged 20–29 were either incarcerated, on parole or on bail awaiting trial." By 1996, the average murderer was under 25 years old (7.5 years younger than in 1965) and already a hardened criminal.

Sixty-one percent of killers average 4.3 prior arrests. More revealing, 70 percent of young convicted murderers on parole were arrested again for new offenses within six years. Indeed, eighteen-year-olds who murder usually resume a career of crime for at least ten more years. "What criminal career research has shown," notes *Science* writer Constance Holden, "is that a small proportion of offenders are responsible for the

TABLE 1 Homicide Rates Worldwide

PLACE	HOMICIDE RATE PER 100,000 PEOPLE	YEAR(S)	NOTES
United States[a]	9.5*	1919	
United States[b]	9.5*	1931–1934	
United States[c]	4.5*	1957–1958	
United States[c]	10.2*	1980	Record rate, 23,040 murders
United States[c]	9.8*	1991	Record number, 24,700 murders
United States[c]	9.3*	1992	23,760 murders
United States[c]	9.5*	1993	24,526 murders
United States[c]	9.0*	1994	23,305 murders
United States[c]	8.2*	1995	21,606 murders
United States[c]	7.4*	1996	19,645 murders
United States[c]	6.7*	1997	Estimated 17,877 murders
Louisiana[c]	17.5*	1996	Highest U.S. state
South Dakota[c]	1.2*	1996	Lowest U.S. state
U.S. cities with population of 250,000–500,000[c]	18.9*	1996	
U.S. cities with population of 10,000 or less[c]	3.2*	1996	
Black males in New York[d]	247.0	1990–1991	Ages 15–24 years old
Juvenile U.S. gang members[c]	463.0	1995	Ages under 18 and of all races
Bodie, California[d]	116.0	1878–1982	Highest for boomtown
Miami, Florida[e]	110.1	1926	Highest ever for U.S. city
Miami, Florida[c]	29.0*	1995	
Washington, D.C.[c]	65.2*	1995	Highest big U.S. city
St. Louis, Missouri[c]	54.9*	1995	Second highest big U.S. city
Compton, California[c]	83.3*	1994	Highest modern U.S. city
Los Angeles, California[c]	24.5*	1995	
New York City[c]	16.1*	1995	
Navajo Reservation[l] (population of 165,000)	25.3	1990–1998	375 murders

TABLE 1 **Homicide Rates Worldwide** (continued)

PLACE	HOMICIDE RATE PER 100,000 PEOPLE	YEAR(S)	NOTES
Hopiland (enclave of 10,000 within Navajoland)[l]	4 murders 1993–1998; 3 committed by Navajos; fourth suspect was Hopi		
Flagstaff, Arizona[l] (population of 73,000)	3.1	1990–1998	21 murders in city near Navajoland
Canada[e]	2.69	1974–1983	
Vancouver, Canada[f]	7.0	1990	
Colombia[f]	49.0	1988	Highest for any nation
El Salvador[f]	40.0	1990	
England and Wales[a]	0.8	1919	Handguns legal
England and Wales[g]	1.35	1984	Handguns banned
Gebusi tribe, New Guinea[h]	568.0	1940–1982	Highest rate anywhere
Ireland[b]	0.6	1984	
Israel[b]	2.3	1984	
Japan[i]	1.20	1988	Likely inaccurately low†
!Kung San Bushmen, Botswana[j]	29.3	1920–1955	
Mexico[f]	20.0	1990	
New South Wales, Australia[e]	1.69	1968–1981	
Spain[b]	4.7	1984	
Switzerland[k]	1.23	1990	

* U.S. "murder" rates may be overestimates because many FBI homicide counts include in with murders justifiable homicides committed in self-defense.

† Japan's true rate is likely higher because when a relative murders a family member and then commits suicide, police record the homicide as a "family suicide."

SOURCES

a: Kleck, G. 1992. *Point Blank: Guns and Violence in America*. New York: Aldine de Gruyter, p. 394; **b:** Reiss, A. J., Jr., and J. A. Roth (eds.). 1993. *Understanding and Preventing Violence*. Washington, D.C.: National Academy Press, pp. 50, 52; **c:** Federal Bureau of Investigation (FBI). 1997. *Crime in the United States, 1996*. Washington, D.C.: U.S. Department of Justice, pp. 13, 62. See also Maguire, K., and A. L. Pastore (eds.). 1997. *Sourcebook of Criminal Justice Statistics, 1996*. Washington, D.C.: U.S. Department of Justice, pp. 306, 327–328. Witkin, G. 1998. The crime bust. *U.S. News & World Report* 124(20):28–34; **d:** Courtwright: D. T. 1996. *Violent Land: Single Men and Social Disorder from the Frontier to the Inner City*. Cambridge: Harvard University Press, pp. 81, 226; **e:** Daly, M., and M. Wilson. 1988. *Homicide*. New York: Aldine de Gruyter, pp. 29, 125, 285; **f:** LaPierre, W. 1994. *Guns, Crime, and Freedom*. Washington, D.C.: Regnery Press, p. 172; **g:** Staff. 1987. Still unsafe on the streets. *Economist*, March 21:56; **h:** Knauft, B. M. 1985. *Good Company and Violence: Sorcery and Social Control in a Lowland New Guinea Society*. Berkeley: University of California Press; **i:** Kopel, D. B. 1992. *The Samurai, the Mountie, and the Cowboy: Should America Adopt the Gun Controls of Other Democracies?* Buffalo: Prometheus, pp. 22, 44; **j:** Lee, R. B. 1984. *The Dobe !Kung*. New York: Holt, Rinehart and Winston, pp. 93–96; **k:** Halbrook, S. P. 1993. Swiss Schuetzenfest. *American Rifleman* 141(5):46–47, 75–76; **l:** Donovan, B. 1999. Major murder drop on the Rez. *Arizona Daily Sun*, Friday, January 22, p. 1. Also: Flagstaff Police Department and Hopi Tribal Police record books.

majority of crimes." An even smaller proportion are responsible for murder: about 70 percent of all violent crime in America is committed by a mere 6 percent of violent criminals.

As with the bulk of known rapists, most murderers are career "bad guys." But what makes them bad?

Why Do Murderers Kill?

SOCIOLOGISTS EXPLAIN MURDER in three ways. Their *cultural-subcultural* explanation says that murder is learned through shared values and behavior (such as on television). Their *structural* explanation says that murder results from racism, poverty, lack of opportunity, and crowding. And their *interactional* explanation says that people murder due to the cultural mores of their interactions during conflict. Although each of these ideas is interesting and touches on a factor implicated in homicide rates, none has been tested scientifically, and none addresses the ultimate roots of homicide. Each of these ideas is shallow, or *proximate;* it refers only to a local mechanism that "triggers" a behavior—in this case, murder. None of these sociological ideas even approaches the *ultimate* deep function of the behavior that the proximate trigger serves. Each of the explanations is like one of the proverbial three blind men who feels only one part of an elephant to tell what it is. One man insists that it is a tree, the second is sure that it is a wall, and the third cries out that it is a snake. Even when combined, these explanations can neither agree with one another nor explain why murder happens.

The big picture of murder particularly eludes those who blame men's violence on *socialization* that encourages boys to be violent and girls to be gentle. This view is based on the notion of the seventeenth-century philosopher John Locke that human nature is a blank slate to be written on by culture. This notion was bolstered by psychologist B. F. Skinner's claim that any human behavior can be conditioned, and it was given even more credibility by Margaret Mead's theory of "cultural relativism," which claims that cultures evolve free of human biology. Hence many non-biologists still believe it.

But as we have seen, human beings are not blank slates. Human nature comes packaged with behavioral contents included. These behaviors serve *ultimate functions for the individual's survival and reproduction.*

The problem with nearly all the sociological paradigms mentioned above is that they are based to a large degree on a phenomenon that does not exist—that human blank slate. This shortcoming is akin to a school of cosmologists who believe that Earth is stationary and non-rotating and exists at the center of the universe. Their basic assumption makes it impossible to explain how nature actually works—except by a lucky accident. As Melvin Konner notes, the biggest problem in studies of violence is that social scientists are ignorant of biology, fear its political misuse, or even suffer a professional "inferiority complex" to biology, which, unlike sociology, is a hard science built on the scientific method. Ultimate causes for human behaviors remain for such sociologists the undiscovered reality that Earth rotates.

In short, sociology identifies symptoms and proximate causes of murder, but it fails to link them to the basic human motivations that offer insights into the ultimate reasons for the murderous design of the human psyche. The key to understanding murder lies beyond these symptoms, in what they may reveal about human biological motivations. Before we proceed, let's round up all the usual, proximate suspects for the crime of murder and see what they confess.

GUNS

Many social scientists say that murder happens for a *structural* reason: easy access to easy-to-use weapons. Many people also blame firearms for emotional reasons. Many fear and detest firearms, and because some of them have been victims of gun crimes or were close to such victims, their emotions are understandable.

But weapons, it turns out, have less to do with murder than do the attitudes of people, and their system of justice, in accepting or rejecting murder (see Chapter 8). The National Academy of Sciences concluded, "Available research does not demonstrate that greater gun availability is linked to greater numbers of violent events or injuries." Rates of murder depend not on numbers of guns, but on *who* possesses them. To reduce murder, the National Academy's Panel on the Understanding and Control of Violent Behavior recommended that "existing laws governing the purchase, ownership, and use of firearms" be enforced.

More data separating guns from murder rates come from Robert J. Mundt's study of homicide rates in twenty-five U.S. cities versus twenty-five similar-size Canadian cities. It revealed that among *non-Hispanic Caucasians,* murder rates were the same, despite the availability of handguns in the United States versus their longtime ban in Canada.

A classic demonstration that ready availability of guns does not, in itself, raise murder rates is a comparison of Switzerland, Japan, and England. Every able-bodied Swiss man is required to keep at home, for life, a fully automatic rifle or a pistol plus ammunition. Yet among 6 million people privately owning 600,000 assault rifles, half a million pistols, and thousands of other guns, murders are extremely rare. Even gun suicides are low. Japan, with no guns, and Switzerland, which is heavily armed, have identical murder rates, 1.20 and 1.23 homicides per 100,000, respectively (less than half the Swiss murders were shootings). England's homicide rate, also with most guns banned, was 1.35 per 100,000. In short, both in America and internationally, the presence of guns does not correlate with the murder rates.

By contrast, a murderer's MO does open a partial window to his or her motive. In cultures whose state-of-the-art weapons of war or hunting are poisoned arrows, clubs, or spears, these are the weapons that murderers choose. Likewise, in the United States by 1996, 54 percent of murderers used handguns, 14 percent used other firearms, and another 14 percent used cutting or stabbing instruments. The rest used "nonweapons": blunt objects, hands, feet, poisons, fire, and so on (usually against infants or children). In the 1990s, one-third of U.S. murders were committed without guns. Indeed, mainstream instruments of murder are sometimes not at all what one would expect. In 1988 in Chicago, for instance, more people were murdered with baseball bats than with guns. And in England, where guns are illegal, 35 percent of murders in 1994 were committed with knives and only 9 percent with guns. The big question here is, Was it the mind of the murderer or the weapon itself that led to murder? The data so far imply that the mind is far more lethal.

The best existing study of hard-core murderers and their weapons is *Armed and Considered Dangerous: A Survey of Felons and Their Firearms* by James D. Wright and Peter H. Rossi. This study was funded to forge new gun control laws. The authors, both sociologists, had neither owned nor fired a gun for years. Indeed, Wright and Rossi admit initially favoring stricter controls on guns. What felons told them, however, tempered this opinion.

Wright and Rossi found that, of all murderers (many of whom kill repeatedly), 61 percent had killed with a handgun and half had murdered with a rifle or shotgun. These gun felons owned twice as many guns illegally (6.6 each) as the average American family does legally. And felons carried guns not because they had grown up with them, but because their peers carried them. To commit crimes, 85 percent of gun "Predators" preferred a handgun. More important, 18 percent of those armed with a gun and 16 percent of those armed otherwise admitted using their weapons not to scare their victims or for self-protection, but to *kill* their victims.

Murderers avoided the cheap, small-caliber, snub-nosed Saturday night specials so frequently cited by the media in the 1980s as causing so many murders. Gun "Predators" themselves considered such cheap pistols useless. Only 15 percent had ever carried one, and the only felons who "preferred" them were men who had *never* owned a gun at all. Instead, Wright and Rossi's "Predators" carried good-quality handguns that they said were accurate, untraceable, well made, easy to shoot, and concealable. In short, they wanted the best tools for killing. In fact, not only did many "Predators" steal these guns from police officers, 8 percent of them said robbing a police officer was the *best* way to rearm after release from prison. Seventy-one percent of the felons carried caliber .38 or .357 or bigger. Wright and Rossi conclude, "Serious criminals prefer serious equipment."

The lesson? *Most murderers use the best weapons they can get, and a large number of them use these weapons with the intent to kill.* Interestingly, most murderers also fear these same weapons in the hands of potential victims.

When I started work on this book, I held the opinion that laws restricting handgun ownership were vital to curbing murder in America. It only makes sense, doesn't it? Not when one knows how men who decide to murder think.

Economist John R. Lott, Jr., surveyed the data on guns and murder from several recent years. He focused on the thirty-one states that have nondiscretionary (also known as "shall-issue") concealed carry weapons (CCW) laws. These states issue to any nonfelon who passes their safety and legal tests a license to carry a concealed handgun. Hundreds of thousands of Americans now legally carry concealed weapons on this basis. Lott examined the records of fifty-four thousand such licenses dating from 1977 to 1994 and analyzed dozens of variables relating to violent crime. He intended his research to answer the question, "Does allowing people to own or carry guns deter violent crime? Or does it simply cause more citizens to harm each other?" The title of his book, *More Guns, Less Crime,* may seem to provide the answer, but that would oversimplify the issue.

Lott found that, contrary to popular notions, even after more than a decade, no CCW permit holder had been convicted of having used her or his gun to murder anyone. Instead, many permit-holding women escaped being murdered (or raped) because of the use of their guns. For example, women who did not resist violent aggressors were injured 2.5 times more than women who used guns to resist them. Further, resistance with a gun led to women being seriously injured only one-quarter as often as did resistance without a gun. Polls reveal that Americans defend themselves with guns between 760,000 and 3.6 million times *yearly!* These figures coincide with a much broader study by Gary Kleck, a professor of criminology who also spent several years researching the effects of guns on enhancing versus preventing violence.

In his book *Point Blank: Guns and Violence in America,* Kleck reports that private citizens in America use guns 783,000 times (handguns 645,000 times) yearly to protect themselves from felonious assaults. That breaks down to once every 48 seconds. Meanwhile, criminals use guns against victims 660,000 times yearly. One-third of randomly polled Americans considered citizens armed with guns to be the best defense against criminals. Roughly half of all gun owners said that their firearms were primarily for protection. Indeed, the FBI reports that between 1992 and 1996, private citizens shot and killed 1,382 violent offenders, a total close to (68 percent of) the 2,035 felons shot by police, to protect themselves. What do the police think of this? Lott cites two major polls showing that more than 93 percent of responding police officers consider private ownership of firearms necessary for the average citizen to protect himself or herself.

Surprisingly, there exists a huge difference in risk to bystanders depending on whether a police officer or a private citizen discharges his or her firearm in self-defense against a felonious assault. Carol Ruth Silver and Donald B. Kates, Jr., found that police shooting at suspects were 5.5 times *more* likely than private citizens to shoot an innocent bystander. By contrast, only about 28 mistaken intruders are shot per year. Many of these shootings result when a gun owner keeps a firearm next to the bed and fires before waking up fully.

Lott explains what the ability to protect oneself means in regard to murder:

> Violent crimes are 81 percent higher in states without nondiscretionary laws. For murder, states that ban the concealed carrying of guns have murder rates 127 percent

> higher than states with the most liberal concealed carry laws.
>
> Overall, my conclusion is that criminals as a group tend to behave rationally—when crime becomes more difficult, less crime is committed. . . .
>
> Guns also appear to be the great equalizer among the sexes. Murder rates decline when either more women or more men carry concealed handguns, but the effect is especially pronounced for women. One additional woman carrying a concealed handgun reduces the murder rate for women by about 3-4 times more than one additional man carrying a concealed handgun reduces the murder rate for men.

Lott reports that studies showing that guns kept in the home lead to more homicides than would otherwise occur were flawed (or fudged). Instead, Lott concludes, a 1 percent increase in gun ownership correlates with a 4.1 percent *drop* in violent crime. He notes that "the passage of nondiscretionary concealed handgun laws in states that did not have them in 1992 would have reduced murder in that year by 1,839; rapes by 3,727; aggravated assaults by 10,900; robberies by 61,064. . . . The total value of this reduction in crime in 1992 dollars would have been $7.6 billion." (Lott also notes that this decline would have been at the cost of perhaps nine additional accidental deaths in all concealed handgun states.) Ultimately, Lott was able to answer—and to successfully defend his answer scientifically against critics—the question his research originally asked. "Will allowing law-abiding citizens to carry concealed handguns save lives? The answer is yes, it will."

This tells us several things about guns and the *structural* reasons why men murder. First, most U.S. murderers prefer handguns. They do so because they intend to murder. Second, many potential murderers also fear guns that might be used against them by a potential victim enough to decide not to murder. Murderers, then, are often rational enough or cold-blooded enough (that is, normally *not* motivated unreasonably by passion) to decide not to attempt a murder when the odds are raised of their being hurt or killed.

Unfortunately, although this is useful information for anyone who needs to protect himself or herself, understanding the tools of murder brings us only slightly closer to understanding why people harbor an intent to kill to begin with.

Low-Performance Individuals (Socioeconomically and Otherwise)

The most thoroughly researched recent book on crime and criminals is James Q. Wilson and Richard Herrnstein's *Crime and Human Nature.* They found that criminals in general are most often men who score low on intelligence tests, score high in impulsivity or extroversion on personality tests, and/or have slow and weak autonomic nervous system responses. Other studies confirm this. "Research consistently places the IQ of convicted lawbreakers at 92," notes science writer Bruce Bower, "some 8 points below the population average and ten points below the population average for law-abiding folks." Long-term research in New Zealand confirms the connection between crime and impulsivity.

Bower reports that boys on the lowest rungs of the neuropsychological scales not only are the most common delinquents, but they also have frequently acted in an aggressive and impulsive manner since age three. Even more telling, reports psychologist Terrie E. Moffitt, is the fact that "results also suggest that poor verbal ability is the 'active ingredient' for delinquency in the [overall] IQ."

Add to this a genetic component. Wilson and Herrnstein found that men who have criminal fathers, even if these men were adopted and could not have known their real fathers, are significantly more likely to be criminals than men whose fathers are not criminals. In addition, an identical twin of a criminal is 2.5 times more likely to be a criminal than is a fraternal twin.

But knowing who most murderers are, how they kill, and that genes play some role does not mean we know why they murder. Do people kill simply because they can get away with it? This is not as flippant as it may sound. FBI reports for 1996 show that only 67 percent of U.S. murders resulted in a suspect being arrested. The most recent Bureau of Justice figures reveal further that fewer than 38 percent of murderers are convicted of murder and that only 97 percent of convicted murderers are sentenced to prison (the other 3 percent are mostly infanticidal mothers; see below). *These figures suggest that about one-third of murderers are sent to prison for murder in America. Moreover, the average prison time they serve is only 10.5 years of their average 22-year sentence.*

But getting away with murder, or being punished only mildly, does not explain why men murder to begin with. Does American culture somehow "create" murderers—by poverty, for example? Poverty often has been blamed as another *structural* cause of murder, but research has shown only a weak correlation between students in poorer neighborhoods or schools and later arrests for violent crimes. The National Academy of Sciences' Panel on the Understanding and Control of Violent Behavior reported, however, that even though few homicide studies have focused specifically on violent crime,

> those [studies] that did continued to find that homicides were disproportionately concentrated in areas of poverty. . . . This pattern held up regardless of which ethnic group occupied [the community]. . . . But finding statistical relationships between community characteristics and violence still does not explain how they are related to poverty and, in turn, how they increase violence. And it does not explain why race differences in violent crime rates tend to disappear when poverty is included as an explanation.

Poverty affects relationships between men and women. Indeed, as we have seen already and will see again later in this chapter, a man's poverty, or his being an economic "loser," drastically affects his ability to attract and hold a woman partner long enough for him to *start* a family with her, let alone raise one. Hence poverty as an "economic" condition in a male is, from a woman's perspective, often a biological signal that the man is not a good bet for a mate. To hide, alleviate, or reverse their lack of financial resources and thus enter the mating arena on more equal footing with other men, many economically poor men rob others of their money or goods. Violence is a chief tool in this process, and murder is a frequent result. But "poverty" as an economic condition is only a superficial manifestation of a much deeper biological reality to a woman: the man will not be a good mate; he will not provide security; he will not be able to afford to raise healthy children; he will ruin my life.

As a *structural* cause for homicide, economic failure by a man seems to create a strong proximate trigger, a biologically predictable predisposition to his using offensive violence to gain more access to desirable women, to gain more exclusive mating "rights" with her, and to sire and even raise his children, which, after all, are the ultimate function of his mating. To some men, rape seems the easiest way to do this. Escalation to robbery and/or murder to gain resources or to eliminate rivals is a far more dangerous tactic, but either act might yield more lasting gains than rape.

Poverty after marriage also affects family social dynamics. Is it these dynamics—*interactional* processes—that sow later violence? Many sociologists think so. I thought so, too. But research suggests otherwise. "Surprisingly," notes sociologist Cathy Spatz Widom, "there is little empirical evidence that abuse leads to abuse." Widom found that only 20 percent of abused children became abusive parents (not significantly more than the rate for unabused children), although abused or neglected black boys were 3 percent more likely than nonneglected black boys to be arrested later for a violent crime. Interestingly, boys who had abusive fathers or who lived in crime-ridden neighborhoods but who refrained from juvenile delinquency often also had higher than average IQs and perhaps better talents to earn money honestly. Overall, Widom concludes, murder is not caused by the family. Although her research did not address murder in all its forms, it does suggest that looking beyond family social dynamics for a cause is vital.

Another intriguing *interactional* and *structural* explanation for homicide rates is offered by David T. Courtwright in *Violent Land: Single Men and Social Disorder from the Frontier to the Inner City.* Courtwright followed homicide rates through history and across America. He found population composition to be a good predictor of murder rate. The highest rates always existed where the local population had the largest proportion of young men and the lowest number of young, marriageable women. Add local competitors from other races (Native Americans, Chinese, Hispanics) to this situation and murder rates jump higher still. Phenomenally high rates, such as those in gold rush towns epitomized by Bodie, California (see Table 1), resembled those in "drug rush" inner-city ghettos. How could these murder rates be cut? By increasing the proportion of young men who are married and engaged in family life, Courtwright concludes. Courtwright's study of American murder is fascinating, but it does not dig any deeper to explain the underlying reasons why men without wives or families kill so readily in the above situations. Is the cause competition for those rare women or for big economic gains that will later help them win women? *Why* these young men kill is still what we most need to know.

Violence on Television

Could television violence be the culprit in raising modern homicide rates? TV programming is a prime *cultural-subcultural* dynamic in the sociological analysis of murder, and for good reason. According to epidemiologist Brandon S. Centerwall, an exhaustive seven-year study by the Centers for Disease Control (CDC) eliminated *every* other sociological cause proposed for homicide (including those above) *except* violence viewed on TV. In *Mayhem,* her exploration of TV violence and its effects, Sissela Bok reports that 98 percent of American households have a TV, most children have one in their bedrooms, and most families have a TV turned on for 7 hours per day. By the end of elementary school, she notes, the average child has watched 8,000 murders. By age eighteen, the average American has witnessed about 18,000 television murders and has watched TV for 15,000 to 22,000 hours (but spent only about 11,000 hours in school). In short, TV is a powerful and persistent companion to most children.

As early as 1972, a ten-year study concluded that "the amount of television violence viewed by boys at age 9 was the best single predictor of juvenile delinquency offenses related to aggression at age 19." A 1986 study, however, concluded that viewing TV predicted violence later in boys only if they had abusive fathers. Suspiciously, however, the homicide rate nearly doubled in Canada after TV was introduced—and it did so with no rise in the number of firearms. More interesting yet, Centerwall cites a poll of U.S. prisoners showing that more than 25 percent of them had committed crimes that were *exact duplicates of crimes they had watched on TV.* The same is almost certainly true of the several copycat juvenile mass murderers who killed their schoolmates in 1997 and 1998. Of the CDC's study mentioned above, Centerwall concludes, "It is estimated that exposure to television is etiologically related to approximately one-half of the homicides committed in the U.S., or approximately 10,000 homicides annually, and to a major portion—perhaps one-half—of rapes, assaults and other forms of interpersonal violence in the U.S."

According to Sissela Bok, one problem with Centerwall's claim is that homicide rates have dropped considerably since his study ended in 1989, yet violence on TV has not decreased. Still, she says, TV violence does contribute greatly to real violence. We just cannot tell *how* much.

How effective can violence on TV or in movies be in shaping human values? U.S. Army psychologist Dave Grossman describes a report on the U.S. government's most sophisticated techniques for conditioning assassins to overcome their reluctance to kill, designed by a naval commander and psychiatrist named Dr. Narut. The conditioning consisted of strapping trainees into a "must watch" TV rig. Then, as in the book/movie *A Clockwork Orange,* these trainees viewed ever more violent and gruesome killings. Unlike *A Clockwork Orange,* however, no revulsion-causing drugs were administered. On the contrary, the goal was to desensitize assassins to violence and to condition them to accept violent killing as matter-of-fact. Whether this information is apocryphal or not, Grossman points out that Americans allow essentially the same conditioning to occur with millions of children who watch episode after episode of lethal violence while munching on "rewarding" junk food or engaging in stimulating physical contact with a boyfriend or girlfriend. He notes,

> We are doing a better job of desensitizing and conditioning our citizens to kill than anything Commander Narut ever dreamed of. If we had a clear-cut objective of raising a generation of assassins and killers who are unrestrained by either authority or the nature of the victim, it is difficult to imagine how we could do a better job.

Do people in general believe that TV conditions people to violence? The watchers themselves think so. In a 1996 nationwide survey of U.S. adults, 92 percent of them agreed that violence in TV entertainment programming impacts real-life violence; 74 percent of these people said it has a "large impact." About half of U.S. teens with an opinion agreed that TV contributes to violence among people their age. Indeed, by the mid-1990s, young people under the age of eighteen were committing 250,000 violent crimes yearly.

The truth is we know beyond any doubt that TV does change human values. Corporate America, for example, is willing to spend a million dollars to air a one-minute commercial during the Super Bowl because TV does this so effectively. How much more effective in altering children's values can thousands of hours of gratuitous violence be—especially for male children in single-parent households lacking a real-life male role model? Without a doubt, we should be worried.

How worried? Sissela Bok reports a chilling conversation presented in a *Frontline* TV documentary.

> A young boy, standing next to his father and mother, is asked what he would say to someone offering him a million dollars on the condition that he never watch the television screen again. His instant reply, "I wouldn't do it." "Not even for a million dollars?" his mother asks incredulously.

> "Not even for a million?" "Why?" Again an instant retort, as if the answer were self-evident: "What [else] would you do?"

And yet, despite a recent *TV Guide* study that counted one hundred violent acts per hour on television programming, and despite an Annenberg School of Communication study counting one violent act every minute even on children's programming, the haunting question remains: why does television violence not incite equally murderous behavior later by women?

As with rape, the keys to unlocking this mystery are simpler than we have been led to believe, but they are deeper than those that sociological inquiry into *cultural-subcultural* versus *structural* versus *interactional* explanations can reveal. As we have seen, 90 percent of murderers are men. Very few are totally random people who lose control. But men do murder most often during arguments or during the commission of a confrontational, face-to-face crime.

Understanding murder requires a leap: we must admit that each murderer has made a conscious decision to kill and is personally responsible for that decision. Of course, personal responsibility alone fails to explain why a murderer makes his decision to kill. But the realization that the murderer himself—not society as a whole—is responsible for his decision is the most important first step toward understanding why men commit murder.

Natural-Born Killers: Gorillas

ONE OF THE MOST IMPRESSIVE—and revealing—products of natural selection is the mountain gorilla, a species whose DNA is 97.5 percent identical to that of humans. Indeed, the charisma of just two hundred mountain gorillas once fueled half the hard-currency economy of Rwanda. Not many tourists realized it, even after viewing gorillas, but male gorillas win the blue ribbon for "most macho primate." Paradoxically, however, their lives seem gentle and idyllic.

Because most gorilla food is common foliage, gorillas can afford to stay together as permanent groups that need to travel only half a mile a day for all of them to eat. Half of a mountain gorilla's day is spent lounging or napping in the sun (if the sun manages to emerge from behind the mountain clouds). The roly-poly juveniles use these long breaks to somersault down the verdant slopes into dog piles that erupt into frantic wrestling matches punctuated by breathless chuckles and goofy-looking play faces. Sharing the day-to-day life of a group of gorillas will convert you to primatology as quickly as a silverback can snap four-inch-thick bamboo.

But as gorilla researcher Dian Fossey and her colleagues found out in the Virungas, as relaxed as such gorillas may seem, they are not at all random aggregations of amicable apes. Instead, they are one-male harems assembled via bloody combat. An average harem holds eight members: four adult females, three immatures, and the adult silverback male who did the gathering. This silverback does not defend a territory, but he fiercely defends his harem against intruding males. In fact, 79 percent of encounters between adult males who are strangers to one another lead to violent threat displays; half lead to combat. Combat between these superaggressive males is so demanding, and often so lethal, that macho sexual selection has pumped them up to a size 237 percent heavier than the females' mere 170 pounds. Indeed, violent death is so common among wild gorillas that their primordial savagery seems to belie their humanlike intelligence and charm.

Interestingly, mating is impossible in any less violent way. When about fourteen years old, most males leave their fathers' harems to roam the forest and assess other harems for females to recruit. No silverback harem master willingly relinquishes females, so these young silverbacks must forcibly drag off many females. Some even kill a female's infant sired by a previous, rival male.

Why would any female bond with such a violent male? The first time a female leaves the harem of her birth, she does so to avoid inbreeding with her father or brother. But once a female leaves, she enters a stable dominance hierarchy based on the order in which she was recruited by her silverback. In these hierarchies, earlier wives dominate latecomers, even to rare cases of killing a latecomer's infant. Consequently, it seems, some low-ranking females "divorce" and "remarry" to raise their status and/or to escape their domineering co-wives.

On the other hand, a few females not only refuse to be recruited by a new male but even defend their silverback against him. Despite these rare cases, most females do not stay with the first male who recruited them. Primatologist David Watts reports one female, Simba, who married into four harems and bore three offspring by three different silverbacks. Simba abandoned one little daughter, Jennie, in Nunkie's Group. Nunkie, a hulking silverback (when I observed him in 1981) and Jennie's father also became her mother. And Jennie became Nunkie's shadow. Wherever he paused in the cold, wet vegetation of the Virungas, she leaned against him to catch his body heat. At night she snuggled with him in his nest.

Nunkie seemed a paragon of gorilla fatherhood. But this is only one side of the male gorilla psyche. A stunning one in seven gorilla infants is lost to infanticide by a male who is not its father. The high divorce rate of females despite this slaughter reveals that a low-ranking female will divorce to seek a higher-status "marriage" (or a superior husband?) even at the cost of one of her infants being killed.

Why do silverbacks brutally murder infants? Physical anthropologist Sarah Blaffer Hrdy carved her niche in primatology by illustrating this point with cold, hard statistics on monkey murder. During her research in India in the early 1970s, Hrdy was puzzled by vanishing infant langurs. (Langurs are leaf-eating monkeys that live in matrilineal groups and mate with the one immigrant adult male in each group.) Over time, Hrdy saw four new adult males combat and usurp four resident males. Each newcomer killed the youngest infants (killing six, three, one, and two infants, respectively), then mated with the mothers. Since the killers could not otherwise have mated with the group's females without waiting from one to three years, their infanticide made reproductive sense for the killers—despite the degree to which it hurt female success. This is because normal nursing by an infant would have stalled each mother in a hormonal state of lactational amenorrhea—an infertile stage of the female reproductive cycle. The infanticide sped up the mother's breeding cycle to instant fertility. When one considers that male tenure averaged only two years, it becomes clear that many new males who did not kill infants would not breed at all. And despite the mothers' heroic and painful defense against these males, they readily mated with the killers. In many cases, as among lionesses, the bereaved mothers rub so languorously against the infanticidal males that nature seems a nightmare.

Overall, new males who did not kill infants but instead waited sired fewer infants than males who did kill them. This is all sexual selection needs to install infanticide as a male reproductive strategy.

Even so, these killings by monkeys were so politically incorrect that many anthropologists simply denied them. (Ironically, even Darwin would have been skeptical. Despite the fact that male infanticide is shaped by sexual selection, Darwin considered infanticide too "perverted" for nature.) As other males in a dozen species of primates were seen to kill infants routinely during group takeovers, however, monkey murder entered the textbooks as vivid testimony to the power of macho male sexual selection. Many species of carnivores—tigers, lions, cougars, cheetahs—do the same thing. Indeed, within-species killing is rampant in nature. Significantly, males kill infants of other males, not their own.

It may seem strange that the female gorillas of a harem forge no lasting bonds with each other that might benefit them in group defense. Even after living together for years, they associate only because a male recruited them. If their male is killed, they normally disband and seem disoriented. In short, female gorilla psychology revolves around being "won" by a male.

Males, if they are to breed, have no choice but to use violence to win or defend mates—despite such violence seeming "bad" for the species itself. The key point missed by many popular natural history books and videos is that natural selection works only through the reproductive success of the *individual,* not the species. Nature's most meaningful maxim is "He or she who produces the most surviving offspring wins," and gorillas seem ruled by this. A male who fights hard enough to recruit and defend, say, four females may sire eight to twenty offspring, at least triple the lifetime reproductive success of an average female. Dian Fossey estimated that one silverback named Beethoven sired at least nineteen offspring.

Recruitment by combat is not the only road to reproductive success for gorillas, but combat is always a necessary reproductive tool. When Beethoven died, for instance, his son Icarus inherited Beethoven's harem (40 percent of harems in the Virungas contained more than one big male, although the silverback leader usually bred exclusively, while his grown son sat by). When Icarus died too, apparently from wounds sustained in combat, another of Beethoven's sons, Zizz, inherited his father's harem of four adult females. Primatologist David Watts describes Zizz as the biggest gorilla he ever saw. Only a few years earlier, Zizz was also the most aggressive blackback (immature) male I ever saw. He even attacked our tracker, Antoine, for example, by running him down, chomping on his arm, wresting his machete away, biting it, and then flinging it into the foliage. As an adult Zizz used his hyperaggressive talents to recruit seven more females, for a total harem size of eleven. Two of these females later shifted to other harems.

The lesson in this is that male gorillas who use combat, even at lethal risk, to build and/or defend a large harem may sire at least three times more offspring than they would by mating monogamously. Indeed, *hyperaggressive males like Beethoven and Zizz may sire even more infants than that—and thus found bloodlines that may last millions of years.* By contrast, losing in combat, or being unwilling to fight to the death, leads to half of all male gorillas failing to sire any offspring at all.

Hence we can imagine each young silverback's dilemma: should he stay in his natal group and wait, or strike out alone and risk combat to recruit a harem? An older silverback faces a similar dilemma: Should he try to recruit yet another female? If so, when, where, and how can he pull it off without losing the ones he already has?

In summary, gorilla society seems gentle and idyllic on the surface, but it actually consists of deceptively impermanent and radically sexist family units forged in blood by xenophobic males who kill to recruit or retain their harems. Unrelated males clash in lethal combat over females, and females themselves commonly divorce males. Mountain gorillas highlight most of the extremes of male violence—including infanticide, bluffing, combat, and homicide—as means to win and monopolize females as breeding resources. Gorillas also show us how macho male sexual selection can shape males into powerful and remorseless fighting machines.

I mention all of this here because, as we now look at who kills whom and when, we must first admit the plausibility that *human murder is no accident. Instead, murder is encoded into the human psyche. People who murder do so deliberately based on their own personal decisions favoring their own ultimate self-interests. They do not murder because they themselves are hapless victims of a society gone haywire.*

Killing Infants

WE WILL START our exploration of human murder with the darkest side of human nature: the killing of blood relatives.

The murder of an infant is perhaps the murder hardest for us to understand. But people, often women, do murder infants, and they do so far more often than most people would guess.

Anthropologists Paul Bugos and Lorraine McCarthy found that young Ayoreo women of Bolivia and Paraguay often "bury" their infants at birth. One Ayoreo woman buried her first six babies before raising her next four. Such infanticide occurs only under one of the following conditions: the mother lacks a husband (Ayoreo women cannot raise children without a husband's support); the infant is deformed; twins are born (not enough mother's milk for both of them); or the baby is born so soon on the heels of a sibling that it imperils the older infant. Mothers are sad when they must let an infant die—which is how they think of it—and in response they often shower the older, surviving infant with extra care.

As cruel as it seems, Ayoreo infanticide prevents lethal competition for limited milk between a newborn and an older, healthier infant. It also leaves intact the unmarried mother's chances of landing a husband. With age and marriage, Ayoreo women become more and more reluctant to bury an infant.

The Ayoreo are not unique. Napoleon Chagnon saw the same, seemingly cold equations among Yanomamo mothers, and Kim Hill and Hillard Caplan saw it among the Ache of Paraguay. Ache parents kill 6.9 percent of their infants due to illness, deformity, close birth spacing, or the infant's being the wrong sex. Another 2.8 percent of all children are killed because their fathers either died or divorced their mothers. In fact, 9.1 percent of 66 children under fifteen years of age whose fathers died are killed, while only 0.6 percent of children whose fathers are still alive are killed. The Ache explain that once the father can no longer hunt and help feed the band, his children will starve and at the same time cause yet other children to get less food.

Does this happen all over the world? Anthropologist Susan Scrimshaw surveyed infanticide worldwide and found that it was common everywhere. In Asia all societies killed infants, in Africa 58 percent did, in North America 65 percent, and in South America 69 percent. Of preindustrial societies, 49 percent were infanticidal, while a mere 12 percent forbade it. In Germany a century ago, for example, firstborn children whose fathers died also often died soon after. In contrast, a widower's children lived longer, usually dying only after the widower remarried (due to the "stepmother-Cinderella" syndrome, in which a new wife eliminates a "rival" child by a previous wife by cutting and redirecting its resources—food, or even oxygen).

But more often it is young women who most harshly murder their own infants when the father is not offering marriage and support. For example, whereas unwed women accounted for only 12 percent of Canadian births, these few mothers (average age 22.7 years) committed 61 percent (88 of 144) of Canada's mother-infant murders. Women in Canada murdered their own children 1.5 times more often than men did. They also showed far less remorse than men. Whereas suicide by infanticidal fathers in Canada was common (up to 43.6 percent), only 1 in 50 infanticidal mothers committed suicide.

The same thing is happening south of the border. Recently, two American teenagers, Amy Grossberg and Brian Peterson, checked into a motel, delivered their baby, killed him, and then discarded his body in a Dumpster. Months later, eighteen-year-old Melissa Drexler went to her high school prom, delivered her baby in a girls' restroom, left him dead in a garbage bin, and returned to the dance floor.

In short, mothers who kill their infants seem to do so not only in cold blood but also simply as a means to a better socioeconomic and reproductive end.

Infanticide somehow seems even more hideous when the victims are no longer newborns. A few years ago in South Carolina, Susan Smith, a young, white mother separated from her husband, claimed that she had been car-jacked by a black man. He had ordered her out of the car and driven off with her two young sons, ages fourteen months and three years, as hostages. This kidnapping launched a highly publicized, ten-day, nationwide manhunt for the car, the kidnapper, and the boys. Meanwhile, my wife insisted, "There's something fishy about her story. No woman who cared about her kids would leave them in the car. I'd have told him, 'No way. Either I stay or I take my kids with me.'" Sure enough, the manhunt ended abruptly when Smith confessed that there had been no car-jacker. Instead, Smith had strapped her two sons into their safety seats, positioned her car on a boat launch ramp, shoved the gear selector to "drive," and launched the car and her helpless boys into the lake and underwater to drown. Why? Smith explained that her new boyfriend did not want someone else's kids.

Shockingly perhaps, Smith is not a uniquely murderous fiend. Instead, she is, in many ways, all too typical. Smith's lethal reproductive logic typifies that of young mothers worldwide who find their long-term reproductive security, especially their marriage prospects, being threatened by an "inconvenient" child. Infanticide seems written into our genes, but it is generally manifested in young, unwed, desperate mothers who murder their neonates. Smith differs from them in murdering her offspring when they had grown into children: talking, walking, hopeful little people. Probably because of this, she was sentenced to life imprisonment. In contrast, forensic psychologist Barbara Kervin reports that of three hundred women in the United States and the United Kingdom charged with killing their newborns, none spent more than one night in jail.

Infanticide runs so deep in many cultures that an infant has no recognized identity until it survives the time of potential infanticide. Among the !Kung, a baby is not considered alive until it is named. This announces that its mother has accepted it as no threat to her older children or her marriage. The Amahuaca of Peru admit that they do not even recognize children as humans until age three.

Infanticide is not, however, merely a "primitive" syndrome. In the United States, for example, infanticide rates rose 46 percent from 1975 to 1992 (during the same period, family murders overall decreased by one-half).

Infanticide is also often sex biased. Eskimo fathers abandon one out of five newborn girls in the snow so that the mothers may sooner bear sons. Hunters are so vital and they die in accidents so much more often that Eskimos must raise more sons than daughters if everyone is to eat. Though awful, this heartless equation is understandable, perhaps unavoidable. Less understandable is infanticide in cultures that are not plagued by a higher male death rate but that kill infant girls only because the culture values sons more highly. In Burma, India, Bangladesh, Jordan, Pakistan, Sabah, Sarawak, Sri Lanka, and Thailand, girls younger than age four mysteriously die significantly more often than boys. "The anthropological evidence seems to indicate that female babies are more likely to be eliminated than male babies," reports Sheila Ryan Johansson. The outlook for female babies can be extremely bleak. In a study of 8,000 fetuses in six abortion clinics in Bombay, for example, 7,999 were female.

Social anthropologists are still trying to make sense of this. Marvin Harris calls infanticide "the most widely used method of population control during much of human history." This idea is supported by the existence of a few groups of people who did kill "extra" infants to curb growth. Among those people are the ancient Greeks, modern Chinese and Japanese, Eskimos, and Yokun and Sakai of Malaya. The Tapirapé of Brazil, for example, allow each family only three children and mandate that extra infants be abandoned in the jungle. These cultures are the exceptions, however. Instead, *most infanticide has been decided by people trying to produce more children in the long term by sacrificing an infant now.*

How can I make such an assertion? Because of the reasons given for infanticide by the sixty societies listed in the *Human Relations Area Files*. Half of all infants murdered in these societies were killed because of circumstances *adverse* to the infant's survival (for example, being born with a twin or too soon after the previous sibling or to a mother lacking a husband); 19 percent were killed because they were deformed or unhealthy; 18 percent were conceived in an adulterous mating; 3 percent were conceived in incest; 3.5 percent were killed because they were females; 2 percent were sacrificed in black magic ceremonies; and 4 percent were killed out of spite, to erase a successor to a hereditary position, or to prolong sexual activity. Only the last 9.5 percent would act to curb population growth (note that the rest were biological investments with a poor prognosis for reproductive success and/or who imperiled the survival of siblings), but *in no case was killing for population control cited as a reason.*

Lest you condemn or discount these practices as something only "heathens" do, bear in mind that Americans also kill infants, and in many cases they do so far more abusively than parents in "primitive" tribes. In the second national family violence survey, sociologist Richard Gelles looked at 6,002 U.S. households to pinpoint the dynamics of child abuse and homicide. He found that single mothers were 71 percent more violent toward their children than married ones, especially when poor. Single fathers, regardless of income, were 420 percent more severely violent than married fathers. In both groups, single parents living with a boyfriend or girlfriend were the most severe and most frequent abusers of their children. These trends not only parallel those among primitive tribes and many nonhuman primates, but they also reveal that when a single parent's marriage prospects are at stake, children are often shoved into a life-and-death arena. In America, where out-and-out infanticide is not only socially unacceptable but a capital crime, remarried adults often abuse infants to death instead of killing them outright.

Of course, some North American adults do kill infants. Human stepfathers often treat stepchildren worse than gorillas do. Canadian data show that an infant under two years old is seventy times more likely to be murdered by a stepparent than by a natural parent. In England, an infant is fifty times more likely to be killed by a stepfather than by a natural father. And in the United States, an infant is one hundred times more likely to be murdered by a stepparent than by a natural one. These heinous murders erupt from the darkest side of the human psyche.

Men kill stepchildren for the same reasons other primate males kill infants: to increase their reproductive success by sweeping the reproductive arena clean of the offspring of male competitors. Such murders free up limited resources for the man's natural children, and they also "push" the mother into a physiological and/or emotional state in which she is ready to breed again.

Hideous as such stepfathers are, they are small potatoes compared to how far the instinct to kill infants takes some men. The Nazis, for example, exterminated at least a million children in their death camps. "A Wehrmacht memorandum of June 1944," writes Mildred Dickemann, "states the purpose of a plan to remove 40,000–50,000 Russian children from controlled areas: 'The operation is planned not only to reduce direct growth of enemy strength, but also to impair its biological strength in the distant future.'" Meanwhile (as mentioned in Chapter 4), Nazi occupation soldiers were mass-raping Russian women.

Infanticide reveals how natural selection can be a brutal sculptor of instincts and also how male and female *Homo sapiens* (and other primate species) are prompted to kill infants by instincts designed to maximize their overall reproductive success.

OCCASIONALLY, THE REVERSE of infanticide occurs. The killers, though, are not infants; they are usually teenagers. Victims are usually abusive fathers or, as Daly and Wilson found, parents who are unwilling to stop breeding (and thereby slicing the family pie too thin). In Canada, for example, sons ages sixteen years or older were 455 percent more likely to kill their fathers than vice versa. And whereas forty-five sons murdered their mothers, only one mother killed her son. Daughters were 238 percent more likely to kill a parent than vice versa.

Bear in mind that most murdered parents had been abusive and that stepfathers were far more likely than fathers to be victims of "patricide." And when all is said and done, children fare far worse in the murder arena than parents.

If infants are murdered for the sake of the overall reproductive success of the adults nearest to them, why do adults kill other adults?

Killing Adults

DECADES AGO, in the 1950s, criminologist Marvin Wolfgang reviewed 560 killers in Philadelphia and identified a dozen motives among them. The most common was an "altercation of relatively trivial origin; insult, curse, jostling, etc." These motives accounted for 37 percent of all homicides. In America in the 1990s, personal conflicts, arguments, and insults are still the most common causes of homicides. They accounted for a whopping 53 percent of all known cases in 1995 and 55 percent in 1996. But these conflicts are light-years away from being "trivial" in view of their lethal consequences. Arguments are the number one reason men murder, and this holds true worldwide.

To dissect the process by which such confrontations escalate to murder, David Luckenbill analyzed homicides in a California county over ten years. In 41 percent of the killings, the victim had insulted the killer verbally (and usually at length), 25 percent had insulted the killer with gestures, and 34 percent had refused to comply with the killer's requests, thus challenging his authority. At least one of these situations had led to every murder.

Most of the killers, however, did not kill readily. Sixty percent of them first questioned the victim or bystanders to confirm that the victim truly had intended to insult or disobey him. Eighty-six percent of the killers challenged the victim verbally or physically, thus allowing him the chance to change his mind by apologizing or ceasing his insult or noncompliance. Fatally, 41 percent of these victims continued to insult or not comply.

Seventy percent of these murders were witnessed, and the witnesses played a role in the murders. In 57 percent of the murders, the onlookers either encouraged the killer or the victim to use violence, blocked outside interference intended to stop violence, and/or provided weapons. Incidentally, 36 percent of the killers possessed a gun or knife at the onset of the confrontation, but only 13 percent of the victims did. Unarmed killers left the scene to get a weapon or else grabbed something nearby (a telephone cord, a kitchen knife) to use as a weapon.

These complex dynamics of victim versus killer led Luckenbill to conclude:

> Criminal homicide does not appear as a one-sided event with an unwitting victim assuming a passive, non-contributory role. Rather, murder is the outcome of a dynamic interchange between an offender, victim, and, in many cases, bystanders. The offender and victim develop lines of action shaped in part by the actions of the other and focused toward saving or maintaining face and reputation and demonstrating character.

Can the desire to save face be important enough to commonly lead to murder? Anthropology offers a startling answer. Studies of "primitive" people from every continent reveal consistently that saving face in connection with the possession of women is a leading motive leading men to homicide.

Consider the Gebusi, who eke out a living in the New Guinea rain forest. They seem to be the gentle natives of anthropological myth, untainted by the dehumanizing influences of civilization. Indeed, their pleasant small talk, exaggerated humor, and generous sharing of food in communal longhouses make Westerners wonder whether we, in our fast-lane, solitary-career-track lives, are missing the point of being alive. But for the Gebusi, the whole point of being alive is to stay that way. Gebusi men murder other Gebusi at a rate nearly a hundred times greater (568 per 100,000 people per year between 1940 and 1982) than Americans do.

Anthropologist Bruce M. Knauft found that murder accounted for the deaths of 35 percent of all Gebusi men and 29 percent of women. The victims were not random: four out of five had been accused by their killers of having used sorcery to engineer other deaths. Incredibly, *most* Gebusi men had murdered at least one "sorcerer." But Knauft exposed the reality behind the "sorcery": most murdered sorcerers were members of families who owed the killer a long-standing debt of a woman in exchange for one they had accepted in marriage. It appears, therefore, that the Gebusi murder to save face against the insult of being cheated on a deal. But "sorcery homicide," Knauft judged, "is ultimately about male control of marriageable women."

This fatal currency of women as prime property was so intense that it even prompted men to kill their own women relatives married out to male in-laws in spiteful revenge against those in-laws for their having failed to provide a woman in exchange.

The !Kung Bushmen also tell a tale of murder, although it took a long time for anthropologists to hear it. In *The Harmless People,* Elizabeth Marshall Thomas's enchanting 1959 ethnography of these hardy hunter-gatherers of the Kalahari, she painted the !Kung as uniquely peaceful.

> It is not in their nature to fight. . . . Bushmen cannot afford to fight with one another and almost never do because their only real weapon is their arrow poison, for which there is no antidote. But even were they to disregard this danger,

> Bushmen would try not to fight because they have no mechanism in their culture for dealing with disagreements other than to remove the causes of their disagreements. . . . The !Kung call themselves *zhu twa si,* the harmless people.

Twenty years later, anthropologist Richard B. Lee's six-year survey showed the !Kung's homicide rate to be 29.3 homicides per 100,000 per year, higher than those in New York City or Los Angeles. Lee notes that *zhu/twasi* does not mean "harmless people"; it means "true or genuine people" or "just folks." More to the point, Lee learned, they are not harmless.

Lee documented twenty-two homicides (nineteen men, three women) between 1920 and 1955 among the !Kung. All twenty-five killers were men, almost all of whom used poisoned arrows. Nearly every killing took place in the context of a fight over a woman accused or caught in the act of adultery. Only seven killings were directly over the women themselves. The other fifteen were retaliations—face-saving—by the kin of a man who had been killed. When Lee asked the !Kung why they used poisoned arrows instead of a less deadly weapon, one man told him, "We shoot poisoned arrows because our hearts are hot and we really want to kill somebody with them."

Melvin Konner, who also lived with the !Kung, notes, "Among the !Kung, it [vendetta killing] is a mainstay of social control." Lee sees the !Kung as an example of a primitive society in which the individual comprises the entire legal system. One band, however, was so angered with a man who had killed three people that they resorted to cooperation. They voted unanimously to execute the killer by ambush. The committee shot him so full of poisoned arrows that "he looked like a porcupine." Even the women stabbed the killer with spears.

The !Kung and Gebusi are not unusual. During his twenty-five-year stint among the Yanomamo of Venezuela's southern rain forest, anthropologist Napoleon Chagnon witnessed many chest-pounding and side-slapping duels over face-saving. Though ferocious, these fights are rigged to show who is the "better man" without leaving the loser dead. In cases of food theft or adultery, however, the foes escalate their duels to club fights or, worse, to ax or spear fights. In a club fight, two men take turns pounding each other on the head with the heavy end of a club eight to ten feet long (say, a roof pole from a house). The first man to drop loses. (Club fights wreak nasty wounds and ugly scars worn as badges of prowess by the survivors.) Once blood flows, however, friends of the contestants may rip poles from houses and leap into the melee. Some sharpen poles at the small end to create spears.

In one club fight between an abused wife's lover and her cuckolded husband, the lover used the sharp end of his pole to stab the husband. The village headman, who had been commanding the people not to escalate the fight, became so enraged at this "cowardice" and failure to obey him that he grabbed a sharpened pole and speared the lover to death. The wife was then given back to her legitimate husband, "who punished her by cutting both of her ears off with his machete."

The headman then ordered the dead man's male kin to leave the village. They left and allied themselves with an enemy village to stage a reprisal raid. Thus spousal abuse led to adultery, which led to nonlethal club fighting, spear fighting and homicide, retaliatory homicide, and ultimately to village fission and full-scale war.

In Australia, 90 percent of lethal conflicts among Northern Territory's Tiwi people also center on face-saving and women. Anthropologists C. W. M. Hart and Arnold Pilling found that only elder Tiwi men can marry, and they marry every young woman. This monopoly leaves young men with no sexual options but adultery or elopement. Seduction of a young woman is a young man's offense against an older man, who, to save face, fights the offender in a public "duel."

The husband carries several spears into the dueling arena. The young man has three choices: he can carry a couple of spears (which is considered insolent), he can carry throwing sticks (less defiant), or he can go unarmed (which shows proper respect). Once inside the dueling circle, the elder harangues the villagers circling them with the history of the accused man. He explains his own kindnesses to him and points out the young man's failure to meet his responsibilities. Then he throws his spears at him.

The young man may dodge the spears, but he must stay in the strike zone. Dodging every spear, however, is bad form. It makes an elder look ridiculous. So, after several minutes, the accused youth normally allows himself to be hit in a leg or arm, thereby bleeding enough to pay his "debt to society." Two-thirds of Tiwi "duels" end this way. But in cases when a young man does not defer, several armed elders suddenly appear beside the accuser. If the accused still does not allow himself to be hit, the newcomers launch their spears simultaneously at him, usually killing him.

Inuit Eskimos of the Arctic Circle, who are chronically short of women due to female infanticide, risk death simply by having a woman. Anthropologist A. Balikci reports,

> A stranger in the camp, particularly if he was traveling with his wife, could become easy prey to the local people. He might be killed by any camp fellow in need of a woman. In ancient times such assassinations led to the formations of revenge parties consisting of the relatives of the victim, resembling war expeditions. The objective of the revenge party was not just to kill the original murderer but members of his kindred as well.

Every example above, as well as a plethora from anthropologist Carlton S. Coon's overview of hunter-gatherers—G/wi Bushmen of Botswana, Akoa Pygmies of Gabon, the Andamanese of the Andaman Islands, Yaghans of Tierra del Fuego, Tasmanians, and the Tlingits of the Pacific Northwest—reveal that men typically commit homicide over face-saving and over women. Moreover, from steaming jungles to frozen tundra, all of these men commit murder without having been exposed to television, drugs, broken families, crumbling communities, racial tension, high school peer pressure, poverty, or lack of equal opportunity. Yet most of their murders, ambushes, and duels match to a disturbing degree the typical murders in California described by Luckenbill. It is like watching "civilized" silverback gorillas. We like to flatter ourselves by thinking that American murder is more complicated than these killings among "primitives," but U.S. murders—and those worldwide—just as often boil down to the twin issues of saving face and which man walks away with the woman.

Why? Because men who win in this murderous game of face-saving do not simply win a single contest. By killing once, men gain or enhance a reputation of ferocity, which later helps them wrest resources from other men without conflict. This is one of the deepest aspects of male human nature, one shared with nearly all other male primates and mammals in general. Women, in contrast, rarely kill adults other than husbands who abuse them.

Certainly, there is a risk of oversimplification here. We do know that many people erupt into violence easily and early. Affronts to self-esteem trigger anger and aggression even in two-year-olds, and this instinct never fades. But far more lurks behind lethal anger than meets the eye. In societies without formal police, note Daly and Wilson, killing the right victim (usually one from outside your kin group) often enhances the killer's status with a fierce reputation.

But to what advantage? Resources everywhere are limited. Thus conflict over their acquisition is guaranteed. And because intimidation is the easiest, most energy-efficient way to win against rivals, a fierce reputation is a vital male trait. Men, incidentally, usually earn their fierce reputations during their early prime—the same age span that encompasses most U.S. rapists and murderers—and it stays with them.

The acid test of the macho male murder hypothesis emerging from Darwin's sexual selection theory would be to see whether the killing of one man by another during a confrontation—as one male gorilla kills another—truly increases his reproductive success. Surprisingly, this has been done.

Anthropologist Napoleon Chagnon documented the legacy of an exceptional Yanomamo headman named Matakuwa (Shinbone), who won his leadership—as do all Yanomamo headmen in the rain forests of southern Venezuela—partly through killing enemies to achieve *unokai* status. By Yanomamo standards, among whom fierceness *(waiteri)* is a prime virtue, Shinbone was legendary. His reputation gained him 11 wives, who bore 43 children, who bore 111 grandchildren, who bore 480 great-grandchildren.

Admittedly, Shinbone was an extreme case. But among the Yanomamo in general, Chagnon found that 44 percent of men over twenty-five years old gained *unokai* status by helping to kill an enemy or by killing a man in a village duel. A stunning 30 percent of all men died from homicide. Fatherhood among *unokais* rose with their fierceness and differed drastically from that of same-age non-*unokais*. On average, *unokais* (137 men, not including Shinbone) had 1.62 wives each and sired 4.91 children; non-*unokais* (243 men) had only 0.63 wives each and 1.59 children. In short, *those Yanomamo men fierce enough to kill other men sired three times more children than the less violent men.*

Not surprisingly, Chagnon's data were attacked bitterly by rival anthropologists who clung to the politically correct idea that homicide could be neither part of *Homo sapiens'* natural history nor something that sexual selection would favor. Chagnon's data, however, speak for themselves. In fact, they scream.

But lest one still be tempted to argue that only "primitive" people kill to win women, Richard Halliburton reported that 90 percent of French prisoners sent to Devil's Island for murder admitted that they had killed over a woman. No matter what one might think of the

French, the male drive to be macho is no accident of culture. Politically incorrect as it may be, macho men are usually top dogs, especially intelligent macho men. My years with back-of-beyond peoples in Uganda, Ethiopia, Kenya, Rwanda, Tanzania, Zaire, Australia, Peru, the Caicos, Palau, Papua New Guinea, Korea, Sumatra, and Turkey—many of whom are little touched by the outside world—have convinced me that people everywhere respect (if not admire) the fierceness of men. Beyond my own meager experience, however, most ethnographies confirm a universal tendency for men to respect other men based on their reputations.

This is especially clear among America's quarter million juvenile gang members. They murder at least a thousand teenagers a year (1,157 in 1995), usually in the context of saving face, building reputations, and protecting territory. This equates to 463 murders per 100,000 gang members per year, a rate rivaling that of the Gebusi.

The counterpart to this macho male respect is that many women are attracted to violent men. "Women are always attracted to power," observes author Gore Vidal. "I do not think there ever could be a conqueror so bloody that most women would not willingly lie with him in the hope of bearing a son who would be every bit as ferocious as the father."

Whatever you may think of Vidal's statement, a man's fierce reputation does cut both ways: it repels challenging males, and it attracts many women. This is what Karen Hill, a middle-class Jewish girl married to mafia wiseguy Henry Hill in Nicholas Pilaggi's *Wiseguy,* has to say about power:

> He [Henry] had something cupped in the palm of his hand. I took it and looked down. It was a gun. It was small, heavy, and gray. I couldn't believe it. It felt so cold. It was a thrill just to hold it. Everything was so wild I began to feel high. . . . In a few minutes Henry comes walking back. The police were waiting. They had spoken to Steve and other people across the street first. [Steve was a neighbor who had tried to rape Karen. Henry had just shoved that gun into Steve's mouth and waggled it like a dinner gong in a threat to kill him if he ever got near Karen again, and Steve had urinated all over himself.] It was the biggest thing anyone had ever seen on our block. I was really excited. I loved that Henry had done all this for me. It made me feel important.

Admittedly, some women, unlike Hill, disdain macho men. But global reports show just how serious this business of earning a fierce reputation via lethal aggression is. Daly and Wilson analyzed thirty-five studies (in the United States, Canada, Mexico, Brazil, Australia, England, Scotland, Iceland, India, Denmark, Germany, Botswana, Nigeria, Zaire, Kenya, and Uganda) to compare the frequencies of men murdering males (outside war) with women killing females. Men committed 95 percent of the same-sex homicides—a ratio of 19.7:1 over women. But even this is misleadingly low. In Mexico, Iceland, India, and Botswana [!Kung], for example, no woman had killed another female. And when infanticides are dropped from this global analysis, the ratio of same-sex killings by men doubles to nearly 40 to 1. Men the world over kill other men to carve out reputations, personal empires, and to gain access to more women. "The difference [in homicide] between the sexes," conclude Daly and Wilson, "is immense, and it is universal."

Robbery-Murder

A SKEPTIC MIGHT POINT OUT that some men kill merely to steal, not to carve out a reputation or win women. In Los Angeles, for example, in one year robbers murdered three victims just to steal their Rolex watches. But the robbers were men. As with murder, arrestees for violent robbery have for decades tended to be men. By the mid-1990s, 90 percent of U.S. robbery arrestees were still men. Sixty-five percent of these men were under twenty-five years old; 58 percent were black. In Canada, 96 percent of robber-murderers were men. Significantly, 80 percent of Canadian robbery-murder victims were also men.

Why are so many male robbery victims killed? The reason is not because men are robbed more often. Department of Justice reports project that a stunning *73 percent* of all U.S. women over the age of twelve eventually will be robbed, raped, or assaulted in some way (domestic violence, school violence, date rape, etc.). But women resist robbery less, more often yielding to their assailants' demands. Men resist more, due to their macho, face-saving psychology—the same instinct that leads to so many murders during so-called "trivial" arguments.

To return to the argument that some men kill "merely" to rob, we should remember that having successfully robbed automatically creates a fierce reputation, money in the pocket, and thus, as we have seen, an increased appeal for many women.

Robbery, moreover, has little to do with culture and nearly everything to do with nature. Biological sociologists Lawrence E. Cohen and Richard Machalek have built an airtight case of fact and logic against calling robbery an aberrant, sociopathic, or pathological result of culture. Such explanations, they argue, are both unjustified and useless. Robbery is not unique to people, civilized or otherwise. Most predators—lions, leopards, tigers, wolverines, bears, badgers, wolves, sharks, eagles, hawks—and most social primates rob prey or other resources from weaker or less-well-armed individuals. Ants do it, birds do it, even strangler figs do it. To call these robberies "sociopathic" misses the point that nature *equipped* these creatures—as it equipped hard-core felons—with the tremendously useful strategy of seizing resources by force or threat of force. The fact that most critical resources are limited not only makes the natural world turn as it does, but it also provides us with the following major insights into male violence: *individuals will defend defensible critical resources, while other individuals will seize critical resources they perceive as poorly defended.*

Just because robberies and murders are natural, of course, does not mean that we must like them any more than we like a mutating flu virus. But to stop such crimes, we must understand them.

Robbery, like murder, is universal. Hence all societies recognize that some people will choose such strategies to get what they want, and they pass laws to deter and punish both crimes. Typical robbers give us an insight into why some men rob violently and some do not. Typically, men who choose to steal via threat or force are those whose ability to bring home the bacon through normal work is low. Because bringing home "enough" bacon is a relative thing, this means almost anyone could become a robber if his or her subjective perceptions allowed him or her to justify it internally. Cohen and Machalek say that each person decides daily, based on risks versus gains, whether or not to steal. But robbery, they add, is most common in societies where the potential gain is high, the robber's perceived need is high, thefts are easy (victims are unarmed or weak), risk of detection and capture is low, and risk of punishment is lower than overall gain. (We could say these same things about a man's decision to rape or murder.)

Nor is it solely the down-and-out who steal or murder, although they are the most likely to do so. Seizing someone else's belongings is such a deep instinct in the human psyche that rampant "inside" theft against companies—by both white- and blue-collar employees—costs every U.S. household an extra "crime tax" of $1,376 per year in higher prices. In total, this represents almost nine times more money than what people lose annually to street-type robberies and burglaries, which themselves account for a whopping $15.3 billion.

Robbery often escalates into murder. The U.S. record for robbery victims murdered (in 1993) was 2,301. Robbery-murders account for a fairly stable 10 percent of all U.S. murder victims annually, and 85 percent of these victims are men. The natural tendency in men to get it while they can, however they can, spawns true sociopaths who kill not just readily but without regret.

An illuminating example comes from psychiatrist Martin Blinder's interview of Ken, a handsome, thirty-year-old, bisexual drifter jailed a dozen times for petty theft. While tending bar, Ken fell in lust with a married woman named Karen. Ken stole the bar's money and was arrested, but he escaped, taking a guard's gun. He fled with a willing Karen in her husband's Lincoln, stopping every hour for sex in the backseat. When they ran out of money, Ken tried to trade the spare tire for gas. When the attendant refused, Ken put the gun to his head, ordered him to empty the till, and then shot him dead. Blinder asked why:

"I don't know. It just seemed like the thing to do. I hadn't intended to kill him. I'd never really hurt anyone before. But Karen was waiting in the car. I wanted to get back to her and it was the quickest way to handle things."

Outside Las Vegas, the Lincoln conked out. Ken and Karen asked a woman for a lift: "As soon as we came to a quiet stretch in the road, I pulled out the gun and made her stop and get out. I shot her in the head, took her wallet, and dragged her into a deep culvert alongside the road."

Blinder asked Ken what he felt when he shot this woman. "Nothing. I didn't feel anything. It was simply an easy, obvious solution to a small problem."

Later Ken stopped to pick up a female hitchhiker carrying a guitar. During a pit stop, Ken shot her too in the head. She was carrying $12.

Months later, Karen turned state's evidence and had Ken arrested.

TO RECAP, most murders by men, and rarely those by women, occur in the context of arguments over saving face, over who gets the sexual partner, and, in robberies, over money to buy illegal drugs or resources vital to attracting sexual partners to mate with them, or to support a family. Even murder of infants is tied to reproduction and reproductive resources. In short, the evolutionary link between sex and violence is exemplified by murder as much as, if not more than, by rape. The murder contexts of sex and reproduction, both direct and indirect, do not constitute all murder circumstances. But, as we will see, they encompass an astonishing majority.

Jealousy Murders

WE HAVE SEEN that women kill to solidify the security of their reproductive situations and that men kill over face-saving and over resources attractive to women. But men are even quicker to kill directly over women themselves due to jealousy, and both jealousy itself and jealousy murder are common. Daly and Wilson cite a study in which only 1 woman in 168 people claimed never to have felt jealous. Jealousy for men is not what it is for women (see Chapter 2). Karen Hill, in Nicholas Pilaggi's *Wiseguy,* explains the typical woman's viewpoint after she found out that Henry was cheating on her.

> But still, when I found out what was going on, it was very tough. I was married to him. I had Judy and the baby to worry about. What am I supposed to do? Throw him away? Throw away somebody I am attracted to and who was a very good provider? He wasn't like most of his friends, who made their wives beg for a five-dollar bill. I always had money. He never counted money with me. If there was anything I wanted, I got it, and it made him happy. Why should I kick him out? Why should I lose him just because he was fooling around? Why should I give him up to someone else? Never! If I was going to kick anybody, it was the person who was trying to take him away from me. Why should she win?

As you might guess, men and women handle their jealousy differently. In 8 studies of 147 "love triangle" homicides, men murdered their rivals in 135 cases. Women killed in only 12. As a motive for murder, jealousy ranks high, from number 1 to number 3, in homicide statistics.

A natural "experiment" created by the mutiny on the *Bounty* shows just how dangerous jealousy is. Two centuries ago, the nine mutineers marooned themselves with six Polynesian men and thirteen women on uninhabited Pitcairn Island. When another ship stumbled upon

Pitcairn eighteen years later, only one man remained alive. A dozen had been murdered, one had committed suicide, and one had died. Meanwhile, only three women had died. According to anthropologists Donald E. Brown and Dana Hotra, "The mutiny [on the *Bounty*] was motivated by the desire of the men not to give up the pleasures of their Polynesian mates. . . . Most of the murderers were motivated by male sexual jealousy."

Although jealousy is often lethal to the rival, it also is often fatal to the jealous one's spouse or lover, especially in situations where the rivals are either unknown or too numerous or dangerous to confront directly. Strangely, marriage itself clarifies why and how jealousy possesses such lethal power. "Women need men to aid in rearing children," notes physical anthropologist Meredith Small, "but men must be assured that his [*sic*] is going into the correct packet of genes. This female need for aid to improve her reproductive success and the male response that he will give only when paternity is clear have dominated the evolution of our species."

Marriage is part of this evolution. It is the ultimate human contract. Men and women in all societies marry in nearly the same way. Marriage is normally a "permanent" mating between a man and a woman whose primary goals (being in love notwithstanding) are procreating via only one husband and then cooperatively rearing their children—with the woman nurturing the infants, while the man supports and defends them. The institution of marriage is older than states, churches, and laws. In fact, "marriage" and mating among thousands of nonhuman species are nearly identical in their economic alliance, division of labor, cooperative rearing, and male defense. But two very big differences exist between animals and people. First, laws make marriages binding. Second, sometimes one spouse in a failing marriage murders the other, and the motive is usually jealousy.

All societies, even polygamous ones, recognize adultery as a married adult having sexual union with a partner to whom he or she is not married. And all societies forbid a wife's adultery by law. These laws show two things: how different men and women are and how intensely concerned men are with paternity. In America, either sex can divorce an adulterous spouse, but as Karen Hill revealed and as Daly and Wilson point out, "men are much more likely than women to feel that spousal adultery actually warrants divorce." Indeed, the first known law classifying adultery by a husband as a crime was not passed until 1852, in Austria.

The term *cuckold* reflects this asymmetry between men and women. It is based on the parasitic breeding habits of the European cuckoo. The female lays her eggs in the nests of other species of birds, so that her chicks are raised by a pair of unwitting avian stepparents—a reproductive strategy incurring minimal investment by the cuckoo. If the host bird dumps out the cuckoo's eggs, the cuckoo returns and destroys the host's own eggs. Similarly, the adulterous man leaves his bastard child in the house of the cuckolded husband to be supported by him unwittingly. When a husband commits adultery, however, his wife faces no such problem of raising someone else's child and pays no such price.

This difference is so significant that many laws worldwide do not just forbid female adultery, but they also allow the cuckolded husband to kill his rival. In Texas, until 1974 it was justifiable homicide for a cuckold to kill his wife's paramour when caught in flagrante delicto. Even where such homicide is illegal, note Daly and Wilson, "many American juries have voted to acquit homicidal cuckolds altogether, on the basis of the 'unwritten law.'" English law holds that a "reasonable man" will rightfully lose his head and kill only when defending himself or a relative—or when finding his wife in flagrante delicto.

All of this leads one to wonder just how common adultery is—and, more to the point evolutionarily, how often married women conceive infants via adulterous relationships. Is adultery and cuckoldry so common that men should be obsessed with it to the point of being driven to murder? Although it may appear that literally every aspect of American life has been poked into by one or another group of researchers, this area is so taboo that it has been left nearly untouched. In the 1940s, a secret study by a Dr. X examined known blood groups of one thousand pairs of parents and their newborns. Dr. X's results revealed that a shocking 10 percent of American babies were fathered not by the mother's husband but by another man. One out of ten husbands were cuckolds! Subsequent studies revealed that between 5 and 30 percent of newborns in the United States and Britain were fathered by men other than the mother's husband. In short, reproductively, it does appear that husbands are validated in their "obsession" over the paternity of the children they are expected to raise. This has likely been true throughout history and prehistory, and sexual selection has likely equipped men with both (1) a suspicious and jealous nature and (2) an instinctive strategy to use extreme violence against sexual rivals and to punish their wandering wives.

Daly and Wilson suggest that *if true motives were known, adultery—real or suspected—would be the number one cause of murder within marriage.*

Jealousy in husbands is far more lethal than in wives. One study found that Canadian women murdered their spouses less than one-third as often (248 cases) as did men (786 cases). Of these cases, 46 percent of convicted husbands were sentenced to prison, but only 12 percent of convicted wives were. This is because most of the wives used a self-defense plea against husbands who had often beaten them and/or their children. In many instances, the beatings occurred because the men had suspected adultery.

Why do men abuse their wives? Such abuse is common worldwide. Surprisingly, however, psychologist Donald G. Dutton found that violence in American marriages is as commonly initiated by wives (11.6 percent of marriages) as by husbands (12.1 percent). (Violence even occurs in 18 percent of lesbian couples.) Dutton found, however, that wife beaters are very *unlike* generally assaultive men (who clashed with other men) in two ways: first, wife beaters "perceive more abandonment" when their wives act independently, and second, they feel more anger about it. How did these men get this way? According to Dutton, "For all men in this study, anger scores [of men who abuse women] were associated with feelings of humiliation and with the prior occurrence of abuse (both physical and verbal) by their mother." These insecure and jealous men, who imagine potential abandonment and cuckoldry in any independent action by their wives, believe that beating their wives is the *best* way to stop them.

Yet humans have no monopoly on spousal abuse. A dark side also exists in relationships between most male and female primates, particularly baboons. Physical anthropologist Barbara Smuts found that a female baboon's special male friend (a male from outside her troop who entered into it as her "partner"), despite his grooming and baby-sitting favors, was also the most likely male to attack her unprovoked. Puzzled at first, Smuts concluded that a male attacks a female as punishment to teach her not to interfere when he mates with other females. "Natural selection has favored a tendency in male baboons to use aggression toward females and infants," notes Smuts, "as a means of increasing their own reproductive opportunities, whenever circumstances permit."

Even stranger, whereas some females avoided males who attacked them, others were attracted to those males. This also puzzled Smuts. She concluded that, to many females, a male who abuses her at times, but who also protects her and their mutual offspring from infanticidal males, may be worth keeping at the price of being abused. Whether this tendency to cling to an abusive partner is a natural trait innate in other primates, including humans, requires more research. But we do know that this behavior is common among humans and that it is often fatal.

The National Academy of Sciences' Panel on Research on Violence Against Women concluded, "At least 2 million women [in America] each year are battered by an intimate partner" and 1,500 are murdered by the same partner—a husband, ex-husband, boyfriend, or ex-boyfriend. And vaguely like baboons, battered women who lack their own financial incomes plan to return to their abusive partners 2.4 times more often than do women who possess their own incomes.

How predictably does domestic violence lead to murder? The Crime Control Institute in Washington, D.C., followed 15,537 reports of domestic battery. Researchers found that only 1 in 33 domestic homicides was connected to prior violence between the same two people. Moreover, of 110 threats to kill, including pointing guns, none led to later injury or death. All of this reveals that many men who decide to kill their wives simply go ahead and kill them without giving as many advance signals as one might expect. These men's decisions to kill, note Daly and Wilson, most often result directly from "the husband's proprietary concern with his wife's fidelity or her intention to quit the marriage." They also note, "The estranged wife, hunted down and murdered, is a common item in police files."

An insight into the incomprehensible (to women) cause-and-effect "package" in the male psyche that transmutes a man's "love" for a woman into his brutal murder of her is embodied in what must be destined to become the most famous quote ever made tying the two together. In 1998, when asked yet again about the murder of his estranged ex-wife, Nicole Simpson, O. J. Simpson said, "Even if I did do this, it would have to have been because I loved her very much, right?"

Everything we know about men tells us that, above all else, they want to, and often do, monopolize the sex lives of "their" women—wives, concubines, and even helpless women they have just met. Men everywhere want this exclusivity of sex, and many use violence, even murder, to gain it. Indeed, in canvassing societies worldwide, Daly, Wilson, and Weghorst found no society in which adultery and men's sexual proprietorship of women had not been a frequent cause of spousal homicide.

Male jealousy is so severe that, of all American women murdered in 1996, 30 percent—a fairly stable figure from year to year—were killed by their husbands or boyfriends. Meanwhile, a mere 3.0 percent—also a stable figure—of male homicides were committed by wives or girlfriends. Canadian women are murdered by their intimate male partners at the same rate as American women (even in the absence of handguns, but often with alcohol as a contributing factor), but Canadian women kill their intimate male partners only half as often as American women do. This problem of homicidal jealousy is so severe that the National Academy of Sciences' Panel on Research on Violence Against Women concluded "scarce resources designated for men's violence against women should be allocated not to 'stranger danger,' but to the problem of intimate mates and acquaintances."

The Upshot of Murder

OVERALL, *most* men can become murderers if given sufficient provocation. In fact, what fascinates us most about murder is that we *know* most murderers are sane, and that nearly all of us recognize that we could be driven to commit homicide given the "wrong" or "worst" set of circumstances affecting our ultimate need to survive or reproduce successfully.

The data in this chapter answer many of the questions posed at the outset—although why men kill so much more than women still needs a clarifying statement on ultimate causation: *Homicide is an instinct coded by sexual selection into the human male psyche in a design that prompts men to kill (1) to* expand *their personal gains leading to reproductive advantage or (2) to keep major gains they have already made. Women, meanwhile, usually kill to* protect *their personal security or reproductive future. Clearly, the implication here is that whereas violence by women is finite and more predictable in perceived self-defense (however twisted), men's violence tends to be more infinite and opportunistic.*

Murder is an ever-present possibility in all cultures because its roots are biological. Murder is coded in our DNA, just as it is in the genes of our close ape cousins. Only by realizing this, and by understanding murder's "greenhouse"—a world of severe reproductive competition, wherein the winners have decided our genetic behavioral heritage for millions of years—can we hope to rise above this genetic predisposition for violence.

• • •

Incredibly, Hawley's shots missed Ray from only a dozen feet away.

Ray fired six shots back at Hawley. Three shots missed altogether. Two passed harmlessly through Hawley's jacket. One bullet finally grazed Hawley's shoulder as he backed away into the open.

Meanwhile, Hawley spun and ran toward the cars parked at the gas pumps in front of the "stop-and-rob." Ray hoped that this fray would end without his having to kill a man. But, as if counting Ray's shots to add up the total fire from a standard police six-shooter, Hawley stopped abruptly after Ray's sixth shot, pivoted, and leveled his gun at Ray in careful aim.

But Ray was not carrying a six-shooter. His Glock had a twenty-round mag.

"This was when I first realized I had to kill him to survive," Ray said.

Ray took careful aim and fired twice. The first bullet entered Hawley's left eye, killing him before he hit the oil-soaked pavement.

A background check on Hawley revealed fourteen prior offenses, including attempted murder. At age fourteen, he had been abandoned by his mother and ignored by his father. He was adopted by a woman who soon became his lover. During his last year, Hawley had been out of prison for five months, during which time he had been arrested in California for being the wheelman in a drive-by shooting by the Aryan Brotherhood, for two armed robberies, and for making pipe bombs. Despite all this, when Hawley had agreed to snitch on the shooter in the drive-by, California's judicial system had released him. The woman in the backseat of Hawley's Buick waiting for him to rob the minimart was his father's girlfriend's sister—and his lover.

CHAPTER SIX

War

If they [the chimpanzees] had had firearms and had been taught to use them, I suspect they would have used them to kill.

— Jane Goodall, 1986

All I want is a Coke," Bill Klingenberg said, "warm, cold, or hot."

After nearly eight weeks in the field, Bill was dog tired. And just today his squad had combat patrolled five miles of rough terrain, then choppered back to Firebase Oasis. He was beat. Hacking and hiking five miles while humping eight hundred rounds of 5.56 mm ammo, two hundred rounds of 7.62 mm machine gun ammo, four fragmentation grenades, a mortar round, and a light antitank weapon (LAW)—plus full field gear—felt more like fifteen miles.

It was late afternoon, December 21, 1966. This was Bill's sixth month with the First Cavalry. He and everyone else was hoping they would be lifted to spend Christmas in base camp, a place he could barely remember. But, as he started off in hopes of finding that warm Coke, First Platoon got the call to saddle up.

"Hustle," Platoon Sergeant Speller commanded, as he dumped cases of C rations on the ground, five meals each. "Just sandbag these C-rats and tie them on your gear—you can repack it right later."

Ah, shit, thought all seventeen men—survivors of what should have been a forty-four-man platoon now whittled down by casualties—as they slogged dry-throated a quarter of a mile to the four Hueys. They sat in the birds for the next hour, waiting for the pilots, who themselves were waiting for their orders.

A forward air controller had spotted North Vietnamese Army (NVA) troops at the base of a hill eight miles away. Now the powers that be were cooking up what they hoped would be a classic hammer-and-anvil mission. Bill's platoon, First Platoon of B Company, Second Battalion of the Fifth Cavalry of the First Cavalry, would be dropped to become the anvil by setting up some sort of fixed ambush site. A squad of a dozen men of the First Battalion of the Ninth Cavalry Blue Team would act as the hammer pushing NVA troops into the ambush site by advancing aggressively against them.

But first, as usual, Bill's platoon had to wait for the air strike on the NVA position to "soften" the enemy. Despite ideal weather, it was difficult from the air to tell whether there were six hundred NVA down there or six. After all was said and done, and despite whatever initial air strikes, First Platoon would simply act as bait (the usual but unstated tactic) to reveal the positions of prey to make them easy targets for U.S. artillery and jets. Trapped in these Hueys, Bill and his fellow grunts had too much time to contemplate this.

Already this war had disproved the conventional military wisdom that ruled Pentagon minds: superior firepower, especially airpower, is what wins wars. Here in Vietnam, it took 80 tons of bombs costing $140,000 to kill 1 NVA infiltrator. This was 25 to 80 times more per enemy killed than during World War II. Half the enemy casualties here resulted from small arms fire—but with 50,000 U.S. bullets fired per casualty. And grunts like First Platoon were the guys who fired those bullets. In retaliation, 10 percent of U.S. casualties resulted from enemy booby traps. More nightmarish yet, nearly twice as many U.S. casualties as that were caused by friendly fire.

Still, the NVA were worse. They came down the Ho Chi Minh Trail 1,500 miles from North Vietnam via nearby Cambodia. The U.S. Air Force could not even close this one network. New NVA recruits loaded Russian munitions onto their backs, then tromped south. The NVA also used bicycles, trucks, even elephants. And despite the current 200 U.S. bombing strikes per day on the Ho Chi Minh Trail—the most intense, concentrated bombing in the history of warfare—the 300,000 North Vietnamese workers and 25,000 NVA guards assigned to the trail kept it open continuously. An average of 60 tons of communist supplies and munitions were being delivered to those nearby Cambodian supply depots daily.

The Hueys finally lifted off and skimmed above the trees. Minutes later, they hovered over a dry rice field in Montagnard country. Bill studied the terrain below. This field was in the Highlands about ten miles from Cambodia, near the Ia Drang Valley, north of Pleiku but south of Kontum. Eerily, First Platoon's current assault was in the neighborhood of the first big battles America had fought in the Vietnam War—the very battles that had turned it from a police action into a war.

Then, only thirteen months earlier, the First Cavalry had lost 234 men in the Ia Drang Valley. That first battle was at Landing Zone (LZ) X-Ray. Colonel Hal Moore commanded 411 Americans of the First Battalion, Seventh Cavalry against 2,000 NVA and Vietcong in a blistering, three-day, close-quarters battle that killed 79 Americans and an astonishing 1,500 NVA. Three days later and four miles away at LZ Albany, an American force twice as big, from the Fifth and Seventh Cavalries, walked into a well-prepared ambush by an equal-size NVA force. Sixteen nightmarish hours of savage face-to-face and hand-to-hand butchery in the tall grass left 155 GIs and 403 NVA dead and even more wounded, in what the Seventh Cavalry feared would be a second Little Bighorn.

Bill and his buddies in their Huey glanced at each other. The entire First Platoon could see that this combat assault was unusual. For one thing, it was late in the day. For another, there had been no fire from below, no strong indication of NVA presence, no artillery prep. Furthermore, their birds were going in with no gunship prep of the landing zone. No one seemed to expect to encounter anything in this zone. Instead, everything felt rushed and unplanned, and at the same time it was starting to look like a wasted effort. So little thought had gone into this operation that Bill and nearly everyone else were thinking the same thing: It's just another useless combat assault. We could be back there drinking Cokes and getting six hours of sleep for a change.

Less than two hours before dark, the four Hueys dropped to a hover at two yards. First Platoon jumped off the skids onto a gently sloping field covered with tree stumps and rice stubble a yard high.

The birds flushed and vanished. First Platoon spread out on a wide line. Bill walked the far left. Looking around, he spotted a crosshatched bamboo mat a foot and a half square sticking up fifty feet off the platoon's line. "Hey, Chico," Bill called to his squad leader, "I see something. I'm gonna check it out."

"Okay," Sergeant Chico Deleon acknowledged.

Bill stepped closer. The mat was sticking out of a hole in the ground. Bill held his M16 with one hand, casually, with the buttstock balanced on an ammo pouch on his belt. His finger was on the trigger, his thumb on the selector switch.

Bill spotted the muzzle of an AK-47 rise up at him from inside the hole. The mat was the cover for a "spider hole," and Bill was about to be cut in half by .30-caliber bullets fired at point-blank range.

Before the NVA fired, Bill flipped his selector switch and pulled the trigger simultaneously. He fired seven rounds before his M16 jammed.

Bill's fire forced the NVA soldier to duck back inside the hole. Even so, the two men were close; the muzzles of their weapons were only three feet apart.

Amazingly, all seven of Bill's rounds had missed.

His M16 now jammed and useless, Bill tried to dive out of the NVA's immediate view. But Bill's web belt unlinked due to his hastily packed, ill-balanced load of C rations. His web gear tangled around the stump and in the limbs of a cut tree and slammed him to the ground about ten feet from the spider hole, with the wind knocked out of him.

The NVA soldier popped up from his hole and began firing his AK in Bill's direction, emptying his thirty-round magazine.

Still tangled and immobilized in his gear, Bill saw the fear-contorted face of the soldier as he opened up on full auto. He was young, younger than Bill's own twenty years. It occurred to Bill that his own expression of primal fear was likely the same—they were both just scared guys who wanted to go home. But only one of them would. In one tug, Bill tore a hand grenade free from his web gear (taped there as per a new regulation), ripping the adhesive tape and metal spoon in half.

The NVA kid's fear-contorted face vibrated with his weapon as its muzzle blasts blew hot gas, dust, tree splinters, and rice chaff into Bill's face, half covering him. Miraculously, the bullets sizzled past, inches above him.

Four other AKs scattered around the field now opened up and caught First Platoon in a cross fire.

Bill knew he was close to dying. Adrenaline pumped through him. He desperately wanted to escape his tangled web gear, but it pinned him to the ground. His fight-or-flight response screamed at him to act, but he could neither fight nor run. Bill yanked so hard on the pull ring of his grenade that, instead of pulling the cotter pin out of the grenade to arm it, he pulled the ring through the eye of the cotter key, breaking it. Now the grenade had two pieces of cotter key stuck in it. It was as useful as a one-pound rock.

The kid had emptied his first magazine and now ducked to reload. Bill frantically tried to extract the two pieces of cotter pin, still keeping an eye on the hole. Whoever was first with his weapon would live.

Whoever came in second would die.

Bill decided to throw the grenade like a hardball if he had to. Just hit the kid with it. He knew it would not explode, but it might stun him. It was a risky tactic, but he had no option other than to lie there and wait to die. Tangled as he was, he could not reach another frag without exposing himself.

The NVA soldier emerged from the spider hole to spray thirty more rounds at Bill from spitting distance. Bill lay entangled on his back and stared again at the blasting muzzle. This was unbelievable. Time moved in agonizing slow motion as Bill waited for those bullets to rip into him. Meanwhile, some other part of his consciousness focused on trying to extract those two pieces of wire. One wire finally pulled loose.

The NVA had now fired sixty rounds toward Bill, yet every shot had whizzed mere inches above Bill's head. The kid ducked back into his hole to load a third magazine.

Bill's mind raced hopelessly: Shit! Shit! Shit! He frantically wiggled that last piece of cotter pin. He knew the NVA would come out blasting again.

The panicked soldier reemerged, firing a third burst at Bill. The muzzle blast of heat and dust blinded him, but, finally, that last piece of broken cotter pin in his grenade moved.

• • •

What War Is; What War Is Not

EVEN IF YOU have never faced being exploded into pieces by driving over a command-detonated mine or being ripped full of holes by automatic weapons in an ambush, it is hard to feel indifferent about war. We all have an opinion, usually a passionate one. Although most of us agree that losing in war is bad, generally we can't even agree on a definition of war. War has been called everything from a great adventure to hell. For example, Prussian military genius Carl von Clausewitz insisted, "War is merely the continuation of policy by other means." In contrast, Colonel David H. Hackworth, America's most decorated and proficient warrior in Korea and Vietnam, asked, "What is war anyway but one big, raging atrocity?"

Because war means so many things to so many people, no useful analysis of this, the ultimate male violence, is possible without a firm definition. So here it is. *War is a conflict between social groups that is resolved by individuals on one or both sides killing those on the opposite side.* The scale of war is relative to the combatants' numbers. Twenty-five warriors from a tiny village can wage a "bigger" war, relatively speaking, than the U.S. Army did during World War II. The offensive goals of war typically include territorial expansion, plunder of property, kidnapping women, seizing other critical resources in short supply, and/or genocide. *Offensive war consists of, and is defined by, these intentions to steal en masse what other men own and to leave them dead. The conduct of war is only a tool to fulfill these intentions.*

In an even broader sense, war is the one male behavior that most fully exposes the fundamental differences between men and women—differences that have fueled male violence since our most distant past.

"War is ancient," notes historian Richard Rhodes. "Nor were mass slaughters ever rare. The Old Testament regularly celebrates their carnage. The history of empires bulk thick with them." Indeed, in the Bible whole civilizations are put to the sword. In Exodus, for example, Moses pulverizes the forces of Og: "So they slew him, his sons, and all his subjects until he did not have a single survivor remaining, and then they occupied his land." Moses took each of Og's sixty walled cities and killed every man, woman, and child in them. Joshua succeeded Moses and led the Israelites to conquer more than thirty kingdoms, again leaving no survivor. Joshua's descendants did likewise in centuries of scorched-earth battles rivaling in casualties those of the twentieth century. For the ancient Hebrews, genocidal war seemed the primary instrument of God.

And so it would seem for all men in all times ever since—and ever before. Recently, in Queensland, for example, I came across a *Geo Australia* article titled "Rock Art Warriors: World's Earliest Paintings of People at War." These Aboriginal paintings—dated at five thousand years old and located in Kakadu, in Australia's Northern Territory—depicted men spearing one another in war. But contrary to the article's title, they are relatively recent. "The earliest actual image of combat," notes army intelligence analyst Robert L. O'Connell, "a Mesolithic cave painting [twenty thousand years old] at Morela la Vella in Spain, depicts men fighting with bows." My point here is not to quibble about the oldest hard evidence of war. Instead it is to emphasize that wars are immensely older than even these paintings. As we will see, wars are older than humanity itself.

Of course, the big question is, Why do men fight wars? Why do *any* animals fight wars? Again, as in our understanding of rape and murder, the clues are in who fights wars and what they gain from them. We know wars were and are fought by men from all backgrounds: farmers, herders, hunters, and the people of industrialized nations, from the sophisticated upper class to ghetto dwellers. And wars are fought often. Since the Napoleonic Wars, for example, men have fought an average of six international wars and six civil wars per decade. Over time, the slaughter has become increasingly more rapid as materiels have evolved from wood to stone and bone to bronze to steel to exotic alloys, plastics, chemicals, isotopes, and particle beams—and as weapons have evolved from spears, to slings and arrows and giant ballistae, to siege weapons, Greek fire, and Chinese rockets, to assault rifles, gas, napalm, helicopters, jets, rockets, virulent biological agents, and nuclear doomsday machines ensuring global death. Tactics also have evolved from routing the enemy to mutual assured destruction. Anthropologists Lionel Tiger and Robin Fox pinpoint the greatest truth in human warfare: "the most dangerous weapon the soldier has is the cerebral cortex under his hat."

War vies with sex for the distinction of being the most significant process in human evolution. Not only have wars shaped geopolitical boundaries and spread national ideologies, but they also have carved the distributions of humanity's religions, cultures, diseases, technologies, and even genetic populations. When the British colonized Tasmania, for example, they used diseases, dogs, horses, rifles, starvation, imprisonment, poison, and bounties of five British pounds per head to eliminate the Tasmanians, who had been isolated there for thirty thousand years. The British murdered thousands, with the last two Tasmanians dying in captivity. The Dutch did the same to the San Bushmen in South Africa; the Spanish killed all the Arawak Indians across the Caribbean; the Germans tried to do the same with the Herero in Namibia; and both the British and the Americans tried to annihilate the North American Indians. In one of the earliest tactics in biological warfare, the British even gave Indians "peace" gifts of blankets deliberately contaminated with smallpox. The U.S. Cavalry in the late nineteenth century was primarily a government instrument of genocide. As directed by Washington, D.C., it nearly extirpated all Plains Indians and replaced them with white Anglo-Saxon Protestant pioneers in little houses on the prairie. By 1864, for example, General Philip Sheridan voiced U.S. policy this way: "The only good Indian I ever saw was dead." This was reworded to become the maxim of the U.S. Army: "The only good Indian is a dead Indian." A more concise formula for genocide would be hard to find.

Genocide, by definition, is the most draconian event in the evolution of a species short of its total extinction. Thus wars have played a major role in human evolution. How important a role? "Probably at least 10 per cent of deaths in modern civilization can be attributed directly or indirectly to war," notes Quincy Wright. Most geneticists would agree that removing 10 percent of a population may well change that population's gene pool.

Despite the claims of some anthropologists and revisionist historians who argue that wars are nothing more than unnatural products of civilization, wars are ubiquitous. *Wars erupt naturally everywhere humans are present*. Although we do not know as much as we would like to about natural primitive war, we possess some sobering chronicles. In 1769, for example, James Cook captained the *Endeavour,* the second ship to visit New Zealand (the first ship, Abel Tasman's, had arrived 127 years earlier). Cook, both skilled at and dedicated to not offending the local people, was attacked by organized Maoris almost everywhere he dropped anchor—just as Tasman before him had been. And not only did these isolated Maoris wage war unprovoked (mostly with each other), but they also ate their enemies.

Likewise, in 1933, gold prospector Michael Leahy and his mates stumbled upon the King Kong of primitive war. During an adventure unparalleled since Henry M. Stanley blazed a trail of blood across Darkest Africa to rescue Emin Pasha, Leahy and his mates marched into the unknown Wahgi Valley of the New Guinea Highlands. They were surprised even to find a valley. Everyone assumed that the mountains of this, the second largest island on Earth, rose up from each side, met in the middle, and then dropped back down. No outsider suspected that New Guinea's central massif guarded huge, elongated valleys. Further, none suspected that in those highlands a million Stone Age farmers speaking hundreds of languages were locked into scores of isolated valleys by an incessant cycle of war. Leahy was shocked to find tribe after tribe. So was the rest of the world. Leahy's first contact with this, the last huge undiscovered population on Earth, made headlines worldwide. More to the point, Leahy and Crain's book *The Land That Time Forgot* describes chronic war among the Chimbu:

> If a village becomes weaker than its neighbors, it will sooner or later be wiped out. If it waxes too strong, it will become overbearing, and in time a coalition of neighboring villages will likewise wipe it out, burning the houses,

> destroying the gardens and ring-barking the trees. Then the women and a few surviving warriors will flee farther back into the hills, there to breed up a new generation of fighting men numerous enough to reconquer their lost heritage. So it goes, the wars being a succession of "pay-back" affairs rooted in long-forgotten feuds.

To test the oft-cited argument that, when left alone, humankind is by nature peaceable, anthropologist Carol Ember probed records of hunter-gatherer groups. She found that 64 percent of such societies waged war at least once every two years, 26 percent warred less often, and only 10 percent rarely or never engaged in warfare. Anthropologist K. F. Otterbein followed suit, again focusing mostly on primitive societies but including some horticulturalists. He found that 92 percent fought wars. Prehistoric American Indians also fought major wars. Even the "harmless" !Kung and !Ko Bushmen defend water holes and foraging areas. Men everywhere have cooperated in war. "No tribes have been adequately described," explains war historian Quincy Wright, "that will not fight as units under certain circumstances, and in most tribes the mores prescribe violent behavior."

We can only conclude that *war is both a significant and a natural state of affairs erupting periodically between social groups of* Homo sapiens.

In view of the huge cost in lives and resources we pay—$1 trillion per year globally—to fight a war or just to be ready to defend ourselves from attack, we should be scrambling to figure out how and why we are a species addicted to war. "If you wish for peace," advised military strategist B. H. Liddell Hart, "understand war."

Understanding the scourge of war is impossible without recognizing first that every war is intended by its aggressors to benefit them. Men cooperate as warriors not because they are sociopathic or stupid, but because they as individuals stand to gain something from winning. Many men who live by the sword often do win something by it. That *something* is the key to understanding why men wage war.

War is *the* quantum leap in nature's arms race. *Homo sapiens* was not the first species to make this leap, however. Nor is human war caused by a backfiring of the killer instinct from our legacy of hunting big game. Killing prey is nothing like killing members of one's own species. Likewise, war is not caused by an aggressive "drive" that builds up in men and demands release. No such drive exists. Nor is war programmed by some sort of internal, for-our-own-collective-good mechanism whose function it is to control our population growth. Evolutionary ecology has revealed that *no such self-regulating mechanism exists in humans "for the good of our species"*—nor, for that matter, does it exist in any other species.

Instead, war is a male reproductive strategy. All that is needed for this strategy to evolve is that aggressors fight and win more often than they lose. But because war poses a lethal risk, we need to ask what spoils could be worth this risk of death?

Natural-Born Warriors: Chimpanzees

Five males from the south silently crested the ridge. They peered north, down into the next valley. They craned their necks and shifted their positions uneasily. A grunt drifted up from below. All five males retracted their lips in nervous grimaces and reached out to touch one another in reassurance. Then, in unison, they quietly hurried down the slope.

In the valley below, a northern male and female were eating figs in the southern periphery of their community's territory. It was the only place where the male could hope to mate with this female away from the competition of his fellow males. More danger lurked here, but that seemed a small price to pay.

Suddenly, the five intruders surrounded the couple's tree. Two males climbed its trunk. Three stayed below to block the escape routes. The female stared in shock at the five males, then shrieked and fled. The southern male guarding her route let her pass, but he tackled her consort as he tried to flee. They rolled across the turf in a tangle of frantically groping limbs.

The other four males converged into a writhing dog pile. One grabbed the northern male's leg, another his arm. A third pummeled and bit him, while the fourth stomped on his back. One ripped two of the victim's fingers off with a popping sound. All five savagely tore at him for ten minutes. Panting and exhausted, they stepped back to study his bloody, inert body. One ran a wobbly circle around him and hooted in a dominance display. Two others joined in. Then, abruptly, they abandoned their dying victim and raced back up the hill to the south.

The female emerged from hiding. Only the faintest hiss of breath from her consort told her that he still lived. She touched him tentatively. He tried to rise, but the five attackers had broken his spine. He sank back to the forest floor. She stayed with him for two days, brushing flies from the wounds where his skin had been torn off and his fingers gnawed and torn from his hands. She tried grooming these nasty wounds, but her efforts were no substitute for surgical repair. Finally, when her consort died, she returned to her adopted community in the north.

• • •

THE ABOVE INCIDENT is a composite of several observations of chimpanzee warfare, but every detail is accurate. Chimps kill chimps. But why?

The recent discovery that chimps are the species most closely related to humans genetically (we share 98.4 percent of the same DNA)—and that we, not gorillas (they share 97.9 percent of the same DNA), are the species most closely related to chimps—makes the answer to this question significant. Our closeness is no big surprise. After all, chimps look so much like us that it is uncanny. Jane Goodall, for example, reports that when Muslim villagers near Gombe found a dead chimp, the first chimp they had ever seen, they were so amazed by its similarity to humans that they gave it a proper Islamic burial.

But chimps share far more with us than DNA and anatomy. Their behavior seems to parody our own. Chimps are the Albert Einsteins of the nonhuman world. They do math, hunt cooperatively, and use medicinal plants. They manufacture and use tools from leaves, stems, wood, and stones. They communicate in the wild via three dozen calls. Captive chimps display both self-awareness and a self-identity. They have learned hundreds of signs in American Sign Language and other languages and used them to speak sentences, to invent other words, and to communicate to self-advantage with people and each other about the present, past, and future. Chimps even mutter to themselves. They also deliberately teach other chimps sign language. This mental evolution of chimps makes their warfare that much more eerie.

On the flip side of the coin, chimps make lifesaving friends. Jane Goodall describes how zookeeper Marc Cusano at Lion Country Safari in Florida painstakingly befriended a captive male, Old Man, who was a tough nut to crack. Zoo authorities warned Cusano that Old Man was a demon due to the years of abuse he had suffered at human hands. Even so, by repeated acts of kindness, Cusano patiently bridged the gap between them a fraction of an inch at a time. One day on Chimp Island,

Cusano's foot slipped as he was setting out food for the chimp colony. He fell near a newborn infant. Its mother attacked him instantly. Two other enraged females joined her. All three piled on Cusano in a three-hundred-pound mass of fists and fangs sinking into his arms, back, and leg. He thought, "This is it; I'm dead." Then Old Man counterattacked. He picked up each female and hurled her away. Old Man continued to hold the other three apes at bay while Cusano painfully crawled to his little boat to escape across the moat. Had Old Man not intervened, those three females would have killed Cusano.

Hero though he may have been, Old Man was not a unique nonhuman primate. Charles Darwin reported a small New World monkey who performed an even more daring rescue. The monkey's keeper had been attacked by a large baboon, who had sunk his canines in the back of the zookeeper's neck. The little monkey, despite being terrified of the baboon, counterattacked and saved his keeper.

Both rescues reveal a deep-seated sense of individual identity and comradeship. Chimps, like humans, make their mark as individual players. Yet paradoxically, also like humans, they abandon their individuality temporarily to cooperate as disciplined warriors.

Today, due to decades of work in Gombe and Mahale Mountains National Parks in Tanzania, as well as shorter projects such as that by Christophe and Hedwige Boesch in Tai National Park in Sierra Leone and mine and others' in Uganda's Kibale National Park, we know more about chimps than about any other wild nonhuman primate. And one thing we know is that the vital ingredient in the formula for chimp warfare is neither intelligence nor individuality nor the capacity to kill (although all of these play a role). Rather, it is their pattern of exogamy, the way genes transfer between groups.

Unlike most social mammals, among whom the rule is for adult males to out-migrate to join a new group, male chimps never out-migrate. Only females do. Genetic analysis of sixty-seven chimps at sites in Africa revealed that males within a community are related on average as half brothers. In contrast, the genetics of females suggests far greater genetic distance between the females themselves and between them and males with whom they live as adults. In short, female chimpanzees come from "out of town." This form of exogamy is rare. Of the more than two hundred species of primates known, we see male retention like this in fewer than ten—one of them being humans.

Breaking this rule opens a Pandora's Box of cooperative male violence. As females emigrate and leave their brothers behind, the resulting community becomes a male kin group. This extended family of brothers, cousins, uncles and nephews, and fathers and sons shares so many genes that it sets the evolutionary stage for exotic male reproductive strategies based on cooperation, even in the face of death.

How this process works has become clear only recently, yet the breakthrough had nothing to do with chimps. It happened because biologist William D. Hamilton was intrigued by a mystery: why do sterile workers and soldiers in ant, bee, and termite colonies slave and sacrifice themselves to defend the queen and her brood? Social insect queens rule the quintessential reproductive aristocracy: only queens and one in a thousand males breed. Why, with no way to reproduce their genes into the future, do these insects work? By what legerdemain could natural selection convince someone to be a slave or, worse, a kamikaze soldier? Why not just cool it and survive?

Hamilton could see that, all else being equal, queens who produce sterile workers who would fight in defense of her brood or work to help raise it would win the reproductive contest against other queens. But how did she trick her sterile workers into being willing slaves?

As Hamilton sat on a park bench at Oxford in the early 1960s and puzzled over this question, he was hit by a flash of insight. The only chance sterile workers and soldiers have to reproduce their genes lies in their helping their mother to raise more queens—regardless of what that help costs them. This is because, by proxy, she and her queen progeny reproduce the very same genes the workers themselves carry. In short, the queen does their breeding for them. The queen is *their* reproductive slave. Helping or defending her is, in fact, self-serving.

Hamilton had identified one of the most important processes of natural selection: *kin selection,* the process whereby family members increase the fitness of certain genes they carry by assisting one or more relatives to breed more than they otherwise would. Before we go any further on this now proven process, however, we should remember that in Darwin's old saw "survival of the fittest," *fittest* refers not to big muscles or marathon endurance or clever wits, but to the most reproductively successful *genes*. Secondarily, *fitness* refers to the fittest *individuals,* defined as those most successful reproductively. Back to the point, we now know that genes code for cooperative behavior and that cooperation increases fitness in many animals. Here's how.

Hamilton found that behavior that seems altruistic, such as a soldier insect suicidally defending his queen, cannot be labeled altruistic if the behavior increases the reproductive success of that soldier's genes.

Instead of altruism, it is nepotism. To analyze behaviors, Hamilton found, one must first evaluate the *coefficient of relatedness* between the actor and the one who benefits from that act. For instance, genetically, our children are our half clones; the coefficient of relatedness between us is one-half.

Hamilton coined the term *inclusive fitness* to define how one relative shares genetically when another breeds. For example, when my child reproduces, my genetic "share" of his or her success—the grandparent's coefficient of relatedness—is one-half of one-half, or one-quarter of a clone. Helping my grandchild yields half the genetic "gain" in inclusive fitness as helping my child at the same age would. The concept of inclusive fitness reveals how a celibate uncle can paradoxically achieve high reproductive success by fostering the births of more nieces and nephews (whose coefficient of relatedness to him equals one-quarter) than his brothers and sisters otherwise would have raised. The classic example is the celibate but nepotistic parish priest who, by directing parish resources to his huge extended family, increases the size of that family and, at the same time, his own inclusive fitness.

"Blood is thicker than water" was dictated by inclusive fitness. In nature's arena, *nepotism* reigns supreme. Many animals possess an inbred talent to recognize their own kin, and they even do instinctive arithmetic before risking themselves for another individual.

Many birds, for example, help raise nestlings that are not their own offspring. This is called nest helping. But nest helpers also do the math. Biologists Stephen T. Emlen and Peter H. Wrege, for example, analyzed nest helping by white-fronted bee-eaters in Kenya. In 88.5 percent of 174 cases in which an adult bird postponed its own breeding to help older adults, the bird helped its own relatives. In 44.8 percent of cases, the helper assisted both parents in raising new, full siblings. In only 19 percent, it helped raise half siblings (offspring of one parent and a new mate). In 10.3 percent of cases, grandparents "nest-sat" for their own adult offspring. Even more convincing, in 115 cases where helpers had to choose between two different pairs of related nesting parents, 94 percent assisted the nestlings most closely related to them. Helpers do raise their parents' reproductive success and thereby increase their own inclusive fitness. Emlen and Wrege's experiments revealed that white-fronted bee-eaters not only learn who their relatives are, but they also discriminate to help their closest ones.

Birds do it, bees do it . . . what about apes? And what about people?

Examples of human inclusive fitness could fill a library. Hunting-gathering peoples reveal some of the clearest examples. The !Kung San of the Kalahari, for example, shared the meat from their kills. Anthropology lecturers love to explain how beneficent these hunters were. What is rarely said, however, is that successful hunters shared first (sometimes only) with their families via very strict rules in which the closest relatives come first.

"Me against my brother," goes the old Arab proverb; "me and my brother against my cousins; me, my brother and my cousins against our nonrelatives; me, my brother, cousins and friends against our enemies in the village, all of these and the whole village against the next village."

If this proverb seems easy to understand, it is because inclusive fitness is so powerful in shaping our behavior—for good or ill. Inclusive fitness, for example, is the architect of nepotism, tribalism, nationalism, and racism, as well as of the tender mother-infant bond we admire. As biologist Richard Dawkins notes, William D. Hamilton's two papers on inclusive fitness in 1964 "are among the most important contributions to social ethology ever written."

Male chimps in a community bond via inclusive fitness. This is obvious in their sharing of mates. According to Jane Goodall, "Analysis of data collected in the group situation at Gombe, from 1976 to 1983, shows that at some point during the four-day POPs [periovulatory periods] of most females, they copulated with most or all of the reproductively mature males of their community."

Why would males share their mates if the process of natural selection works through the reproductive success of the individual? Actually, despite the power of kin selection and inclusive fitness among males, their sharing can be begrudging. Some males try to exclude other males from a female. Some also try to entice, or coerce by force, a female to go off into the hinterlands of the community's territory "on safari" so that he can breed with her exclusively. Many of these safaris, if composed of a human couple instead of apes, would be categorized as rapes. Despite these male stratagems, however, a female chimp usually manages to mate with most of the males in her community, copulating on average 135 times before she conceives.

What, if anything, does a female gain by mating with every male in her community? A female's promiscuity leads each male into suspecting that he is her infant's father. This then leads to each male protecting all the chimps born in the core of his territory as his own. This protection consists mainly of keeping alien males at bay. Otherwise, male chimps at best make dilatory fathers. But why would such dominant males allow females to mate with every male? Because, as cousins, brothers, uncles, and so on, every male gains—through inclusive fitness—when an infant is conceived. Moreover, pragmatically, except for during safaris, male chimps have a hard time stopping females from mating with other males.

The key point here is this: the kin-selected male community that shares its females sets the stage for exotic adaptations in the macho male sexual selection arms race for any trait that increases the males' reproductive success. These traits emerge in the chimps' sexist lifestyles. Males at Gombe, for example, take male bonding to a whole new level. They travel most often with one another, and they cover 66 percent more territory per day than the females do. These males scout or patrol their territorial fringes against aliens. Normally loud, the males metamorphose into platoons of stealthy guerrillas. "When the chimpanzees go out on patrol," notes biologist Christopher Boehm, "they often maintain a strict silence for hours, during which they manage, nevertheless, to behave much as a coordinated group."

How coordinated and stealthy chimp patrols are must be seen to be understood. Primatologist Mark Leighton, who worked at Gombe, told me of an episode during which a young male was traveling "with the big boys" on their territorial fringe. All the adult males were so tense that the young male started whimpering in fear. Instantly, an old male nearby clamped a hand over the youngster's mouth to seal off his whimper before it could betray their presence to alien males.

Males stay on the move because they are always in search of females in heat. Females, by contrast, neither patrol nor travel beyond the area that satisfies their simple need to forage. The males I observed in Kibale Forest not only foraged longer than the females, but they also rested less and preferred one another, rather than the females, as grooming and traveling partners. Meanwhile, the females preferred one another over males. In studies at Kibale, Gombe, Mahale, and Tai, male patterns of activity differed from female patterns in this way. Solidarity among male chimps also was much higher than among females.

Male chimps and bonobos are unusual in their abilities to strike off from their groups to travel in small parties or on their own and yet still remain in touch. Males often vocalize across miles of forest, revealing their identities and directions of travel, thus allowing other apes to join or avoid them. Indeed, males often "dial" friends and family every few hours. The "phone" they use is a long call Jane Goodall labeled *pant-*

hooting, a call unique to each chimp. And when these apes reunite, they hug, kiss, pat, groom, vocalize, and puff up in dominance.

One big reason for those chimps' community to fission into smaller parties is ecology. Sixty percent of a chimp's diet consists of ripe fruit. Yet fruit is often so hard to find that wild chimps are drastically underweight compared to captive ones. Nor do enough huge fruit trees exist for all fifty or so chimps of a community to travel together and still get enough to eat. In any tree, the least dominant chimps, females in particular, lose in competition over what little fruit exists. Here again, however, males place solidarity ahead of calories. Despite the importance of a square meal, when approaching big fruit trees, males at Gombe and Kibale—but not females—have been observed pant-hooting loudly and drumming tree buttresses with their feet in a wild tattoo resounding through the rain forest for up to a mile. This bedlam attracts other chimps, who share the food of the calling males. This cooperative "food calling" pays off in three selfish ways for the males who called: by facilitating mutual grooming to rid them of parasites, by adding more male companions for safer territorial patrols, and by being able to mate with a female arrival. It also pays off in inclusive fitness by helping all relatives within earshot to achieve better nutrition. All of this, incidentally, is gained at a low cost because males usually call at trees big enough to feed all comers. By contrast, a female would gain nothing by food calling, because males habitually usurp the best feeding spots. And, to add insult to injury, she would be cheated by arriving males, who would not groom her after she groomed them.

Chimps typically travel in groups of two to six adults, but scarcity of food often forces them to go it alone. That they travel together anyway whenever they can leads us to ask the biggest question in social behavior: why do they bother to be social at the cost of not getting enough to eat?

Pieces of this puzzle fell into place in the early 1970s. The process began after Jane Goodall stopped her eight-year program of giving Gombe chimps six hundred bananas a day to habituate them to human observers and to keep them nearby. Her study community split into two factions. The biggest, the Kasakela community of thirty-five apes, stayed in the north. The Kahama faction of fewer than fifteen chimps went south. Within a year or two, the Kasakela males forayed south to the Kahama Valley in sortie after sortie, during which they killed at least five of the seven Kahama males (the last two vanished due to causes unknown). They likely also killed two of the old females. These gang killings were at least as brutal as the one described at the beginning of this section. Males stomped on, twisted, bit, yanked, dragged, gouged, pounded, dismembered, and threw boulders at their outnumbered opponents with such fierce and deliberately lethal aggression that Goodall admitted, "If they had had firearms and had been taught to use them, I suspect they would have used them to kill."

But these killer chimps were neither an anomaly nor unique to Gombe. Chimps in the Mahale Mountains (less than one hundred miles south of Gombe) launched a war a decade later. Toshisada Nishida and his colleagues concluded that males of Nishida's huge M-Group (more than eighty chimps) systematically stalked and murdered the six adult males of the smaller, neighboring K-Group, which contained twenty-two chimps at the onset of hostilities. Their violence, too, was shockingly brutal, premeditated, and deliberately lethal.

In Gombe and Mahale, after all the adult males in each vanquished community had been killed and all the adolescent males had sickened and died, seemingly from depression, the young females shifted their allegiance and home ranges and mated with the victorious males. The victors instantly expanded their territories to include part (Gombe) or most (Mahale) of the territories of the dead males. Both defeated communities ceased to exist, having been wiped out by genocidal warfare. Tanzanian chimps, like Hitler's storm troopers, had fought for lebensraum.

My own fieldwork on chimps in Kibale Forest was spurred directly by these Gombe killings. In the 1970s, one primate killing another of the same species was the hottest issue in primatology. My mission? To find out whether these warlike killings were normal or an artifact of the researchers having provisioned the chimps with food to observe them. This was an important question, because if the chimps had killed only because humans had severely disrupted their normal lives, it would shed far less light on the natural origins and functions of warfare than if their killings were normal.

In Kibale I found that wild males (none of whom were ever fed anything by anyone) were very tightly allied with each other. They traveled together and preferred one another as companions over females in virtually all activities except mating. But I saw no lethal clashes between the males of the Ngogo community (my primary, and larger, study community) and the Kanyawara community (my secondary community) ten miles north. Indeed, I never witnessed any male from one community meet or be near any male known to be of the other community. In short, I saw no occasion for border clashes. But the solidarity with which males from both communities "cemented" themselves to their buddies convinced me that they were tight for a reason. So, with a boldness that I knew would receive the scorn of some social scientists, in my book *East of the Mountains of the Moon* I predicted that the male chimps I observed in Uganda were naturally organized socially, as were those in Gombe and Mahale, and that this male solidarity was a specific adaptation to survive (at least) or win (at best) in intercommunity relations shaped primarily by war. I went out on a limb and said that Kibale male chimps were warlike—as warlike as any chimps anywhere.

After my fieldwork ended in 1981, years passed before my conclusion was proven accurate. In 1988, 1992, and 1994, an adult male of the Kanyawara community, then under study by Dr. Gilbert Isabirye Basuta, was killed, apparently by males of the Wantabu community. I had only suspected the existence of the Wantabu community, located between Kanyawara and Ngogo. The Wantabu chimps range immediately south of Kanyawara and north of Ngogo, where I had spent most of my time. Significantly, all three Kanyawara males were killed in combat, all three died in the same border region between the two communities' territories, and at least two died within hours of prolonged territorial pant-hooting and displaying by males of both communities along this border.

No reasonable doubt exists today that the natural strategy of common chimpanzees is to establish, maintain, defend, or expand a kin group territory via lethal warfare. Clearly their aggressive expansion is most likely when one community outnumbers a neighboring one. Between equal-size groups, active (as opposed to latent) war seems very rare, if not nonexistent.

Annihilation on the scale observed in Gombe and Mahale may seem extreme outside human war. A stunning 30 percent of Gombe's adult males died at the hands of other chimps. But these killer apes simply acted out the logic of reproductive advantage. They gained more of the two resources that limited their reproduction most: females and the territory needed to raise more offspring. Unlike their hyperaggressive but noncooperative cousins, the gorillas and orangutans, *these chimps won in warfare via extreme cooperation and solidarity among kin-related males.* (The endangered pygmy chimps, or bonobos, register lower on the violence curve than chimps, but veteran Japanese researchers have witnessed several "severe" fights between adult males of opposing communities. Thus the wishful claim that the "peaceful" bonobo should give us hope is likely premature, as it was with chimps until 1974.) In addition, male chimp solidarity was not merely a martial fervor that erupted briefly into lethal aggression and then abated. Instead, it was a permanent state, evident further as males shared each other's females.

What is the significance of chimp war? It is important in what it shows us about cooperative violence, which usually erupts only as enlightened self-interest. Group aggression confers such a huge winning edge against single competitors that, once it entered the arms race of sexual selection, kin selection instantly forged it into the most serious weapon in any male's behavioral arsenal. No male on his own could ever compete and win against a martial system like that of wild chimpanzees.

Eerie parallels also exist between warfare by juvenile street gangs and warfare by chimps. First, gang homicide, though driven by the illegal drug trade (instead of by fruit trees and females), is far more often related to control and defense of territory than to drugs. Second, street gang death tolls are horrific. In the 1990s, U.S. juvenile gang killings have averaged 1,000 per year, and these killings are often the result of tit-for-tat retaliation. Third, as violent as gang members are, they commit few rapes. Instead, enough women are attracted to gang members to satisfy them. Fourth, homicides in individual gangs cycle in bursts and lulls that varied 660 percent in annual death tolls. Like chimp wars, gang wars run hot and cold.

It should be noted that male chimps from Mahale and Gombe (and likely Kibale as well) waged war on a neighboring community only when it, the "enemy," was a lot smaller and weaker than their own community, containing half or fewer adult males. It is no overstatement to say that chimps are Machiavellian—or, to put it another way, politically devious and violent men are chimpanzee-like!

These studies of wild chimps tell us that solidarity in aggression among a community's male kin is their standard strategy to reproduce and that this strategy has been around for a long time. How do we know? Despite their fierce and violent competition, male chimps are only 123 percent the weight of females, evidence that winning against other males no longer hinges on the more primitive orangutan or gorilla strategy of being a huge and formidable individual. Instead, winning depends on group size of male kin who cooperate as an army. Were this a recent evolutionary development, male chimps would be both big and cooperative.

Although chimps teach us what the law of the jungle really means, they also teach us what being social is all about. Sociability is for individual advantage. Within the chimps' fusion-fission society, each ape's decisions as to whether to socialize and with whom are based solely on how to best enhance his or her own reproductive success. Thus *chimpanzee social structure—violent and otherwise—owes its form ultimately to each individual's reproductive strategies and, by extension, to each one's individual decisions.* Warfare is simply the social version of combat.

Chimp social structure would be unique were it not for humans acting similarly. This is no coincidence. By most taxonomic criteria, chimps and humans are *sibling species.* Overall, chimp society is not only extremely sexist—with all adult males dominant over females—but also xenophobic to the extent of killing all alien males, many infants, and some old females who enter their territory. To some readers, my use of the word *war* may seem too strong to describe what male kin groups do. But systematic, protracted, deliberate, and cooperative brutal killings of every male in a neighboring community, plus genocidal and frequent cannibalistic murder of many of their offspring, followed by usurpation of the males' mates and annexation of part or all of the losers' territory, matches or exceeds the worst that humans do when they wage war.

Wild chimps reveal the natural contexts of territoriality, war, male cooperation, solidarity and sharing, nepotism, sexism, xenophobia, infanticide, murder, cannibalism, polygyny, and mating competition between kin groups of males—behaviors that have evolved through sexual selection. Also significant is the fact that none of these apes learned these violent behaviors by watching TV or by being victims of socioeconomic handicaps—poor schools, broken homes, bad fathers, illegal drugs, easy weapons, or any other sociological condition. Nor were these apes spurred to war by any political, religious, or economic ideology or by the rhetoric of an insane demagogue. They also were not seeking an "identity" or buckling under peer pressure. Instead, they were obeying instincts, coded in the male psyche, dictating that they must win against other males.

Nuclear physicist Freeman Dyson warns, "If we are to avoid destruction, we must first of all understand the human and historical context out of which destruction arises. We must understand what it is in human nature that makes war so damnably attractive."

The great apes, especially chimpanzees, are the best living mirrors of primeval humankind. It is up to us to look into that mirror (before we have destroyed all their tropical forests and killed them all) and identify what it is in the human male psyche that makes violence so "damnably attractive." Now that we have seen the Machiavellian nature of martial chimpanzees, it is time to revisit *Homo sapiens*.

Are Human Warriors Natural-Born Killers?

IN ALL WARS ever fought by men, some men have killed, while others have avoided doing so. This inconsistency has spurred many idealists to deny that men are instinctive warriors. Instead, they insist, killing must be pounded into each of us against the grain. According to journalist Alessandra Stanley, "Yes, boys have a primitive urge to fight, an easily tapped aggression. But killing is not instinctive; it is an acquired taste, something that grownups must pass on." Historian Gwynne Dyer likewise claims, "Aggression is certainly part of our genetic makeup, and necessarily so, but the normal human being's quota of aggression will not even cause him to kill acquaintances, let alone wage war against strangers from a different country."

Both are wrong. People's "quotas of aggression" are all too often high enough to kill both acquaintances and strangers. During the Vietnam War, for example, Ho Chi Minh's forces killed fifty-eight thousand Americans. But during the same time span, Americans murdered far more Americans at home—and most of those murdered were acquaintances (or even more intimate). According to political scientists Paul Seabury and Angelo Codevilla, "An ineluctable fact is that human intercourse all too naturally produces circumstances in which reasonable people regard kill or be killed as the best option available."

This reality is so obvious to biologists that, despite his personal aversion to believing that men are innate killers, German ethologist Irenäus Eibl-Eibesfeldt found himself listing the universal traits of men the world over that are vital to war: loyalty to group members; readiness to react aggressively to outside threats; motivation to fight, dominate, and act territorially; universal fear of strangers; and intolerance of those who deviate from group norms.

Despite overwhelming evidence to the contrary, many recent books on war—*War* by Gwynne Dyer, *Aggression and War: Their Biological and Social Bases* by Jo Groebel and Robert A. Hinde, *On Killing: The Psychological Cost of Learning to Kill in War and Society* by Dave

Grossman, *Of Arms and Men* by Robert L. O'Connell, and *Blood Rites* by Barbara Ehrenreich—insist, in unabashed wishful thinking, that killing is an acquired proclivity that society must inculcate into men. Men, they say, do not and could not possess an instinct to kill other men, because that would be bad for the species. These books all were written by people who understood little or no biology—or who simply ignored or denied its findings.

The most ingenious example of this thinking is found in Ehrenreich's *Blood Rites*. To make her explanation work, she ignores all data on the predatory nature of primates in general and on violence between male primates. She also ignores evidence showing that human males are violent by nature toward male rivals, unprotected young women, and any creature that is big enough to eat or is capable of competing with them. She suggests instead that, late in human evolution, men abruptly quit being the prey of big cats and, in a turnabout, became the hunters of those big cats' other prey. Men thus carved out for themselves a role "superior" to that of women, who remained mere scavengers and gatherers of plants. But when prey herds dwindled and agriculture replaced hunting as a lifestyle, men found themselves with no useful role. To avoid becoming obsolete through unemployment, they invented war. War not only kept men busy, but it also helped them regain a predatory role superior to that of women. (Besides, Ehrenreich says, stabbing enemies with spears and other weapons was a lot like sex.) War, Ehrenreich explains, "is too complex and collective an activity to be accounted for by a single warlike instinct lurking within the individual psyche." Instead, war ritually "reenact[s] the human transition from prey to predator." As time passed, warriors became a class unto themselves, serving functions that changed with societal convenience. And war itself became a parasitic, self-replicating entity separate from and above any individual people. War became so much bigger than all of us, Ehrenreich says, that we are now its slaves. We must have the courage to free ourselves of war, Ehrenreich advises: "What we are called to [wage against war] is, in fact, a kind of war."

Almost everything about Ehrenreich's explanation is, as we will see, wrong despite its selective use of reliable facts. I mention all this because the magnitude of what we suffer due to men's violence is too serious to pussyfoot around by granting that everyone is entitled to an opinion. Of course, everyone *is* entitled to an opinion, but anyone insisting that men do not possess an instinct to kill other men in certain conditions is in factual error.

The central "truth" of sociologists is that nature, especially that of humankind, is nice and that people are designed to do things that, all in all, favor the survival of their species. Hence people could never be equipped by nature with instincts to kill other people. This idea comes from the Bambi school of biology, a Disneyesque vision of nature as a collection of moralistic and altruistic creatures. It admires nature for its harmony and beauty of form and for its apparent "balance" or even cooperativeness. It admires the deer for its beauty and fleetness, and it grudgingly admires the lion for its power and nobility of form. If anything is really wrong with us, it explains, it is a sociocultural problem that we can fix by resocializing people. It is not a biological problem.

Nature, however, is actually a dynamic state of recurring strife—of relentless competition, dedicated predators and parasites, and selfish defense. The deer owes its beauty and fleetness to predators such as mountain lions, which kill the clumsiest and slowest deer first; to competitors for food; and to competition between males to mate. Without predators, deer would not only lack fleetness; they would lack legs altogether. They would be slugs oozing from one plant to another. Yet even if these deer-slugs were the only animals out there, natural selection would favor the evolution of faster and more aggressive deer-slugs and would favor any other trait that made them superior competitors against each other. This would include the killing of one deer-slug by another in situations where it boiled down to kill or die.

Moreover, the power and noble visage of the lion (or of the family cat or dog, for that matter) rest entirely on natural selection having shaped not only a fleet predator and efficient killing machine but also a very violent competitor against its own kind in situations where the options were narrowed to exclude or kill, or else fail to survive or reproduce.

Since the first pair of amoebas vied over a tidbit of organic detritus, conflict has been rampant in nature. Robert L. O'Connell explains:

> Weapons are truly ancient, far older than man—perhaps nearly as old as life itself. It is among the stingers of colonial invertebrates and the body armor of Paleozoic crustaceans that the genesis of weaponry is to be found. . . . Too frequently, weapons development is viewed as fundamentally unnatural, a particular curse of mankind that sets him at cross-purposes with the mechanisms of his environment.

> This is far from true. The world of nature is essentially a violent one.

This violent world of nature, including human nature, multiplied by millions of competing species, is one that biology has revealed to us. It is the world of Darwinian natural selection and its variants of kin and sexual selection. These important processes shaped the behaviors of the life forms on this planet (and on any other planet where life may exist). All behavior has been shaped for the maximum survival and reproductive success of the genes *of the individual and/or its close kin* (not of species). And although natural selection produced the beauty we admire in nature's form and function, not everything it has shaped is pretty. Much of it is selfish, ugly, or violent—some aspects of human nature included.

To pretend otherwise is not just self-indulgently ignorant. When it comes to our desperate need to understand human violence, it is dangerously, misleadingly ignorant, possibly criminally so.

Trying to explain human behavior without the insights of Darwinian biology is akin to trying to explain how the solar system works based on the belief that the earth is stationary and the universe revolves around it. Explanations are possible, and some may be poetic, beautiful, reassuring, or otherwise appealing. But none will offer us an understanding of reality.

That most men are resistant, reluctant, or personally afraid to kill does not equate to men being inhibited by nature from killing. Everyone—even our warlike sibling species, the chimpanzees—knows that killing is a very serious and dangerous business, and almost everyone is reluctant to kill. But in war, all that most men need to know to prompt them to kill is that their opponent is a true enemy—someone who is either trying to kill them or has seized something vital from them—and that the odds are good of winning. Of course, many soldiers in politically motivated wars, as opposed to tribal or community-based wars, have refused to kill opponents. But their reluctance is normally the result of their being unconvinced that the opposing men are true enemies who deserve death and/or are worth risking their own lives to kill. Significantly, the more powerful the weapon at hand and/or the greater the distance between a soldier and his opponents, the more willing he is to kill.

Most combat veterans know how easy it is to kill a true enemy without remorse, although they rarely speak of this because of the appalled looks they would get. Archconservationist David Brower, for example, felt such remorse over shooting a bird and a rabbit that he never hunted again. But later, as an artillery officer in World War II, he found it easy to call in barrages on German troops. "There's nothing more exciting than killing a man," one of my best friends explained after combat in Vietnam, "not exciting as recreation, but exciting physiologically." This man also never hunts, believing that it is cruel. Consider Colonel John George's reaction to World War II combat on Guadalcanal: "I cannot recall the least bit of thinking about having just killed a man for the first time. All I remember is a feeling of intense excitement." And Sergeant John Fulcher, a U.S. Army sniper in Europe, explains: "When you go into combat, you revert to the most vicious kind of animal that ever walked the earth. You become a predator. I got to where it hurt me more to kill a good dog than a human being."

In my research for this chapter, I found pages of similar confessions by male combat veterans, but none by women. This is due in part, at least, to the reality that no nation or tribe has ever relied heavily on women combatants (not Israel, not North Vietnam, and not even the Soviets during World War II). This male monopoly on war urges us to identify what it is about men that makes them so different from women and so predisposed to war.

Male Bonding in War

"[IN] THE INTIMACY of life in infantry battalions," notes Philip Caputo, "the communication between men is as profound as between any lovers. Actually it is more so. . . . It is, unlike marriage, a bond that cannot be broken by a word, by boredom or divorce, or by anything other than death. Sometimes that is not even strong enough."

When circumstances force unrelated men to rely on one another in combat against a common enemy, they often develop camaraderie. They act like kin and even call themselves "brothers in arms." William Manchester, for example, though horrified by World War II in the Pacific, said of his compatriots:

> It was an act of love. Those men on the line were my family, my home. They were closer to me than I can say, closer than any friends had been or ever would be. They had never let me down, and I couldn't do it to them. I had to be with them, rather than let them die and me live with the

> knowledge that I might have saved them. Men, I now knew, do not fight for flag or country, for the Marine Corps or glory or any other abstraction. They fight for one another.

Clearly, a key ingredient of war is male bonding within a fighting force. Psychologist Drury Sherrod explains how bonding differs between men and women:

> For most men, most of the time, the dimension of intimacy in friendships with other men is irrelevant to their lives. According to the research, men seek not intimacy but companionship, not disclosure but commitment. Men's friendships involve unquestioned acceptance rather than unrestricted affirmation.

Male bonding, however, especially in the context of violence, has been targeted by some feminists as a sick side effect of a sick society. Dorothy Hammond and Alta Jablow call it the "myth" of "epic male friendship." Using the movie *Butch Cassidy and the Sundance Kid* as their archetype, they say that this myth is promulgated unrealistically in heroic movies and literature.

> Friends are described as male peers cooperating in a hazardous venture, whether it is robbing a bank or slaying monsters. They support one another throughout a life of adventure and danger, in which they exhibit great courage and fortitude. Mundane affairs such as marriage, making a living, or having a family, which make up the lives of other men, do not concern them. The emotional bonds between them are stronger than any other, even ties to wives, children, or kinsmen.

Hammond and Jablow not only insist that such male bonding is strictly cultural, but they also condemn it as irrelevant and unhealthy today, because it is socially irresponsible in its emphasis on aggressiveness and combat. Yet, ironically, Hammond and Jablow also admit being unable to understand why each normal boy's search for his epic friend begins so early on the grammar school playground.

"By the second grade," notes psychologist Perry Treadwell, "boys begin to bond and become more disruptive." Treadwell suggests that this is a hormonal phenomenon. Levels of testosterone, he explains as an example, have been found to soar in Norwegian and Finnish men who master military skills in parachute school, but they plunge in recruits who fail. These men are like Sapolsky's baboons from Chapter 1: confidence spurs an aggressive attitude, which spurs testosterone levels to rise, which reinforces an aggressive reality, which spurs yet more testosterone, and on and on. The same thing happens to boys by second grade.

Not only does this *not* happen in women, but anthropology also reveals that women do not bond as men do. Two inadvertent real-world "experiments"—one American and the other Israeli—yielded fuzzy results in this regard. The American "experiment" is described below. The Israeli "experiment" involved the use of women in combat in 1948. It was a disaster, due in part to out-of-control reactions by male Israeli soldiers to having women killed and/or raped. Afterward, Israel limited women to support units. Whether Israeli women combatants bonded into do-or-die units is unclear.

Illuminating this question is Janet Lever's classic study of children's games. In 1972, Lever observed games played by boys and girls. Girls' games, she found, were shorter and less competitive than boys' games, and they involved fewer participants. Lever attributed this to girls' lower ability to resolve disputes. She described the major differences between boys and girls: "Boys were seen quarreling all the time, but not once was a game terminated because of a quarrel, and no game was interrupted for more than seven minutes." Meanwhile, "most girls claimed that when a quarrel begins, the game breaks up, and little effort is made to resolve the problem."

Whether girls' lack of bonding (which seems similar to that among female gorillas and chimpanzees) persists into adulthood or not is unclear. But after the Vietnam War, the U.S. military began its own experiment with women and weapons. At that time, the military had too few male volunteers intelligent enough to handle its high-tech weapons. So it enlisted women (who scored higher on intelligence tests) to perform high-tech support roles, which could expose them to enemy fire. To keep these recruits, the U.S. military passed women in physical tests at much lower levels of fitness than those imposed on men. And when women failed to meet even these standards, as Arthur Hadley reports in his book *The Straw Giant: Triumph and Failure: America's Armed Forces,* officers often lied to cover for them. The results were not only that most women recruits were too weak to carry a full combat load (ninety-plus pounds), but also that many U.S. officers doubted that women could cope psychologically with combat or killing. The acid test of units of women engaged in unrestricted combat remains to be conducted. But even without this test, history reveals no armies of women bonded into the do-or-die military units vital to win in combat.

History and anthropology reveal instead that the "mythical" power of male bonding to strike deep emotional chords in boys and men is universal. Men in every culture bond, primarily in anticipation of danger. Today danger remains as common as it was in the primeval jungles and savannas of our past. Perhaps a more important question is, How is it that men, who in most of today's wars are no longer close kin, bond so tightly?

This is a critical question, because war would be impossible if the individuals who waged it did not decide to bond and cooperate by fighting against other men. *Without such decisions, wars could not happen.* These decisions to bond and fight do not come easily. Take, for example, Colonel John George's description of the terror faced by combat veterans during World War II on Guadalcanal before every attack on Japanese jungle defenses: "This terror is born of the memory of horrible sights [they have] seen before—visions of splintered, jagged ends of bone, with bloody flesh and marrow; of gaping wounds with intestinal juices and bits of semi-digested food gurgling out of a man who is still alive and feeling it all."

Likewise, during the Vietnam War, men bonded in battle and performed unbelievable heroics. Lieutenant General Harold G. Moore, who commanded the First Battalion of the Seventh Cavalry, explains the chemistry wrought in a horrific three-day battle in 1965 in the Ia Drang Valley, where Moore's 411 men fought more than 2,000 NVA and Vietcong. Moore explains that as the First/Seventh suffered 200 casualties, including 79 men killed, while killing 1,500 enemy soldiers, "We discovered in that depressing, hellish place, where death was our constant companion, that we loved each other. We killed for each other, we died for each other, and we wept for each other. And in time we came to love each other as brothers."

The same bonding occurred in the Korean War. Colonel David H. Hackworth, decorated with 110 medals for his service in both wars, tells why, while in Korea, he committed himself to becoming a superlative warrior:

> Sure, I was fighting for America, for all that was "right" and "true," for the flag, the national anthem, and Mom's

> Apple Pie. But all that came second to the fact that the reason I fought was for my friends. My platoon. And as I walked on, I concluded that that was why most other soldiers fought, too. The incredible bonding that occurred through shared danger; the implicit trust in the phrase "cover me"—these were the things that kept me going, kept me fighting here in Korea. . . . [T]he most important thing was that I knew with other troopers' respect came their trust: they knew that I wouldn't let them down. And to the best of my ability, I never would.

Such bonding occurs worldwide. Consider, for example, anthropology's "nonviolent" Semai of Malaysia. Drafted by the British after World War II to fight communists who had killed some Semai villagers, the Semai men eagerly trained and even more eagerly attacked. They slaughtered their foes in a frenzy of "blood drunkenness" so intense that some of them even *drank* communist blood. Most were so eager to kill yet more communists that they forgot to loot the bodies. Once home, however, they reverted to their typical nonviolent behavior. The Semai sent anthropologists back to the drawing board.

To uncover what predisposes men psychologically to fight together, Colonel S. L. A. Marshall interviewed soldiers in four hundred U.S. infantry companies battling the Japanese and Germans in World War II. In this war, during which 300,000 bullets were fired for each man killed, many men never even fired their rifles. "On average," Marshall found, "no more than 15 percent of the men [and, adding in casualties who could not be interviewed but did likely fight, no more than 25 percent] had actually fired at the enemy positions or personnel with their weapons during the course of an entire engagement." Since World War II, more appropriate training (by having trainees fire at human silhouettes instead of circular targets, for example) has increased the proportion of U.S. infantry soldiers firing in battle to 55 percent in Korea and 90 to 95 percent in Vietnam. But the 15 percent ratio during World War II astounded officers. Who were these 15 percent?

A further survey revealed that, on average, those 15 percent who fired and advanced on the enemy—battle leaders of their own initiative—were better educated and scored higher on intelligence tests than did men who did not fire. A third survey showed that 24 percent of World War II's most outstanding enlisted combat performers had attended college. This seems to indicate that men who decide to fight in combat are smarter than average—or at least not stupider (in contrast to rapists and murderers, who are less intelligent than average). Is this because true warriors are smart enough to know that their success in combat will earn them rewards and, conversely, that a poor showing may cause them to lose everything? Probably both. But to be more certain, we must dig into the lives of some warriors to see what makes them tick. Consider the Maasai warrior Tepilit Ole Saitoti:

> I remember from my own experience as a warrior how self-confidence takes over the whole being, along with pride and a feeling of ease, as if you yourself and all those around you were thinking, "everything will be all right as long as the warriors are here." We were supposed to be brave, brilliant, great lovers, fearless, athletic, arrogant, wise, and above all, concerned with the well-being of our comrades and of the Maasai community as a whole. We realized that we were totally trusted by our community for protection, and we tried to live up to their expectations. . . . Maasai warriors share practically everything, from food to women.

For Saitoti, at least, bonding as warriors reaps major rewards.

Meanwhile, noncombatants are horrified by war. These same noncombatants, however, revile cowards and deeply admire selfless and successful warriors. A classic example, known and remembered by nearly every American citizen, would be twenty-one-year-old American Revolutionary Nathan Hale, a freedom-fighter who declared before the British hanged him, "I only regret that I have but one life to lose for my country." Winston Churchill expressed the sentiment of all his countrymen when he extolled British fighter pilots before the House of Commons after the Battle of Britain: "Never in the field of human conflict was so much owed by so many to so few."

The bottom line is that male bonding during war is universal. We now even understand exactly why it happens.

Male Bonding and Reciprocal Altruism

How such do-or-die trust developed between distant male kin, and even between nonrelatives, was yet another evolutionary mystery until 1971, when Robert Trivers used his evolutionary tool kit to solve it.

Trust and mutual aid by those not closely related, Trivers saw, developed via a process he called *reciprocal altruism*. Though powerful enough to prompt many males to willingly risk death, this process is fragile. It works only if the following conditions hold: the cost to the actor of his risky act is less than the benefit to the receiver; the actor can expect the receiver, or someone else, to reciprocate during his lifetime; and the actor can recognize other members of his social group so that he can remember who owes him a good turn versus who is a cheater.

With all these restrictions, it seems that reciprocal altruism is more likely to fail than to succeed, especially in modern social groups of millions of people who are commonly anonymous in their cheating. And if recipients of good deeds too often cheat by not reciprocating, reciprocal altruism can never get off the ground. In fact, for natural selection to favor reciprocal altruism, the reciprocal altruists must net more benefits than the cheaters do. In short, *reciprocal altruism must be enlightened self-interest, or it will go extinct in a flash*. True selfless altruism, of course, is always doomed because it can never survive when pitted against any form of selfish self-interest.

Despite the fragility of reciprocal altruism based on enlightened self-interest, dozens of field studies have found that it exists. Cleaner fish in coral reef ecosystems are a classic example. These tiny fish (about four dozen species are known) set up permanent cleaning "stations." Their clients are huge fish of other species that could swallow six cleaner fish in one gulp. But they never swallow even one. The tiny cleaner fish roam over a client, nibbling away at its ectoparasites. Many cleaners even enter the client's mouth—and exit unscathed. Finally, the client leaves, often to go hunting for non–cleaner fish. Another, more macabre example is found among vampire bats. Pairs of these unrelated bats develop a tight buddy system, a lifesaving relationship in which one bat will regurgitate blood to a specific buddy in danger of starving. Bat buddies who fail to reciprocate when they can (yes, bats *can* tell this) are ostracized.

Although these two examples are intriguing, it is African lions that show us the role of reciprocal altruism in war. In lion society, only females remain in their natal pride, defending their turf against other females, sometimes in lethal battles. Males, however, live even more violent lives. Two- or three-year-old males are forced out of the pride by the adult males. When these young males emigrate, they do so in a kin group of littermates, half brothers, and cousins, and they all stay together.

Eventually, these young males fight together as a "combat squad" to usurp or kill the adult males of a different pride. Then they kill the pride's young cubs. An astounding one out of four cubs born in the Serengeti is brutally murdered by an alien male. Why? Because an average kin group of males controls a pride for only thirty-three months. Like the male monkeys discussed in Chapter 5, lions who kill the young cubs sired by previous males sire more cubs of their own—and sire them soon enough that their own cubs will have their best chance of surviving to adolescence under their fathers' brief protection.

Success in these pride takeovers requires the cooperation that kin selection can provide, because superior numbers are vital to winning in this lethal warfare. The payoff is not just survival; it is increased inclusive fitness between male littermates and cousins after victory—normally. But litters are sometimes decimated by starvation or by predation by hyenas or alien lions, and only one or two male cubs survive. At adulthood, these survivors are too few to hope to win in combat against a larger pride, so they make a giant leap in strategy: they form alliance coalitions, based on reciprocal altruism, with other unrelated males who have been caught in the same predicament.

The males in such a coalition would otherwise be deadly enemies, but as coalition allies, they are nearly as successful as kin groups in usurping prides from other males. Field biologists Anne Pusey and Craig Packer found that 44 percent of male "kin groups" in the Serengeti were not kin groups at all. Instead, they were coalitions of unrelated males using reciprocal altruism to the hilt. Nor are lions unique in this. Other social carnivores, such as cheetahs and wolves, do the same thing to gain a numerical advantage in lethal combat.

But reciprocal altruism has its limits. Coalitions of unrelated male lions never number more than three. Larger groups are always true kin groups. Three is the magic limit for male coalitions because rarely do more than three females in a pride ever come into heat at the same time. Remember, these males bond into a coalition so that they can breed, but because they are unrelated, they can gain reproductively only if each one of them gets to mate. Moreover, their coalition would instantly disintegrate if two of these allies were pushed into fighting over a female to the point of injury or death. Male lions seem to know their arithmetic too. In stark contrast, in a pride controlled by a large, true kin group of many males, when only one or two lionesses come into heat and only one or two males get to mate, all the other males gain reproductively via inclusive fitness.

Nearly all human relationships are also colored by reciprocal altruism. They are often cemented by exchanging gifts or favors. Indeed, along with kin selection, reciprocal altruism rules meat sharing by hunters and gatherers. Trivers identifies the emotions and attitudes of friendship, gratitude, sympathy, trust, and integrity—versus those of dislike, moralistic aggression, indignation, suspicion, guilt, dishonesty, and hypocrisy—as psychological adaptations fostered by natural selection to keep our reciprocal altruism relationships on track. Think, for example, how quickly you feel wronged by someone who fails to return a favor. Conversely, think how quickly you feel good about one who has done you a favor he or she did not even owe you. The rules of reciprocal altruism are emblazoned into our psyches.

In this look at male bonding via kin selection and also via reciprocal altruism, we should bear in mind that men bond and cooperate in war only because their ancestors who did so left more descendants than the ones who failed to bond. Yet we also know that too often the rewards to warriors in the twentieth century have been so diluted by political-religious dogma that no rational man should give his life for them. Japan's desperate gambit in World War II, for example, was to convert thousands of willing soldiers into human bombs. These kamikaze pilots flew suicide missions that killed 5,000 Americans by sinking 34 U.S. ships and damaging 288 others. Consider Japan's *First Order to the Kamikazes:*

> It is absolutely out of the question for you to return alive. Your mission involves certain death. Your bodies will be dead, but not your spirits. The death of a single one of you will be the birth of a million others. Neglect nothing that may affect your training or your health. You must not leave behind you any cause for regret, which would follow you into eternity. And, lastly: do not be in too much of a hurry to die. If you cannot find your target, turn back; next time you may find a more favorable opportunity. Choose a death which brings about a maximum result.

That half of Japan's 2,363 kamikaze planes crewed by about 5,000 kamikazes completed their missions reveals how deeply the instinctive power of male bonding runs. That this instinct in men could be led so far astray—based on and enhanced by kin-selection, promises of racial dominance, and spiritual-racial unity—also tells us that it must indeed be one of the most powerful impulses in the male psyche. Overall, the evidence of men at war reveals that *men worldwide bond together to increase the odds that their warfare will be successful.*

But what is it, precisely, that triggers *Homo sapiens* to launch a war to begin with?

The Dynamic Elements of War

WAR ANALYST Stanislav Andreski concluded that the trigger for most wars is hunger, or even "a mere drop from the customary standard of living." Anthropologists Carol and Melvin Ember spent six years studying war in the late 1980s among 186 preindustrial societies. They focused on precontact times in hopes of collecting the "cleanest, least distorted" data. Andreski, it seems, was right. War's most common cause, the Embers found, was fear of deprivation. The victors in the wars they studied almost always took territory, food, and/or other critical resources from their enemies. Moreover, unpredictable disasters—droughts, blights, floods, and freezes—which led to severe hardships, spurred more wars than did chronic shortages.

This also holds true among modern nations. In 1993, political scientists Thomas F. Homer-Dixon, Jeffrey H. Boutwell, and George W. Rathjens examined the roots of recent global conflicts and concluded, "There are significant causal links between scarcities of renewable resources and violence."

In short, many wars seem to be a mass, communal robbery of another social group's life-support resources. This conclusion, however, as enlightening as it may seem, does not explain why men fight wars while women stay home and worry about wars. Is there some instinct of war logic programmed into the male psyche that renders men unable to resist starting a war when the odds of winning look good? All wars are initiated by someone who decides to launch an offensive. What makes this happen?

We know that war is ruled by an unforgiving logic—one that compels and rewards wise action with victory but punishes unwise action with defeat. This logic can force the decision to wage war merely by placing too high a price on *not* fighting. Indeed, all analysis of war relies on a game theory format. In war each player, willing or not, uses strategies to maximize his own gains and to minimize his losses. Of course, the victor takes the biggest chunk of the prize, although wars may shrink this prize overall. Even so, men who start wars always expect to end up better off. Paradoxically, most of the major wars of the twentieth century were initiated by the side that lost.

So what are these dynamics that rule the logic and psychology of war? Biologists John Maynard Smith and G. R. Price analyzed and categorized war "players" into all possible strategic types. The three main types include *hawks,* who fight any player regardless of consequences; *doves,* who are pacifists and never fight; and *bourgeois,* who fight to hold their territory but never to steal someone else's. Bourgeois strategy conveys both rationality and irrationality to an opponent by saying, "We are reasonable; we don't attack without provocation, but we do hold sacred principles for which we will fight to the death." Thus bourgeois are "conditional strategists" whose acts depend on those of their opponents. When facing doves, bourgeois act like doves. When attacked by hawks, bourgeois fight like hawks. Other conditional strategists are *bullies,* who act like hawks until attacked in retaliation, at which time they flee, and *prober-retaliators,* who act like bourgeois when attacked by hawks but who, when an opportunity presents itself, attack doves.

Maynard Smith and Price found that among all players of equal ability, bourgeois consistently win. This is because bourgeois fight on home turf—with the home court advantage—and thus use the strategy with the least risk. Bourgeois also occupy the moral high ground: they are "in the right." Hence acting bourgeois is *the* "evolutionarily stable strategy" in the universe of war. In contrast, doves (pacifists) always lose, except against each other.

The ultimate self-defense nation employing a classic bourgeois stance is Switzerland. Spurred by William Tell into revolt against Austria's Hapsburg Empire in 1291, the Swiss launched a two-century war for independence. By 1499, they had won. By 1848, the Swiss had shaped the most democratic government on Earth and voted to guard their independence by always arming their citizens with state-of-the-art weapons. Every able-bodied adult male today spends a year in active military training. At age twenty, each is issued an assault rifle to keep for life—along with any handguns, machine guns, and even howitzers he may care to own privately. "Indeed the militia is virtually synonymous with the nation," note David Kopel and Stephen D'Andrilli. "'The Swiss do not have an army, they are the army,' says one government publication." This has saved their lives against Russia, France, and even Hitler's Nazis (who had prematurely drawn Switzerland onto their map of the Third Reich in hopes of emptying its coffers of gold). Today tiny Switzerland can mobilize 650,000 well-armed, well-trained citizen soldiers in under twenty-four hours! In nature, most animals use a bourgeois strategy like the Swiss.

Prober-retaliators are the next most successful strategists. Hawks attack bourgeois or probers who have been weakened for some reason, or they attack doves. They usually win and thereby keep offensive war as a permanent event on the calendar of the human condition.

Many political leaders understand these realities instinctively—as do chimpanzees—even if they are unable to explain them. "Wars are won or lost," note political scientists Paul Seabury and Angelo Codevilla, "nations live or die, primarily by the people's willingness to fight, their ability to impose discipline on themselves, and the readiness to subordinate themselves to chiefs who know what they are doing."

Beyond this logic, however, real war is universally nasty, because while seizing someone else's resources or territory by force would pay big if the owners did not resist, the owners *do* resist. This is because no one can afford to lose and because no player who is a pacifist can ever win. Hence it is easy to create a world in which most social groups adopt a war stance. As anthropologist Andrew Bard Schmookler explains, *if ten tribes exist and nine crave peace, it takes only number ten to start a war—and thereby induce a defensive readiness for war in each of the other nine. They simply cannot afford not to muster a self-defense.*

The only reality that keeps war from erupting even more frequently is its high cost, which, even among wild chimpanzees, often deters both sides from attacking unless the other side seems weak.

The Yanomamo of Venezuela are a prime example of a people trapped on the "logical" treadmill of war. The Yanomamo believe that all deaths not due to a direct attack by people or animals can only be due to the sorcery of rivals in other villages. This prompts villages to resort to a practice called *nomohoni:* they invite those they suspect to a feast at which they treacherously murder them while their guard is down. The killers always abduct the slain men's women—and thus start a war that may last for twenty years. The Yanomamo recognize that it is their theft of women that ultimately starts all their wars. But once war begins, not to avenge themselves by launching sorties is unthinkable. Otherwise, they say, their enemies would soon overwhelm them.

Always suspicious of Machiavellian treachery by "allies," the Yanomamo weigh the strengths of their alliances with other villages against those of their enemies before deciding on a raid or a *nomohoni*. If a raid, the raiders target *waiteri* (fierce ones) or the enemy headman himself for an ambush. Such strategic assassinations often win a quick victory and may even spur wholesale flight of an enemy village. As with all military sorties, surprise is the most important advantage. Yanomamo wars seesaw in tit-for-tat revenge raids aimed at killing one or more of the enemy, then escaping undiscovered. Because any loss among the raiders means failure, no matter how many enemies they kill, they use extreme stealth and kill by ambush with curare-tipped arrows. But if the targeted victim is unavailable in enemy territory, any enemy will do.

Despite all this stealth and ceremony, about 30 percent of Yanomamo men die violently. This is about the same risk of violent death as among Gombe chimpanzees.

If living like this seems crazy, remember the level of American fear that prevailed from 1949 to 1991 due to the cold war with the Soviet Union and the nuclear arms race it fueled. Like the cold war, this cycle of raid and counterraid and tit-for-tat assassinations persists among the Yanomamo because it seems their best option. Indeed, if the ranking man of a village whose headman had been murdered "failed to put on a show of ferocity and vindictiveness," explains Napoleon Chagnon, "it would not be long before his friends in allied villages would be taking even greater liberties and demanding more women. Thus, the system . . . demanded that he be fierce."

That this warfare is rough on women is also clear. At least 17 percent of all wives are abducted by *waiteri,* usually during *nomohoni* treachery. And most of these women are raped.

All these horrors persist due to the implacable logic of war and the power of male bonding. Even so, not every man in a nation at war is a warrior. Some men adopt what evolutionary biologists call a "sneaky" strategy of appearing to fight or of supporting men who do. The larger and more complex the social group, the easier cheating—or being a pacifist—is. Further, successful cheaters and pacifists can try to prosper by taking none of the warrior's risks but then sharing part of his prize (reproductively) during victory. Some theoreticians who view war as human nature admit that cheaters could nearly eliminate any genes that exist for war, because eventually warriors would kill one another off to the point of whittling down the population to almost nothing but cheaters—and a few surviving victorious warriors. For the last warriors to be eliminated, cheaters and pacifists would have to outbreed warriors (which, as we saw among the Yanomamo in Chapter 5, they do not). But even if nonwarriors did manage to eliminate all the warriors among them, eventually that population of nonwarriors would be attacked and wiped out by another group with warriors. And even if all social groups on Earth ended up composed of nothing but cheaters and pacifists, once one warrior arose through mutation, he would reproduce like crazy at the expense (and death) of nonwarriors, who, unprotected by warriors, lose all wars. This tells us that if war is in our genes, the frequency of these genes will oscillate up and down in cycles of war so that most groups will contain some men who wage war and others who cheat. In reality, this is what we see in all tribes and nations.

The bottom line is that once the forcible seizure of another group's resources, territory, or women is possible, the threat of war is here to stay.

Analysts agree that the only immunity to war—whether fought with poisoned arrows or nuclear bombs—is deterrence. *Deterrence depends utterly on communicating one's willingness to retaliate massively, no matter what the cost, if the other side strikes first*. Indeed, the stated policy of the world's most successful bourgeois nation, Switzerland, is "prevention of war by willingness to defend ourselves," and their standing strategy is to inflict on invaders the costliest and bloodiest price possible. When both sides use this strategy of being ready and willing to retaliate massively, "stable deterrence" develops, as in the Soviet–American cold war. Although the slogan "Better Dead Than Red" may have seemed crazy, it actually prevented war as long as the threat of total retaliation was credible to the Soviets. This credibility, however, was possible only if the Soviets were convinced that America believed its greatest loss would result from *not* retaliating. "Most people today," note Paul Seabury and Angelo Codevilla, "still agree that military strength, backed by the credible will to use it, is the guarantor of world peace."

But what happens to this logic of deterrence if one side is truly crazy (Hitler, Hideki Tojo, Ho Chi Minh, Pol Pot, Saddam Hussein) and invades regardless of the cost? The invaded side knows that resistance will be so expensive that it may seem more prudent not to fight and to cut its losses by simply relinquishing some of its resources. But by doing so, it would be committing suicide rather than being prudent, because it would be signaling to all neighbors that it is a pushover. If history tells us anything, it teaches us that these neighbors would then carve the "rational" one to pieces. The lesson here is that *the irrational militarist who communicates a willingness to fight coerces the other side into a dilemma wherein their best solution is also an irrational willingness to fight, even at a huge cost.*

This cost can be endless war. The Dani of the huge Beliem River valley of Dutch New Guinea (now Irian Jaya), for example, live in perpetual war due to this logic and to the apparent irrationality of retaliating to create deterrence. Anthropologist Karl Heider explains:

> Every Dani alliance [numbering about five thousand people] was constantly at war with at least one of its neighboring alliances. Every individual Dani was touched by war. People who lived near a frontier, like those of the Dugum Neighborhood, could see and hear battles, and whenever they went to their sweet potato gardens, they had to be alert for enemy raids. Every Dani had seen friends and acquaintances dead or dying from spear wounds or arrows. Most men had helped to kill an enemy in that way, and everyone had often attended "fresh blood" funerals.

Overall, 28.5 percent of adult male deaths and 2.4 percent of female deaths among the Dani result from war. Despite this carnage, the Dani believe that revenge is vital to appease the ghosts of the slain, but they also know that they would be overrun if they failed to retaliate. Hence, like the Yanomamo, they continue their tit-for-tat killings until a hotter war draws them to a new theater. In battle, two armies of one hundred men or more face each other in broad lines on open terrain thirty to sixty feet apart and surge back and forth. They launch spears, arrows, and insults at their adversaries and dodge those launched in return. Boys as young as six years old face off along their own lines of skirmish. Coached by old men, they shoot arrows at each other.

Riskier and also more lethal than these flashy but controlled open battles are surprise raids. A dozen raiders must navigate undetected in daylight through swamps teeming with skittish ducks whose flight is a dead giveaway. The raiders must slink through a frontier scrutinized by sentinels stationed in guard towers. These sentries have memorized every detail of their terrain. Detection will lead to the raiders being killed in a counterambush.

As might be expected, the most hideous phase of Dani warfare is the first attack. In 1966, for example, an aspiring Dani big man of the Getulu clan named Mabel (New Guinea social dynamics are not ruled by hereditary chiefs but by "big men" who lead via charisma and persuasion) led a dawn raid by hundreds of Getulu men against the dozen nearest compounds that were firmly tied to some insulting big men of the Dani Wilihiman-Walalua alliance. The Getulu torched the enemy's homes, and as the unsuspecting men, women, and children fled the fires, the Getulu massacred 125 of them. The Getulu regained face against the insulting Wilihiman-Walalua alliance—and also looted hundreds of pigs (a Dani's most valuable commodity). The Wilihiman-Walalua alliance swiftly counterattacked but failed. The gardens between the new enemies became a blackened no-man's-land.

These horrors of perpetual war, as among the Dani and the Yanomamo, have spurred war strategists to agree most on one thing: the ideal strategy is to win without fighting. War's most timeless analyst was Sun Tzu, a Chinese philosopher turned brilliant general who lived 2,500 years ago. In *The Art of War,* he prescribed the rules of war as they are still understood today. For instance, he was the first to write, "Supreme excellence [in war] consists of breaking the enemy's resistance without fighting."

Sun Tzu insisted further that this can be achieved only by a good commander. *Behavior* is the number one weapon in any arsenal, and it is the commander who decides how his forces will behave. Brilliant commanders, however, are hard to find. Some of the most amazing include Alexander the Great, Hannibal, Kublai Khan, Hernando Cortés, George Washington, T. E. Lawrence, John R. E. Chard, Paul von Lettow Vorbeck, Erwin Rommel, Douglas MacArthur, Dwight D. Eisenhower, and David Hackworth, all of whom managed their men and resources to win victories against huge odds. George Armstrong Custer also is legendary, but for the opposite reason.

Many elements—violent instincts coded in our genes, reciprocal altruism, male bonding, sex, hunger, logic, opportunism, genocide, strategy, self-defense, and leadership—blend in natural selection's recipe for war. But, let's face it, the phenomenon of war produced by this recipe evolved in the lineage Hominoidea as a very risky adaptation for individual advantage worth the risk. What exactly is this advantage?

Why War?

MOST OF THIS CHAPTER focuses on the male *psychology* and *logic* of war. What we really need to understand, however, is the ultimate *impetus* behind that psychology. What is it that men seek in war that is worth risking death for and that women do not likewise seek? If it is territory, which is nearly always a goal of war, there must be something vitally important that males want from territorial war and women do not.

The answer lies in male versus female biology. Due to the endless reproductive contest between men, and due to how much less men must invest in children compared to what women do, sexual selection and kin selection have designed human males—*compelled* human males—to wage war as a strategy to cooperatively seize the territory, resources, and women of other men and to use them reproductively. Indeed, when in 1993 cognitive psychologist Leda Cosmides posed the question "Why would anyone be so stupid as to initiate a war?" the data were so clear that they allowed her to answer it with no oversimplification, "To get women."

War is typically men's ultimate reproductive gamble. Many researchers agree that the goal most worth the lethal risk of war is women or the resources that attract or support more women and their offspring. That *something* men—or male apes—seek through war is selfishly expanding or securing their own families. It is macho male sexual selection in its ultimate manifestation as the highest-stakes reproductive risk that males can take.

"If the earliest cause of fighting [wars] was to obtain and retain the means wherewith to support women and children," explains Rear Admiral Bradley A. Fiske, "it probably has been the fundamental cause ever since, even though this fundamental cause has been overgrown with others more apparent."

In contrast, a woman can gain nothing by making war that she is then capable of defending or retaining and that is also worth her risking death. This is because, according to the primeval conditions under which war evolved, a man could accrue more wives through war and thus raise his reproductive success by an order of magnitude. A woman, no matter how successful she might be in war, could barely improve her reproductive success at all with more husbands, being limited most by her own body. Worse, she would unnecessarily face death for no or little gain.

All of this leads us again to one of the most important conclusions on the instinct for violence in the male versus the female psyche: *women use violence only to defend their reproductive interests; men use violence far beyond this to expand their reproductive interests.*

It is no coincidence that Yanomamo *unokais* have three times more wives and children than non-*unokais* (see Chapter 5) or that Kublai Khan sired forty-seven sons. All men—Aka Pygmies, Ifaluks, Yomut Turkmen, Dani, Eskimos, Tartars, Akoa Pygmies in Gabon, the Ona and Yaghans of Tierra del Fuego, Australian Aborigines, the Nootka of the Pacific Northwest, the British in Tasmania, whites in America, and the Nazis and Japanese in World War II—compete most for reproductive advantage. And they do so most glaringly in war.

Denying men's communal reproductive imperative for war is as effective as a physician denying that AIDS is caused by a retrovirus, insisting instead that it is due to poverty, drugs, politics, or TV. The doctor may get away with it, but the patient will die.

• • •

Abruptly, after seventy-five rounds, the AK-47 blasting at Bill's face fell silent. (It had jammed because the kid's third magazine was half full of mud, as were the rest of the mags in his pouch.) But now M16s also fired at Bill and at the young NVA. First Platoon, sure that Bill was dead, was returning fire.

Chico approached the spider hole in a John Wayne crouch. He fired at the hole one round at a time in slow cadence. The NVA ducked inside. *Bang, bang, bang, bang,* Chico's M16 blasted.

Chico reached the spider hole as he fired the eighteenth (and last) round in his magazine. He reached down into the hole and yanked the kid out.

Bill finally worked the last chunk of the pin out of his grenade. Still holding its striker down with his thumb, adrenaline vibrated him into a time warp. He watched Chico wrestle with the kid. The NVA soldier snatched Chico's hunting knife from his belt. Still entangled, Bill watched as Chico wrestled the knife back in seeming slow motion and stabbed the thin young man in khaki. Chico knelt, and his arm raised in slow motion again and again to stab the NVA.

The NVA soldier stopped struggling. Chico dropped his knife and picked up his M16. He ejected the spent mag and loaded a new one. He looked up just as the NVA struggled back to his feet. Chico fired once, striking him in the right eye. The NVA collapsed in a heap, instantly dead. His terrified young face suddenly became peaceful.

Sporadic incoming AK-47 fire continued. First Platoon returned fire with M79 grenades, but fired them short to avoid the First/Ninth, the "hammer" to First Platoon's anvil approaching from beyond. Still holding the frag, Bill yelled, "Fire in the hole!" and hurled it toward one of the AKs. It exploded four seconds later. Bill finally untangled himself from his web gear and from the limbs growing from that tree stump, and unjammed his M16.

He stared across the rice field as the First/Ninth arrived, drawn by the firing, which abruptly ceased.

First Platoon dusted itself off. No casualties. The rest of B Company flew in. They set up a perimeter and dug in for the night.

Bill was shaken by this very close brush with death. Months earlier, on patrol, he had heard a click and spun to see an NVA soldier twenty feet away—having emerged from nowhere and aiming an AK at him. Somewhere a friendly M16 had rattled, and the NVA had crumpled into the foliage. The NVA had pulled the trigger of his AK, but he had forgotten to chamber a round. Otherwise, Bill would have been cut in half. And there had been other times. Right now it seemed like too many other times.

After helping two other grunts dig their communal foxhole, Bill could not sleep. He walked in the dark to the dinky command post (a slightly larger hole dug in the ground). He demanded the captured AK—it had worked, and his M16 had failed yet again, nearly getting him killed. The lieutenant, sergeant, and medic would not let Bill near it. Its sound signature would get him killed by friendly fire. Bill already knew that, but he was obsessed. He tried to push past them. After his third try, the medic, Jack Dempsey, wrestled Bill to the ground, grabbed his throat with both hands, and strangled him until he almost blacked out.

Bill tapped him on the arm to get him to stop. Gasping, he said, "I'm OK, I'm OK. It's all right. Thanks."

Dempsey now talked with Bill to make sure he really was all right. Still shaky, Bill returned to his post. He pulled guard all night while his buddies slept.

At dawn two NVA soldiers, suffering concussion wounds from yesterday's air strike, surrendered to Bill's unit. The "Charlie Charlie" (the colonel's chopper) dropped battalion commander Lieutenant Colonel Robert Stevenson and a dog handler, then lifted both prisoners out.

With First Platoon on point, B Company hiked along a footpath toward the river to recon the air strike. They pushed past bushes and trees, past the Montagnards' pepper bushes and marijuana plants, and past abandoned NVA rucksacks loaded with rice that the NVA had stolen. Specialist-Four Pratt walked backup behind the dog handler on point. Bill walked sweep, behind the colonel, at First Platoon's rear.

The dog alerted. The handler fired a round from his M14. Pratt shoved the handler aside and opened fire with his M16, emptying his twenty-round magazine on full auto. All the men dived right and left off the trail except Colonel Stevenson, who stood watching. Bill reached up and grabbed the colonel by his belt, yanking him to the ground.

"Sorry, Sir," Bill apologized.

"That's all right, soldier."

One of Pratt's bullets had creased the head of an NVA machine gunner waiting in ambush and knocked him backward into the creek. This gunner also had been wounded in the previous day's air strike. His unit had abandoned him with four hundred rounds and an RPD machine gun. First Platoon now took him prisoner. Charlie Charlie touched down again to evac Colonel Stevenson and the new prisoner. The NVA unit, First Platoon learned, had been there to steal Montagnard rice but had vanished to avoid being pounded between First Cav's anvil and hammer. By now they were probably already back in Cambodia.

Cambodia, Bill reflected. *That's* where all this should be happening. As things were, we were fighting a war against an enemy who had his own convenient safety zone to scuttle back to, where Americans were forbidden to follow. Everyone knew that the NVA had cities in the rain forest across the border that were stocked with enough munitions and supplies to invade most of South Vietnam. It was insane to let them get away with it.

But General Westmoreland had forbidden, by order, his commanders in the field even to mention to anyone that Cambodia was the NVA supply route, their logistical sanctuary, and their staging point for attacks into South Vietnam. (His commanders would later insist that forbidding U.S. attacks into Cambodia was the reason America lost the war.) Westmoreland's basic war strategy was to fight a war of attrition that his arithmetic told him would exhaust North Vietnam's army before his own.

Barred from Cambodia, B Company reconned the area all day. But instead of B Company finding an NVA unit, the NVA turned the tables and shot down Charlie Charlie.

Now Bill and eleven other men of First Platoon made a night insertion by chopper to guard the colonel's machine for extrication the next morning.

"Jump! Jump!" the door gunner of the hovering Huey screamed at Bill as he jammed his thumb downward out the door.

Bill looked down at the twenty-five-foot drop. Tall elephant grass waved in the searchlight, but the ground was invisible.

"No! Lower, lower!" Bill jammed *his* thumb downward emphatically and shook his head.

The door gunner could not hear, but he understood. They argued silently in the roar of the chopper. Elephant grass convulsed ghostlike below them.

For many U.S. combat troops, this war had already degenerated into pure survival psychology. Every man's first concern was counting down his 365 days toward DEROS (Date of Eligible Return from OverSeas). Going back crippled because some nervous pilot wanted to kick them out of his ship twenty-five feet above invisible ground was not an acceptable option.

Bill shook his head again and jammed his thumb downward. The pilot finally lowered to fifteen feet. He would not budge any lower.

Bill and the others stripped their web gear and dropped their sixty-pound loads into the air. Now at least they could see what they thought might be the ground. Then the men jumped with their M16s. A trooper on the next bird, whose pilot refused to drop to fifteen feet, broke his ankle. A chopper medevaced him. The eleven guards spent a restless night waiting for the NVA to try to blow Charlie Charlie to high heaven.

But the NVA never tried. After dawn, a Chinook CH-47 slung out Charlie Charlie for repair. Bill and the ten others hiked to a hilltop, where Hueys lifted them to LZ DucCo (a Special Forces camp) to link up with the rest of B Company, dug in along the airstrip. Here the powers that be dropped B Company their Christmas packages (much of which they had to jettison hours later due to weight-mobility requirements). On December 24, B Company flew back to base camp at An Khe for a Christmas stand-down and cease-fire—and for their first shower and clean jungle fatigues in eight weeks.

At An Khe, Bill finally got that warm Coke.

• • •

IN ITS OWN TEPID WAY, that Coke was an allegory for the way America's undeclared Vietnam War was strategized by Lyndon B. Johnson and then eventually ended via withdrawal by Richard M. Nixon.

"Our mission was not to win terrain or seize positions," explains

Marine Lieutenant Philip Caputo, "but simply to kill: to kill Communists and to kill as many of them as possible. Stack 'em like cordwood. Victory was a high body count, defeat a low kill ratio, war a matter of arithmetic." Morale in Caputo's regiment sank so low as early as 1966 (U.S. combat troops would persist in Vietnam for six more dismal years) that Caputo's company commander announced a new policy: "From now on, any marine in the company who killed a confirmed Vietcong would be given an extra beer ration and the time to drink it. Because our men were so exhausted, we knew the promise of time off would be as great an inducement. . . . This is the level to which we had sunk from the lofty idealism of a year before. We were going to kill people for a few cans of beer."

America's Vietnam War was "war" all right, but not war in the clear-cut context of one social-political group (tribe, nation, religion, kingdom, or clan) vying in genocidal conflict with another over possession of territory, resources, women, trade rights, and the like. Americans at home really did not care about, covet, need, depend on, claim rights to, have a stake in, or even admire anything in Vietnam. The Vietnam War, however, was not an exception to normal war. Instead, it was an extreme lesson of war as a corrupt and brutal government tactic in international posturing as part of a strategy aimed at establishing recognized world leadership as the dominant (alpha) nation. In short, the Vietnam War was what war is ultimately all about—primary control of critical and scarce resources on the widest scale possible: globally. It was also about trying to defeat an enemy (the Soviets) vying for the same thing. Significantly, however, the Soviet enemy was absent from the field of battle.

The lesson here lies in that the scale, location, and conduct of the war were so "unnatural" that they failed to engage the instinctive warrior psychology—or patriotism—of most Americans sent to fight it.

True, this was not the sort of war that Bill and the 3 million other Americans shipped to Vietnam understood from watching John Wayne movies on TV while growing up. Why we draftees entered U.S. military service in this era so willingly is a sobering lesson in what a "totalitarian" government can create. When Bill and I and our nineteen-year-old peers were conscripted in 1966, few of us questioned the legitimacy or outcome of the Vietnam War. We were ignorant due to a propagandized and simpleminded system of public education. First, we had been told repeatedly that the communists were murderers who sought to enslave the world (which they were). Second, most of us were loyal Americans who trusted our president enough to accept the idea that Americans who refuse to fight when their country calls are traitors. Third, America always won. In short, we fought out of naïveté and out of loyalty and love for our country. Some even fought out of a desire for adventure. Most fought to avoid disgrace—and being sentenced to prison (hence my "totalitarian" jibe at U.S. conscription).

All this gives us a final huge—and vital—lesson in war. The Vietnam War was an inadvertent "experiment" revealing the natural limits of the male psyche to "agree" to kill in a warlike situation. In short, as strange as it may sound, this war was not a "war" for most American combat troops. GIs were shipped to Southeast Asia to shoot the bad guys—as defined by some rather hazy principles—in order to help the good guys—who turned out to be one, then another political regime too corrupt even to help themselves. Few, if any, Americans actually believed that anything of personal importance was at stake in Vietnam. Yet U.S. troops were ordered to kill NVA and Vietcong—or, again, go to jail. Under these conditions, U.S. troops did kill, but in so doing *they exceeded their natural limits of acceptable violence in the context of what Vietnam meant to them, which was close to nothing.* From the perspective of most American combatants, the war was unnecessary.

"We're the unwilling led by the unqualified doing the unnecessary for the ungrateful." This was the lament, a sad one indeed, of most U.S. troops in the field in Vietnam. But even this vastly understates the effects on the "guinea pigs" used in President Johnson's "experiment" in meaningless "war."

"Our rule of thumb," writes Matthew Brennan, veteran of thirty-nine months of combat with the First Air Cavalry's "Headhunters," "was that the typical new recruit had about six months before he was killed, wounded, or pushed to the edge of insanity." It was this "unnecessary" killing that drove many U.S. combat troops crazy. Yet the psychological damage these men sustained was not due just to having followed insane orders to kill nearly a million NVA, Vietcong, and civilians, or even from seeing their own brothers in arms die horribly. Instead, it was because the killings and deaths and maimings on both sides were for nothing—"nothing" because no strategy or plan even existed in the Pentagon to win the war (and U.S. troops *knew* it), and "nothing" because there was nothing important at stake in Vietnam to any average U.S. combat soldier, except his own survival, which he could have more easily ensured simply by staying at home. In short, the lethal violence of U.S. troops made no sense at all in that delicate natural computer, the human male psyche.

One major lesson here is this: although young men can be convinced initially via propaganda and coercion to kill opponents (especially from other racial groups), unless these men *believe* their opponents to be true enemies, they will ultimately disobey orders, mutiny, or go crazy. The Vietnam War revealed this in spades. It revealed the instinctive limit in the human male psyche for killing: *all killing must be fully "justified" in the mind of the killer, or the killer's sanity slips away.*

The structure of governments today fosters self-growth. Bigger government demands higher taxes—either in money or conscription. To justify higher taxation, some political leaders abuse their vested powers by "inventing" big enemies (drugs, poverty, guns, communists) from which we little citizens on our own cannot protect ourselves. If we fail to pay, politicians warn, we are doomed. This is *political extortion.* It leads to massive abuse of power.

Johnson's political war was a massive misuse of men's lives and spirits, one that scorched the patriotism of the entire baby boom generation to cold ash and replaced it with a supreme cynicism of U.S. leadership. Lieutenant Colonel Charles F. Parker explains:

> A great nation, one that is conceived in liberty with a government of the people, by the people, and for the people, does not have the right to abuse the trust, courage, endurance, and sacrifice of its soldiers—who are its own sons and daughters—this way.
>
> It is true that soldiers don't really fight for King and Country. They fight, first of all, to survive, and second of all, not to let their comrades down. Sometimes those priorities are reversed. But somewhere in the back of the American soldier's mind is a childlike faith that somehow this horror is worth it to the nation. To betray that faith, like the Johnson administration did in Vietnam, is contemptible. And that betrayal has probably killed that childlike faith of the nation's soldiers forever.

• • •

Bill now drinks *cold* Cokes—in therapy meetings with fellow Vietnam combat vets suffering from post-traumatic stress disorder. These meetings, he tells me, are sometimes worse than that damned war.

CHAPTER SEVEN

Genocide

It seems, then, that the attacks are an expression of the hatred that is roused in the chimpanzees of one community by the sight of a member of another. Strangers of either sex may trigger this hostility, but the unthreatening females are attacked far more often. In this way, males dissuade them from moving into their territory—if, indeed, they survive—and food resources in the community range are protected for their own females and young.

— Jane Goodall, 1990

Nuclear, chemical, and biological warfare continue to be perils, particularly in the hands of nationalists and religious extremists trying to preserve intact their antiquated ideas and their already hopelessly contaminated gene pools.

— Christopher Wills, 1998

If only I could blink, thought U.S. Gunnery Sergeant Lloyd Paul Blanchard as he lay spread-eagled on his back. He tried to stare away from the sun. With his eyelids taped open, this was impossible. His struggles only delayed the inevitable. The Japanese would blind him and kill him here. Knowing that this was the end, Blanchard's mind wandered back to the twists of fate that had landed him in this hellish prison camp.

Japanese soldiers storming the Philippines had used suicide tactics at Bataan in early 1942. They had charged American machine gun emplacements and had thrown their bodies on the barbed wire defense entanglements to serve as a writhing, groaning bridge over which yet more Japanese infantry charged. Blanchard had watched in stunned fascination as the Japanese had charged through minefields and past sharpened bamboo stakes despite appalling losses. Some Japanese had managed again and again to reach the Philippine and American riflemen and machine gunners—only to be shot by Blanchard and the rest and then collapse into grotesque heaps of death that became partial bulwarks. To Blanchard's fellow soldiers firing from shallow foxholes scooped out of the rocky soil with a helmet, even the stench of death had seemed a fair trade for this added protection from Japanese fire.

This battle for the nearly useless, tortured landscape of Bataan dragged on for day after day and month after month of heroism, cunning, heavy losses, and terror on both sides. Blanchard feared it would never end.

Finally, on February 22, President Franklin D. Roosevelt had ordered General Douglas MacArthur to abandon his Philippine command on Luzon and retreat to Australia. The trapped command had then fallen to General Jonathan Wainwright, whom Blanchard considered the best general he ever knew.

Now a bleak situation became hopeless. Rations in Bataan had been cut in half as early as January. Now, in April, the pack mules, water buffalo, and dogs had long since been eaten. Even the wild pigs, monkeys, crocodiles, iguanas, and pythons of the surrounding, mountainous jungle were scarce. Epidemic starvation had convinced Wainwright that his men could no longer stay put and expect to hold Bataan's mile-high volcanoes. So, on April 4, he ordered a counterattack on the 80,000 Japanese veterans assaulting them. His 12,000 starved and diseased GIs and 66,000 Filipinos ferociously counterattacked, regaining up to five miles a day. But finally, knowing that no U.S. support in food or ammunition was available, and that the Japanese soldiers were getting fatter daily by looting Manila, MacArthur ordered Wainwright to explode his own remaining ammunition and surrender. Wainwright did so, then he and a few GIs escaped, retreating to the island fortress of Corregidor. Sergeant Blanchard was not among them.

On April 9, along with 75,000 other starving Philippine and American defenders, Blanchard had been forced to surrender to Japanese general Homma. This surrender was intended by MacArthur and Wainwright to avoid a slaughter, but most of these gaunt defenders may have done as well by having fought Alamo-style to the last man. On April 10, General Homma ordered the 75,000 prisoners to march nearly seventy miles to a railway junction. The GI prisoners—all of them starving, many without water to drink, many wounded seriously with limbs amputated, and a third of them racked by malaria, dengue, or other diseases—reeled and stumbled in the tropical heat. Sadistic Japanese soldiers stole their hats and precious canteens of water. They bayoneted those who collapsed. They forced Filipinos to bury them alive.

Blanchard drove himself to focus on walking an orderly four men abreast. He watched the Japanese shoot or bayonet man after man who faltered. Blanchard and his mates encouraged one another to keep walking. He knew that many of his comrades would not survive this forced march, and that even those who did would live only long enough to endure the nightmare of a slower death in a prison camp. (Indeed, the Japanese would murder about 10,000 men during the Bataan Death March, and yet another 22,000 men would die of starvation and disease over the following two months in the prison at Camp O'Donnell.)

Two thoughts dominated every man's mind: escape and water. Blanchard planned to slip into the jungle at the first opportunity. It came on the third night of the Death March. Philippine guerrillas crept into the column and sliced Blanchard's bonds with their jungle knives. Blanchard and a few comrades melted silently into the night. They left the Bataan Death March behind.

The guerrillas helped Blanchard and the other escapees to a small boat, which took them to Wainwright's force on Corregidor. This old Spanish fortress—situated on a four-mile-long, pollywog-shaped island—would soon become the last American stand in the Philippines. This stand would last only one month, but it may have prevented Japan from taking northern Australia. At the fortress, Blanchard was assigned a gun. Despite the deep tunnels drilled into Corregidor, this gun was, like all the others, exposed to enemy fire. Even so, Blanchard explained, "if we had had the ammunition, we could have shot Japs [aircraft] all day long."

But they did not have the ammunition. The situation there was a replay of Bataan. Corregidor's 13,000 defenders ran low on ammunition and ran out of food. On May 4, Japanese bombardment reached the rate of 1 shell every five seconds (16,000 shells per twenty-four hours), tallying one hit for every five-by-five-yard area of the island. This destroyed many of Corregidor's big guns. Finally, on May 6, Wainwright was ordered to surrender again, but only after destroying his last antiaircraft guns. This surrender would launch years of horror so abysmal that Blanchard would later say, "Some of our men wouldn't surrender. When the Japs took us, they kept fighting, and the Japs tracked them down and killed them. If I had known then what I know now, that is the way I would have done."

The Japanese victors herded many of Corregidor's defenders onto the central Luzon Plain and drove them to Cabanatuan Prison Camp, located at a gigantic U.S. depot holding 50 million bushels of rice. Ironically, during the first four months there, 4,000 Americans died of starvation, scurvy, beriberi, and malaria. Beriberi tortured most of the men. It paralyzed the legs of many, including Blanchard.

In Cabanatuan, the Japanese were even more sadistic than during the Death March, if such a thing is imaginable. They practiced calculatedly cruel genocide. They beat Allied soldiers for no infraction, and although the prisoners were forced to work in their gardens, they were not allowed to eat any of the food they grew. When the Japanese caught a GI eating a vegetable, they broke his arms.

The Japanese assigned every prisoner to a "group of ten." When any man escaped, the remaining nine in his group were executed. One British soldier, undeterred, tried to escape anyway. He was caught. Blanchard and the rest were forced to watch him dig his own grave. The prison commander then ordered him shot.

A Japanese soldier shot him. The Brit fell into his grave, then crawled back out. The Japanese shot him again. Again the Brit crawled out. The commander ordered him shot again. But after a third bullet wound, the Brit climbed out again. Again the Japanese guard shot him. Again he crawled back out. Blanchard and his fellow prisoners stood and watched as the fifth bullet finally stopped the Brit from climbing out of his grave.

This defiant Brit haunted the camp commander. Every night for two weeks after this execution, Blanchard says, the commander went to the Brit's grave, excoriated him in Japanese, and whipped the grave with his "vitamin stick" (a length of split bamboo that "gets you with the program"). Eventually, this commander was replaced with one even more sadistic.

Blanchard watched another Brit, a leg amputee, sharpen a scrap of bone at night to secretly carve a wooden leg. Not only was he going to survive, he said, he was going to walk out of this prison camp. But the Japanese found his wooden leg and burned it. Undaunted, the Brit carved another, this time hiding it better. A miracle of Cabanatuan—

and ultimately of a U.S. Army Ranger operation—he did walk to freedom on it.

One of Blanchard's hometown buddies, Luke Mondello, also fought back. He became a Japanese nightmare. Mondello slipped carefully outside the barracks at night, stalked a Japanese guard, and then strangled and left him dead as a Philippine guerrilla infiltrator would have done. Mondello killed three or four guards this way.

These revenge killings helped keep up morale. In fact, the Americans remained unreasonably positive despite daily rations consisting of only a teaspoon of rice, which Blanchard and the rest called "R, R, and R" (rocks, rat shit, and rice), and scraps of maggot-ridden fish; despite beatings and ghastly, random tortures; and despite the theft of the GIs' only medicine from the only two Red Cross boxes they received in three years. (In a turnabout, a few merciful Japanese, who had been American educated, actually stole medicine back and used it to keep some of the prisoners alive.)

During his three years in Cabanatuan, Blanchard dreamed of home, of Port Arthur, Texas, and of his wife (who, unbeknownst to him, had divorced him and remarried). Meanwhile, the Japanese, seeking secret information, pulled out Blanchard's fingernails, one by one. But he was only a sergeant; he had no secrets.

The worst torture was the sun cure. One day Japanese guards dragged Blanchard into the open yard and staked him to the ground on his back, spread-eagled, facing the sky. They taped his eyelids open. For no apparent reason, the Japanese were burning his eyes out of his head.

• • •

Men and Genocide

TYPICAL EVIDENCE to soldiers that they are winning a war is the dead bodies of their enemies. The ultimate evidence is the ownership of those enemies' territory, resources, and/or women, all now to be converted into reproductive advantage by the victors. War is a reproductive gamble that in the primeval, preindustrial past, when natural selection shaped men's psyches, often paid off for the aggressors. *Whom* men kill in war, however, is far from random. They usually kill other men of other races, languages, tribes, and religions. Brother against brother is the rarest of circumstances. Far more often, war has been—and still is—aimed at genocide.

Today small wars rage nonstop. In the first forty-five years after World War II, dozens of dirty, expensive, murderous, and wasteful little wars have been waged. And 17 million people have been killed in them in the name of some ideology (often Marxism) but actually for genocidal goals and territorial gains. During the far more recent past, from 1994 to 1997 alone, Western aid groups tried to help victims in more than fifty significant military conflicts. These recent conflicts led to 23 million international refugees and 25 million more people displaced within their nations. War is so much a part of human nature that at least one genocidal conflict will be happening as you read this, no matter *when* you read this. The wars may change, but nearly all of them will be fought with genocide as a major, though unspoken, motivation.

Wars of genocide rage now in Burundi, in the southern Sudan, in Sri Lanka (against the Tamil Tigers), in Liberia, in Congo (formerly Zaire), in Chechnya, and in Kosovo (between Serbians and ethnic Albanians). In the Sudan, for example, northern Arabs have been at war with southern black Africans for thirty-two of the years since Sudan became independent in 1956. Currently, 2.4 million Africans in the south are facing starvation. One and one-half million have died since 1983, untold tens of thousands of them just recently. Why? Because northern Arab troops have displaced southern black villagers from their farms, crops, herds, and grazing lands. And worse, because rebel soldiers of the brutal Lord's Resistance Army in the south even steal United Nations relief food from their own starving people.

Such wars also simmer elsewhere. The most relentless colonial-type war seethes in Indonesia. The Indonesian government, based on the central island of Java, has been relocating the prodigious surpluses of Java's 100-million-person reproductive machine to Irian Jaya (Western New Guinea), East Timor, Sumatra, and several other of Indonesia's thousands of "outer islands." This translocation program moves millions of Javanese people each year. But to make room for this many people, non-Javanese people already living on the outer islands often must vacate their own land. Indeed, relocating natives in Irian Jaya, as well as government-sponsored logging and mining operations there, has led to whole villages—and the villagers themselves—vanishing.

Although Indonesia denies it, these practices are genocidal. Euphemisms for genocide, such as "translocation" and "modernization," are more politically correct. Unfortunately, however, as history records all too often, the human psyche easily distinguishes between "us" and "them" to permit genocide.

Two human instincts contribute to this: *ethnocentrism,* focusing on one's own culture as the "right" one, and *xenophobia,* fearing outsiders or strangers. Studies worldwide show not only that ethnocentrism and xenophobia are universal in people, but also that they start in infancy. Journalist David Gelman explains:

> There is more than satire in that lament: human beings love to hate. Having enemies fulfills an important human need, as evidenced by children forming rival packs in a playground or nations stock-piling nuclear weapons. Psychologists say that nothing promotes the cohesion of a social, ethnic or national group as surely as a common object of loathing. . . . There is no "us" without . . . a corresponding "them" to oppose.

Carl von Clausewitz stated this even more simply in 1830: "Two motives lead men to war, instinctive hostility and hostile intention; even the most civilized nations may burn with passionate hatred of one another." Universal or not, ethnocentrism and xenophobia are so powerful that when a social group grows too large for the men in it to recognize each other, these men feel impelled to flock to smaller subgroups—clubs, religions, associations, armies, underground cliques, baseball teams, political parties, or militias—in which each man can belong and yet still recognize the others *individually* as allies.

The human male psyche seems compelled to categorize other men as "us" or "them" and to be biased toward "us" and label "them"—those with whom "we" share the fewest genes and least culture—as enemies. *Xenophobia and ethnocentrism are not just essential ingredients to war. Because they instinctively tell men precisely whom to bond with versus whom to fight against, they are the most dangerously manipulable facets of war psychology that promote genocide. Indeed, genocide itself has become a potent force in human evolution.*

Ethnocentrism and xenophobia can be evoked in humans with miraculous speed. In a chilling example, Muzafer Sherif created "us" and "them" identities in twelve-year-old boys new to summer camp and to each other. Sherif divided the boys into two teams, the Bulldogs and the Red Devils. In just five days, these boys were so loyal to members of their own "tribe" and so hostile to boys of the opposing "tribe" that violence between them spiraled out of control. Fearing that one of the boys would be killed or maimed, Sherif aborted the experiment. Apparently, no one has tried it since.

Grown men are equally susceptible to these instincts. Indeed, xenophobia is one of the biggest tricks in the politician's bag. Politicians exhort men to war, notes political scientist Gary R. Johnson, by using terms of kinship—by appealing to the "motherland," "fatherland," or their "brotherhood" to instill patriotism. Politicians remind men of their genetic unity to convince them to risk death for their fellow geneholders (aka the citizens of their "nation").

To ensure this process, all governments and minority groups try to create customs, languages, symbols, and sentiments to enhance ethnocentrism. According to Quincy Wright, "Propaganda and opinion control have become the most important methods for integrating social and political groups." Wright notes too, "Our unity is promoted by identifying the enemy as the source of all grievances of our people."

But we cannot blame politicians for inventing xenophobia. They did not. Primitive societies are equally prejudiced, often going so far as to label their enemies as subhuman. The Eipo of Irian Jaya, for example, refer to their enemies as dung flies, lizards, or worms. Melvin Konner notes that although "the San lack the manpower to mount even the simplest war, when they talk about other tribes, even other groups of San, they make it clear that they are not above prejudice." If war were logistically feasible for the !Kung San, he adds, "they would probably be capable of the requisite emotions."

What exactly are these requisite emotions? Vietnam veteran Michael Decker revealed "The Emotion" when his drug-smuggling employer asked him if he could handle assassinations. "I didn't know what the job would be but I knew it couldn't be anything dirtier than Vietnam," said Decker, "and I had done that. So I knew I could do it. I had already formed in my mind that it would be the same thing, like an enemy, like Vietnam. If I had to kill someone, it would be an enemy."

Again, to kill as a warrior, men need only to believe that their foe is an enemy. Enemies always look different, talk differently, and hold allegiance to the wrong leaders or principles or gods.

Most colonial powers practiced genocide to some degree. The extremes were Britain's complete annihilation of the Tasmanians, Spain's complete annihilation of the Arawak Indians across the Caribbean, and the near extermination by the Dutch of the !Kung San Bushmen in South Africa (see Chapter 6). Amazingly, into the 1950s, there still existed bounty hunters in northern Australia whose job it was to shoot Aborigines. But it was America that amassed a history of xenophobia and genocide that impressed even Adolf Hitler. Estimates vary greatly, but between 5 million and 40 million Indians lived in North America before 1492. If 40 million, 150 years later nearly 90 percent of those Native Americans were gone, mostly due to diseases brought by the white invaders. After 1650, white Americans resorted to more aggressive genocide. Aside from the dozens of major anti-Indian wars sponsored by the U.S. government (remember General Phil Sheridan's "The only good Indian I ever saw was dead"?), American citizens in general, especially males, often went out of their way to slaughter Indians. Innocent Indians. Indian women and children asleep at night in their tipis. Even Indians under flags of truce. Sociologist David T. Courtwright explains:

> The weapons and tactics employed against Indians bespeak Anglo-Americans' indiscriminate hatred and contempt: poisoned meat and drink, smallpox-infected blankets, booby-trapped bodies, cannon charged with slugs, dogs unleashed on captives, and the execution of the wounded, women, and children. California Indians found "naked or wild" were gunned down without pretense or hesitation while Indian women with children were dispatched with no more compunction than stray dogs. Some white men in California, wrote a disgusted French missionary, Father Edmond Venisse, "kill Indians just to try their pistols."

In South America, the same thing happened between Portuguese colonists and indigenous people. One estimate holds that 11 million Indians lived in what is now Brazil (which encompasses more land than America's lower forty-eight states) in the year 1500. Today only about 300,000 survive. As Portuguese colonists arrived, millions of Indians died due to war, slavery, starvation, and disease.

North American xenophobia has not been restricted just to hating the original owners of the continent. Whites in America have been lethally prejudiced against other races as well. (Today in U.S. prisons, blacks, Hispanics, and whites still instantly self-segregate, organize, and begin vying for control of whatever they can control.) Racism is rampant. Courtwright describes, for example, atrocities that whites committed against the Chinese during California's gold rush:

> As with the Indians, to whom whites often compared the Chinese, the way such killings were carried out revealed a deep, almost feral hatred. Chinese men were scalped, mutilated, burned, branded, decapitated, dismembered, and hanged from gutter spouts. One Chinese miner's penis and testicles were cut off and then toasted in a nearby saloon as a "trophy of the hunt."

As hideous as the wholesale murder or annihilation of another social group is, genocide itself can be committed in more ways than one. The clearest possible examples include eliminating most, or all, of a rival group's members, and thus all of its genes. Ultraviolent British colonization, for example, ensured that no Tasmanian exists today—nor any living Tasmanian DNA, however diluted. Only the descendants of the victors now live in Tasmania.

Genocide, however, can be more "subtle." As we saw in Chapter 4, genocide can involve killing males only and then mass-raping surviving fertile females. As mentioned earlier, men everywhere seem more likely to rape women who have lost their defenders, whether in Russia, the Congo, Rwanda, occupied Germany, Bengal, Bosnia, Algeria, or Indonesia. In 1937 in Nanking, for example, Japanese soldiers raped at least a thousand Chinese women (most of them sixteen to twenty-nine years old) each night, murdering many but certainly not all. During a nine-month period in 1971, West Pakistani soldiers raped thousands of Bengali women, siring an estimated 25,000 or more children.

In short, mass rape is not only a massive reproductive victory in the sense of eliminating genetically different male competitors from the mating arena, but it is also a victory in the sense of using the "extra" women conquered as additional reproductive machines for one's own genes. Looked at evolutionarily, mass rape "swamps" the gene pool of the conquered women with the genes of the victorious men. Granted, this is a less "pure" approach to genocide than the complete annihilation favored by racial supremacists. But it offers the victorious males up to an order of magnitude advantage in reproducing their own DNA above what they might otherwise have achieved outside the context of war and invasion. If each man's DNA were calling the shots (in the sense of each human male being nothing more than a "gene machine"), mass rape in warfare is exactly what it would code for as a behavior to increase itself in a strictly Darwinian arena.

So it is no surprise that journalist Barbara Crossette concluded in 1998, "The new style of warfare is often aimed specifically at women and is defined by a view of premeditated, organized sexual assault as a tactic . . . achieving forced pregnancy and thus poisoning the womb of the enemy."

There is nothing new, however, about this style of warfare. It happened when Rome fell. Hence, as I concluded in Chapter 4, rape during war may be an instinctive reproductive strategy of human males—and, for the same reason, it may be an instinctive form of genocide.

Recognition of this has been a long time in coming. In September 1998, for example, a United Nations court handed down the first-ever international conviction of genocide. It found a mayor of a small town in Rwanda guilty of genocide via mass rape.

Unfortunately, every race, ethnic group, and tribe has its prejudices. Nearly all have led to atrocities, many lethal, often including full-scale war. The message here is that the human psyche has been equipped by kin selection to urge men to eliminate genetic competitors—males first, females second—when such killing can be safely accomplished. War itself, declared or otherwise, is often motivated by these instinctive genocidal goals. I believe that this happens because men are born ethnocentric and xenophobic by nature. Men bond along kin lines and/or via reciprocal altruism to fight and kill other men genetically more distant from them in genocidal wars aimed at seizing or usurping what those other men possess, including the reproductive potential of their women. All of this is clear enough in human history. But the ultimate causation for genocide is made even clearer by our nearest cousins.

Natural-Born Genocidal Maniacs

> Cannibalism has been observed twice in East African chimpanzee populations. This behaviour is rarely seen among wild mammals and is hitherto unrecorded in nonhuman primates. . . . In both instances, a group of adult males somehow acquired an infant chimpanzee and began to eat it alive.

As lurid as this description sounds, it comes from the first two paragraphs of David Bygott's classic report on chimp infanticide and cannibalism, published in the prestigious British journal *Nature* on August 18, 1972. Bygott is referring to his own observations in Gombe and to Akira Suzuki's in Uganda's Budongo Forest. At the time, these two reports shattered long-cherished beliefs about the innate "goodness" and "only-if-good-for-the-species" behaviors of chimpanzees like a hand grenade in a phone booth.

Indeed, Bygott's observations were so bizarre that they almost defied belief. The infant chimp he observed, one he did not recognize, had been handed off from one male to another six times, for a total possession time of six hours. Had this infant been a colobus monkey kill, it would have been devoured in a fraction of that time down to skin and bones. But this infant's corpse was nearly intact after six hours and six males. Only its legs, one hand, and the genital region had been eaten.

Bygott's description of chimp infant cannibalism, much like Suzuki's, includes a horrifying juxtaposition of behaviors. Every Gombe male who held the little corpse nibbled at it. This was bad enough. But nearly all of them also did things they never did with monkey prey. They examined it curiously, often poking it or prodding it on the chest, as if to see whether it might spring back to life. Many of these males also groomed the corpse, as if in some hideous mockery of concern. It seemed as though it was too sick to be real—or at least to be natural.

But we eventually learned that this was just business as usual. As observers became more sophisticated and followed chimps into their own terrain and away from feeding stations at human researchers' camps, they watched male chimps kill many infants of their own species. Nearly all of these infants belonged to females from *outside* the males' territory. A few belonged to females residing in the periphery of the killers' territory but adjacent to that of alien males. Hence every infant probably was, or could have been, sired by an alien male. On top of this, most of the murdered infants were males. More to the point, cannibalism was clearly not the primary object of these murders. Murder was. Genocide was. And, ultimately, a reproductive victory of the killers' DNA was.

We now know, as we saw in Chapter 6, that our nearest primate relatives, chimps and gorillas, are wildly infanticidal, but in a strictly genocidal way. That chimps wage genocidal war against adults, too, only deepens the meaning of this knowledge.

These killings, in fact, were what originally impelled me to plan and design a new chimp project in Gombe, working under Jane Goodall's direction but in an untouched region to the north of her main study area. Had it not been for Laurent Kabila's attempt to kidnap Goodall in 1975 for a large ransom to fund his revolution in Zaire, I would have worked in Gombe instead of Kibale Forest, Uganda, where I eventually studied the wild chimp community at Ngogo.

My point here is twofold. First, field research on apes and humans has revealed to us that our own dark side is a deeply embedded set of instincts to kill, rape, or rob when it looks as if doing so might safely pay off reproductively or in our own survival. Second, much of human history is written around conflicts between "gene pools" that have decided, as a social group, to use these primitive and immensely selfish strategies to the hilt. Even so, many people continue to deny this, as if living with the ongoing consequences of an incomprehensible reality of global violence is preferable to moving forward and away from such violence armed with an understanding of the ultimate dark side of man.

To learn how important an understanding of our genocidal dark side is, we will now look at some relatively recent examples of human genocide. The best example, of course, is the lethal competition between gene pools that exploded during World War II.

World War II: Genocide Unleashed

AS MY OPENING CHRONICLE of Lloyd Paul Blanchard's hideous ordeal in the Philippines shows, World War II revealed the entire dark side of the human male psyche. This was not just because it was humanity's biggest war, but because it was humankind's most blatant clash of gene pools.

World War II lasted for six years (1939–1945) during which time 70 million people from all but six of the world's nations donned military uniforms. Roughly 70 million people were killed in the war; 16 million of them were soldiers, 54 million civilians. This last figure reveals the war's monumental genocide. Indeed, World War II is such a telling example of the deep instinctive nature of human war precisely because it was flagrant genocide executed on a mass scale never before imagined. Its dynamics in Europe have been chronicled in hundreds of books, but William L. Shirer's chilling *The Rise and Fall of the Third Reich* remains the most revealing.

The German offensive in World War II occurred because of one man, Adolf Hitler. Hitler's vision of racial purity and "Aryan" destiny as

Europe's master race impelled a nation to wholesale genocidal atrocities. Hitler wrote in *Mein Kampf:*

> All human culture, all the results of art, science and technology that we see before us today, are almost exclusively the creative product of the Aryan. This very fact admits of the not unfounded inference that [he] alone was the founder of all higher humanity, therefore representing the prototype of all that we understand by the word "man." . . . No boy or girl must leave school without having been led to an ultimate realization of the necessity and essence of blood purity.

To Hitler, Jews, Slavs, Gypsies, and even Russians were *Untermenschen,* "subhumans," to be annihilated from the face of the Earth. Although Hitler was a madman whose twisted male psyche went berserk, German documents show that a sobering number of German scientists agreed with his racist rhetoric.

To achieve Aryan dominance over Europe and to ensure blood purity within it, Hitler banned private ownership of guns, committed 40 percent of German men between the ages of eighteen and forty-five to the military (3.5 million of them were killed), and transformed the resources and population of Germany into a racial war machine. His genocidal "cleansing of blood" (aka murder of political enemies and "non-Aryans") often took place in hellish death camps.

American tank commander Arthur Hadley recalls liberating the prisoners in a major concentration camp in Magdeburg, Germany: "When I came round the corner of that pine forest lane and saw the human skeletons hanging from the barbed wire enclosing the camp, I thought, how barbarous of men to string up corpses. Then some of the corpses moved slightly and I realized I was looking at the starved living. There was a horror beyond the horror of all the dying I had seen."

These Nazi murder camps were so ghastly that some people today refuse to believe they existed. But under Gestapo boss Heinrich Himmler, Hitler's SS *(Schutzstaffel)* murdered 5.7 million (of their targeted 11 million) Jews in Europe. The SS also murdered hundreds of thousands of Gypsies, Catholics, and political prisoners and another 3.5 million Russians, Poles, and other civilian and military prisoners in more than thirty *Vernichtungslager* (extermination camps). The SS also systematically robbed their victims of everything they owned, down to their gold teeth. In conquered nations, after the army had moved on, Himmler's SS murdered civilians in a "housecleaning of Jews, intelligentsia, clergy and the nobility."

In *Mein Kampf,* Hitler justified his genocidal plan to eliminate non-Germans and to conquer all of Europe for the Aryan master race because of the need for lebensraum, additional "living space" for the German people.

To me, *The Rise and Fall of the Third Reich* is the most chilling book of all time because it reveals, step by step, how one twisted xenophobic male dictator galvanized totalitarian support within a nation whose citizens obeyed him based on his ethnocentric rhetoric of racist destiny. Shirer's book clearly reveals that Hitler's sole purpose was to conquer by territorial war and transform through genocide a huge region of the world into an enslaved reproductive machine for the gene pool of his master race. And if not for a few fatal mistakes, he would have pulled this off. Hitler epitomized the biological logic of ethnocentrism run amok. But Nazi Germany also teaches us that when governments decide that the rights of the state exceed those of the individual, this is the result. Hitler's genocide will happen again. As mentioned above, it already has.

As Carl von Clausewitz noted (see page 211), racist hate fuels nearly all international wars. The knight in shining armor who (to "rescue" France, England, and Russia) slayed the twin dragons of World War II, Japan and Germany, was the United States. During this war, 12 percent of all Americans donned uniforms, at a cost of 39.1 percent of the U.S. gross national product. There is no question that the United States was committed, but how pure was this knight?

In *War Without Mercy,* historian John W. Dower documents the blatant, deliberately racist attitudes of both the Japanese and the Americans during World War II that shaped genocidal military and political policies. The Japanese, Dower notes, believed that they were the world's leading race *(Shido minzoku),* based on "filial piety and loyalty as expressed under the influence of the divinely descended Japanese imperial line" and on their racial "purity." They held all other races, Asians and Westerners alike, as so inferior and barbaric that they forbade interbreeding with them. "Intermarriage," they insisted, "would destroy the psychic solidarity of the Yamato race."

Even more telling is how Japanese philosophers of the Kyoto school looked at war. War, they said, was an "eternal . . . creative and constructive . . . purifying" exercise of "unique racial power." Japan's racist dogma went beyond being merely blatantly genetic to become a lofty spiritual mission of genocide. Japan's war plans called for conquering all of Asia west to Turkey (nations west of Turkey would be absorbed by Germany and Italy). And Japan's stated goal in these plans, which by 1942 had already resulted in the defeat of one-seventh of the globe, was to create a new genetic order. This would be a pan-Asian society ruled by the destined overlord, Japan, which by increasing each conquered nation's population by 10 percent with pure Japanese, would secure "the living space of the Yamato race." Japan, too, waged its genocidal war for lebensraum. Adolf Hitler had chosen soul mates for an ally.

Racial hatred in the Pacific was thick enough to slice with a bayonet. For example, whereas 4 percent of U.S. prisoners died under the Germans, 27 percent died in Japanese hands. But Japan's racial hatred extended in all directions. I have already mentioned how the Japanese mass-raped thousands of Chinese women in Nanking and murdered about thirty thousand people during and following the Bataan Death March. But the Japanese committed even worse atrocities in the name of racial dominance. Attacking Pearl Harbor as their ambassadors were promising the U.S. government a treaty was only a minor treachery. They routinely maltreated and killed prisoners via torture, forced labor, and "medical experiments." These experiments included lining up and shooting prisoners with different types of bullets to evaluate the relative damage they inflicted. In 1937, Japanese newspapers reported that two Japanese officers engaged in a "friendly contest" to see who would be the first to cut down 150 unarmed Chinese with his sword.

The Japanese frequently massacred innocent noncombatants. As William Manchester notes, nuns in Hong Kong were raped in the street and then murdered. European colonial officials were forced to dig their own graves and then shot. In Papua New Guinea, Japanese secret police yanked the fingernails out of so many natives' hands that the word for *manicure* instilled immediate fear. In the Gilbert Islands, Japanese soldiers forced natives to defecate on their Christian altars and forced women to perform obscene acts with crucifixes. In an orgy of murderous retaliation against U.S. General Douglas MacArthur's triumphant 1945 return to the Philippines, Japanese troops bayoneted nearly 100,000 of Manila's civilians. They strapped patients to their hospital beds and set them afire. Even more hideous, they gouged out infants' eyeballs and smeared them on the wall like jelly. "Probably in all our history," notes Pulitzer Prize winner Allan Nevins, "no foe has been so detested as were the Japanese."

America vented its own hatred by tossing 112,353 Japanese Americans into detention camps (while ignoring Italian and German Americans), an act that constituted one of the most serious violations of U.S. civil rights in American history. In public opinion polls, 10 to 13 percent of Americans said that Japan should not just be defeated, but the Japanese should be exterminated as a race. At the same time, American psychologists and anthropologists dreamed up "scientific theories" justifying why American genocide of Japanese was not a bad idea scientifically. The Japanese race, they explained, was mentally ill; it was composed of mediocre minds, of excellent copiers but unoriginal thinkers; its adults were actually stunted mentally and/or physically to the level of children; and, as a race, it was inferior due to thousands of years less evolution. The Japanese were primitive and compulsive, the experts pronounced, a true "Yellow Peril" bent on ruling the world.

In 1943, General Sir Thomas Blamey, commander of MacArthur's ground forces, spoke for all Allied combatants while inciting his Australian troops in New Guinea with blatant genocidal rhetoric:

> You have taught the world that you are infinitely superior to this inhuman foe against whom you were pitted. Your enemy is a curious race—a cross between the human being and the ape. And like the ape, when he is cornered he knows how to die. But he is inferior to you, and you know it, and the knowledge will help you to victory. . . . You know that we have to exterminate these vermin if we and our families are to live. We must go on to the end if civilization is to survive. We must exterminate the Japanese.

Just how important are racial (genetic) differences in men's willingness to kill during combat? An exhaustive study during World War II by S. A. Stouffer and his colleagues notes that 44 percent of U.S. soldiers said that they would "really like to kill a Japanese soldier." Only 6 percent said the same about killing a German soldier.

Meanwhile, the Japanese government sheltered its people from the army's atrocities, while encouraging the belief that Americans were preoccupied with sex, comfort, and conquest. Japan's xenophobic propaganda was so effective that when GIs took Saipan, thousands of Japanese civilians leaped off cliffs to escape being raped, tortured, and murdered by the Americans. Similarly, in the earlier, bloody battle for Guadalcanal, no Japanese officer surrendered. Surveys revealed that 84 percent of Japanese POWs expected death or torture—and many received both at American hands.

World War II was genocide run amok. But thanks to the truly heroic self-sacrifice of American GIs and their allies—and to the physicists of the Manhattan Project (see below)—that hellish genocidal nightmare ended a long time ago. Things are different now. Aren't they?

Genocide Today

A BIT AFTER 6:00 A.M. every morning, I would reluctantly roll out of bed in our thatched hut on Kenya's Athi-Kapiti Plains. In the dim light of impending dawn , I would put on my cross-country trainers and try to ignore what my lower back was telling me. Leaving Crystal and Cliff sound asleep in their bunk beds, my wife, Connie, and I would jog out onto the savanna.

After the first hill, I usually pulled ahead. We had incompatible paces, and we knew it. Zebras and wildebeests sprinted off the path ahead to the east, then stopped sixty-five yards away to stare at me. They ought to be used to this by now, I often thought.

Minutes later, the sun would appear before me as a huge red ball emerging from the horizon north of Kilimanjaro. The sky blazed crimson. Layers of clouds belted that huge orange ball. Elegant oryx stood unicorn-like, backlit, and stared at my approach. Thank God for this run, I would think. I was field director of a wildlife research school offering full semester credit. Every day I faced demanding people and challenging tasks to fix already-established but poor operational designs that might never be resolved. Running was the best part of my day.

On this particular day in April 1994, I glanced down at the decomposed granite. Small flakes of obsidian scattered ahead spoke of a thousand millennia of previous mammalian bipeds here. Only those guys, *erectus* and later models, were hunting these antelope then—or maybe they were hunting australopithecines in genocidal warfare? Either way, they were not jogging between antelope like an actor in a Kenyan wilderness ad for Perrier.

I wondered again, as I did every dawn, whether a leopard or lion had slunk through one of those holes in the forty-plus miles of fence surrounding this game ranch. Buffalo got through. If lions had, I was doing the exact worst thing I could be doing: running. Worse yet, I was running alone.

A couple of Wakamba herdsmen had been killed and eaten by lions just outside the fence. This made these morning runs slightly less idyllic. I knew that even my serious hunting knife would not offer much protection against a lion. Man-eating lions are the most terrifying predators that exist on land. Except for one other.

"Did you hear the news on the radio last night?" Otieno asked me twenty minutes later.

I admitted I hadn't.

"In Rwanda they are killing each other again," he said. Otieno was our overall ranch liaison man. His moment of glory had come as an oxcart driver in the movie *Out of Africa* with Robert Redford and Meryl Streep. Over my attempts to make a mug of tea as sweat dripped off my nose, he told me the news.

Six months earlier, genocidal intertribal warfare in neighboring Burundi had resulted in the murder of 50,000 to 100,000 people. Burundi's two tribes—the short, dark, horticultural Hutus (84 percent) and the tall, brown, warrior-herder Tutsis (15 percent)—had repeatedly slaughtered each other in major bouts of genocide—in 1965, 1972, 1987, and the previous year, 1993, when newly elected president Melchior Ndadaye (a Hutu) had been assassinated. In each purge, the minority but highly militant Tutsis had ultimately emerged on top. Now war was raging again in Burundi. (Nor would it stop. This tiny nation of 5.5 million is, as John Heminway put it in 1997, "a place where hundreds, perhaps thousands, are hacked to death each month, where genocide is an everyday matter.")

Now Rwanda was at war, too. Burundi and Rwanda, ancient Tutsi kingdoms, share a border. They also share the same two tribes, Hutus and Tutsis, and the same problem: severe overpopulation combined with resource depletion. Finally, the presidents of both countries occasionally share the same jet.

The day before my conversation with Otieno (April 1994), Rwanda's longtime president Juvenal Habyarimana, and Burundi's president, Cyprien Ntarymira, were shot out of the sky by a surface-to-air missile just outside Kigali airport. Yet another genocidal nightmare ensued. Rwanda was a tiny land teeming with an incredible 9 million people backed up against the ecological wall of no more croplands. The majority Hutus took this chaotic opportunity to slaughter thousands of innocent people—both among their ancient enemies, the Tutsis, and among other Hutu enemies to "resolve" feuds. Many Hutu assassins raped and/or hacked and shot to death the Rwandan middle class because they had "soft hands."

The Tutsis instantly reciprocated in reverse slaughter, killing thousands of Hutus and driving 2.4 million of them into exile across the borders into Tanzania and Zaire. There hundreds of thousands died of disease, starvation, or murder in hellish refugee camps west of the Virungas. In the first few months, more than a million Rwandans, mostly Hutus, died. In 1997, tens of thousands more refugees vanished mysteriously. Meanwhile, as Hutu-controlled Milles Collines radio continued to broadcast "Kill the Tutsis," Tutsi refugees from earlier Hutu-led genocidal purges returned to Rwanda after exiles dating back to 1959.

The plot soon thickened as both Ugandan- and Rwandan-supported Tutsi troops invaded Zaire to destroy those original Hutu militants now hiding out and sporadically raiding both countries—and even spitefully killing mountain gorillas. These Hutu guerrillas had been given sanctuary—if not a green light to raise hell—by longtime Zairean strongman Mobutu Sese Seko. By 1997, however, Mobutu faced what was to him an unthinkable retaliation. Zairean rebels enlisted both Rwandan Tutsi soldiers and Ugandan soldiers into their ranks in a savage jockeying to usurp Mobutu's dictatorship. Led by Laurent Kabila, this army wiped out many of the Hutu refugees from Rwanda, as well as Hutu rebels, in yet another genocidal purge. Then it invaded Zaire.

That same year Mobutu secured his alleged $10 billion in loot and forsook Zaire after a reign of thirty-one years. He died soon after of prostate cancer on the French Riviera.

Ultimately, Laurent Kabila's rebel army—the Alliance of Democratic Forces for the Liberation of Congo-Zaire—seized power and tried to restore some sort of national identity in this vast jungle land of Zaire. Kabila had a long history of Marxist revolutionary struggle. He was briefly, in 1965, a "pathetic" student of Che Guevara. In 1975, he was the guerrilla mastermind of the attempted kidnapping of Jane Goodall, the event that eventually sent me to Uganda instead of Tanzania to study chimps. By mid-1997, Kabila had finally won. He renamed the nearly 1 million square miles of Zaire the Democratic Republic of Congo and declared himself president of its 40 million people and 250 tribes. Next he outlawed all opposing political parties and named himself defense minister. Why? Because Congo-Zaire is in the midst of yet another civil war, and the "new" rebels—backed again by Uganda and Rwanda—by late 1998 were closing in, forcing Kabila to retreat south.

In 1998, Rwanda was still in the early stages of attempting to pull itself together into a functioning nation. At last report, it was relying on United Nations aid and had posted teenagers with assault rifles on most street corners in Kigali.

Lest anyone equate Africa with genocide simply because of its persistent tribalism, remember North America, South America, Europe, Tasmania, Australia, and the South Pacific. Genocide is a human male problem, not a regional one, and it is the most reprehensible of all forms of instinctive male violence. Although it can be accomplished with the simplest weapons—machetes in Burundi and Rwanda, for example—more advanced weaponry makes genocide a far greater threat than it ever was during the past 10 million years of genocidal wars.

Pandora's Box

SIXTY YEARS AGO, as World War II loomed on the horizon, a science-fictional yet real-life melodrama began that would alter the course of this large-scale, genocidal war. And because of the unimaginable superweapon that the arms race of this war would yield, the tragic formula of men, weapons, and genocide would never again compute the same way.

Inventing such a destructive superweapon—or, more to the point, *desiring* it—ultimately opened a Pandora's box. Once the ancient xenophobia, emerging from the darkest side of the ancestral male psyche and urging men to wage genocidal war, was mated with an intellect capable of nuclear physics, it inspired men to build a doomsday weapon. And even though atomic bombs were invented in self-defense, they instantly became *the* case study of a problem in human violence with no solution beyond expanding the problem.

It began in 1933 with Leo Szilard. Szilard was a thirty-five-year-old Hungarian theoretical physicist out of work in London. He had just escaped from Germany in the first wave of the Jewish hegira from Hitler's Third Reich. Szilard had cut his academic teeth on Albert Einstein's mass-energy equation, $E=mc^2$. Despite this tumult in his life, Szilard could not help mulling over a claim made by Lord Ernest Rutherford in London's *Times:* "Anyone who looked for a source of power in the transformation of the atoms was talking moonshine." As Szilard walked the streets of London, Rutherford's myopia and dogmatism piqued him. When he stopped for a red light, he abruptly knew that Rutherford had to be wrong.

The light turned green. Szilard started walking across Southampton Row. Like a flash, it hit him. "It . . . suddenly occurred to me," Szilard explained, "that if we could find an element which is split by neutrons and which would emit two neutrons when it absorbs one neutron, such an element, if assembled in sufficiently large mass, could sustain a nuclear reaction . . . [to] liberate energy on an industrial scale, and construct atomic bombs."

This was no trivial insight. An atomic chain reaction was the secret power of the universe. It was the stuff of science fiction. Indeed, Szilard dreamed of harnessing this power in the atom to drive spaceships beyond Earth to explore the solar system and the galaxy beyond. Ironically, however, instead of using the atom to catapult humankind to the stars, nine years later, on December 2, 1942, Szilard and Einstein convinced President Franklin D. Roosevelt that atomic fission could yield a superweapon more powerful than anything then known—one that could vaporize cities. The Germans, they added, and likely the Japanese as well, were already working on such a superweapon. (They were, but fortunately, neither Axis power was visionary enough to invest in the nuclear reactors necessary for research into atomic fission.)

But instead of feeling elated when Roosevelt decided to launch the top secret Manhattan Project, Szilard felt a chill: "I thought this day would go down as a black day in the history of mankind."

Nevertheless, Szilard went to work on the Manhattan Project with dozens of other top nuclear physicists. They were supervised by Robert Oppenheimer and commanded by General Leslie Groves, who imagined spies everywhere. The future would prove Groves's suspicions well-founded, but he also suspected Szilard and even Oppenheimer, on whom he ordered surveillance. In these two men, Groves was wrong.

Not surprisingly, Szilard and Groves despised one another. Despite this obstacle, the team worked together to produce a uranium bomb ("Little Boy") and a plutonium bomb ("Fat Man"). "[Hans] Bethe's group completed the elaborate calculations, but nobody would believe them, not even the people who performed them," notes Edward Teller, a physicist on the project. "In the end a much simpler design was adopted for the plutonium bomb used against Japan. The calculations needed for it could almost have been carried out on the back of an envelope."

Those calculations had to wait for chemist Glenn T. Seaborg's team to chemically extract enough transmuted fissionable plutonium, one microgram at a time, before they were put to the test. This took years. Then "moonshine" suddenly became blinding atomic radiation.

In April 1945, Roosevelt died. Harry Truman succeeded him as president. Within 24 hours, James Francis Byrnes, director of economic stabilization and also of war mobilization, revealed to him the top secret project. "Jimmy Byrnes . . . told me a few details," Truman reported, "though with great solemnity he said that we were perfecting an explosive great enough to destroy the whole world . . . [and] that in his belief the bomb might well put us in a position to dictate our own terms at the end of the war."

Truman was puzzled. The whole idea seemed unbelievable, especially when Admiral William D. Leahy, Roosevelt's principal military adviser, warned him, "This is the biggest fool thing we have ever done. The bomb will never go off, and I speak as an expert in explosives."

But it did. And when the Los Alamos crew tested their experimental atomic bomb, Oppenheimer stared at the gigantic explosion and quoted Vishnu: "Now I am become Death, the destroyer of worlds."

At 8:16 A.M. on August 6, 1945, the *Enola Gay* (pilot Colonel Paul Tibbets named this B-29 after his mother) dropped the first military atomic bomb, "Little Boy," on Hiroshima. The 12.5-kiloton bomb detonated at 1,900 feet. Its explosion erased 4 square miles of the city. Tail gunner Robert Caron had the best view of the world's first atomic bomb unleashed on an enemy target:

> The mushroom itself was a spectacular sight, a bubbling mass of purple-gray smoke and you could see it had a red core in it and everything was burning inside. As we got farther away, we could see the base of the mushroom and below we could see what looked like a hundred-foot layer of debris and smoke and what have you. I was trying to describe the mushroom, this turbulent mass. I saw fires springing up in different places, like flames shooting up on a bed of coals. I was asked to count them. I said, "Count them?" Hell, I gave up when there were about fifteen, they were coming too fast to count. I can still see it—that mushroom and that turbulent mass—it looked like lava or molasses covering the whole city, and it seemed to flow

> upward into the foothills, where the little valleys would come into the plain, with fires starting up all over, so pretty soon it was hard to see anything because of the smoke.

Of 76,000 buildings in Hiroshima, "Little Boy" destroyed 48,000 and damaged 22,000. Of its 330,000 people, 54 percent were incinerated, killed instantly, or, with their charred and blistered skin peeling and hanging off them like rags, died soon after. The rest were injured, some horribly, many thousands of them dying within the next few years.

Even so, Japan did not surrender.

Three days later, "Fat Man," the 22-kiloton plutonium bomb, exploded over Nagasaki and killed 140,000 more Japanese. Truman warned Japan to surrender instantly and unconditionally, or he would drop more atomic bombs. Meanwhile, Truman ordered a thousand B-29 bombers to drop 12 million pounds of high-explosive and incendiary bombs on Japan. (Despite their shocking destructiveness, both atomic bombs accounted for a mere 3 percent of America's bombing destruction of Japan's industrial centers.)

Five days after "Fat Man" destroyed Nagasaki, the Japanese Supreme War Council surrendered—despite Japan's 2.6 million home troops, equipped with more than 500 combat planes, 1,400 kamikaze planes, and mountains of munitions and aided by 32 million civilian militia, many of whom were prepared to fight to the death. Indeed, Japan's home defense was expected to kill up to 1 million of the 5 million Allied troops being prepared to invade Japan had "Fat Man" and "Little Boy" failed.

Impressed by two lethal demonstrations of the most devastating weapon ever imagined, Emperor Hirohito broadcast (his first broadcast ever) Japan's surrender on August 15, 1945. World War II ended, and this planet has not been safe since.

Again, atomic bombs instantly became *the* case study of a problem in human violence with no solution beyond expanding the problem. And expand it did, beyond science fiction. Back in 1941, Enrico Fermi had shared with Edward Teller his curiosity about whether an atomic bomb could be used to heat deuterium hot enough to spark a thermonuclear reaction. Richard Rhodes explains the outcome in *Dark Sun:*

> Each gram of deuterium converted to helium should release energy equivalent to about 150 tons of TNT, 100 million

> times as much as a gram of ordinary chemical explosive and eight times as much as a gram of U235; theoretically, twelve kilograms of liquid deuterium ignited by one atomic bomb would explode with a force equivalent to one million tons of TNT—one megaton; a cubic meter of liquid deuterium would yield ten megatons.

Teller spent the next decade pitching, then pushing the H-bomb beyond curiosity and into reality. This, paradoxically, was despite Teller's own nightmare that just one of these bombs might ignite the Earth's atmosphere and roast the planet.

On November 1, 1952, on the South Pacific island of Elugelab, the United States detonated "Mike," the first full-scale and, at 10.4 megatons, vastly more destructive hydrogen bomb. Elugelab vanished. Everything alive for miles around was fried instantly. A hundred-mile-wide cloud billowed into the sky. Eighty million tons of fallout rained down on the Pacific.

By 1954, the United States had developed the lithium-deuteride-fueled successors to "Mike," thermonuclear bombs deliverable by aircraft. Russian spies played midwife to the Soviet H-bomb less than two years later. This spurred the nuclear arms race into overdrive.

Today the atom bomb used at Hiroshima (0.0125 megaton) would be a puny weapon good only for a surgical tactical strike. By the 1980s, the world's nuclear arsenals held nearly 10,000 megatons of thermonuclear hydrogen bombs, each operating on the principle of solar fusion and possessing a thousand times the megatonnage of an atomic bomb. Their total capacity equaled nearly 1 million Hiroshima bombs, enough to kill every human being on Earth twenty times over. Human weaponry had not only reached frighteningly convenient genocidal levels, but it had exceeded them to the point of threatening the extinction of all humans and many thousands of other species as well.

This threat was so obvious at the onset of thermonuclear bombs that, even then, how to break the cycle of escalation to a level of global suicide became the biggest challenge on Earth. As early as 1953, between the testing of "Mike" and the development of air-deliverable H-bombs, President Dwight D. Eisenhower foresaw the most logical escape from the dilemma of retaining U.S. nuclear superiority:

> This [H-bomb] would be a deterrent—but if the contest to maintain this relative position should have to come indefi-

> nitely, the cost would either drive us to war—or into some form of dictatorial government. In such circumstances, we would be forced to consider whether or not our duty to future generations did not require us to *initiate* war at the most propitious moment that we could designate.

In short, duty dictated blowing the Soviet Union off the map before the Soviets permanently jeopardized the planet.

Instead of leveling the Soviet Union before it could retaliate, however, the United States built thousands of nuclear warheads. It installed most of them in missiles that can hit distant targets a mere twenty-five to thirty minutes after launch (only eight minutes if launched from a submarine). The Soviets built a similar arsenal. The problem was that these weapons added up to a doomsday machine, the ultimate genocidal tool.

"The splitting of the atom," said Albert Einstein, "has changed everything but man's mode of thinking, thus we drift toward unparalleled catastrophe."

To avert this catastrophe, the Americans and Soviets reluctantly cut their megatonnages by about 80 percent. Full nuclear disarmament, however, butts into the biological stone wall of xenophobic mistrust: "we" choose a rational selfishness for "us"—at the cost of imperiling the planet and all the people on it—instead of choosing altruism, which would save the planet but imperil "our" security against "them." The critical question here is, Can we ever really trust the other side, or is our xenophobia too entrenched? The answer is that we cannot trust each other—both sides have cheated (the only "rational" decision for any superpower *is* to cheat)—because we *are* too xenophobic.

Does any solution exist to the problem of combining catastrophic weapons—nuclear, chemical, and biological—with a Pleistocene psyche whose dark side urges some men to commit genocide even in the face of possible global destruction? Most analysts say, realistically, no. One positive sign, however, is America's signing of the Biological Weapons Convention in 1972, banning the offensive use of biological weapons. Another is the 1989 senatorial prohibition against "knowingly" developing, producing, stockpiling, or possessing a biological agent or delivery system as a weapon. Despite these gestures, madmen—and slightly mad men—still rise to lead nations. Neither logic nor morality nor external legalities are their forte. In 1990, Daniel E. Koshland, Jr., former editor of *Science,* summoned an unfortunately realistic vision of the ghost of the nuclear future: "A world in which there are many relatively smaller powers, each with its own nuclear weapons, may in a few years look even scarier than two big superpowers eyeing each other's overwhelming arsenals with caution."

We now stand on the threshold of that future—one at least as scary as Koshland predicted because of the likelihood that Soviet thermonuclear weapons may already have been sold to black market buyers with terrorism in mind. The best scenario might seem to be that we eliminate all nuclear weapons. But two critical problems would remain even if this very unlikely scenario came to pass. First, the knowledge of how to build new ones is impossible to destroy—even if all the books were burned, all the computers erased, all the physics departments outlawed, and all the physicists beheaded. The invention of nuclear weapons opened a Pandora's box for the duration of human existence. Indeed, as Richard Rhodes notes, "every nation that has attempted to build an atomic weapon in the half-century since the discovery of nuclear fission has succeeded in the first try." Second, nuclear weapons are first an *idea* in our minds—a deadly one and, unfortunately, a desirable one—emerging from the imperative in the human male psyche to seek such weapons as an advantage for "us" to annihilate "them" and then usurp their salvageable territory. This is precisely why they are here to stay. Our genocidal imperative is the real problem. Doomsday weapons are merely a symptom of men's instinctive dark side, which today increasingly slip-slides away from direct confrontations and resorts to the more elusive political tactics of terrorism.

The new face of war at the end of the twentieth century is one of terrorism. Modern terrorism is often executed via guerrilla warfare, with goals—though couched in terms of combating or correcting perceived political inequities—of genocide.

On August 4, 1998, after climbing Kilimanjaro, I drove past the U.S. embassy in Nairobi. I'd been inside it with my wife, though not for a few years. This time, ironically, my only thought as I looked at it was: That has to be the most solidly built structure in Nairobi.

Two and one-half days later, this embassy and the U.S. embassy in Tanzania were simultaneously exploded into rubble by car bombs. These bombs were detonated during peak working hours to ensure maximum casualties. Adjacent buildings were reduced to rubble. About 265 people (nearly all in Kenya) were killed by the blasts; a dozen were Americans. More than 5,500 other people, most of them Kenyan, were injured, many horribly.

No one claimed responsibility. (Terrorists in general quit claiming responsibility for many of their heinous acts after President Ronald Reagan launched a retaliatory raid against President Mu'ammar Gadhafi in Libya.) But the evidence points to Osama bin Laden, an extreme Islamic fundamentalist exiled in 1994 from his native Saudi Arabia, which revoked his citizenship. He controls a fortune of about $250 million and a network of about three thousand people (his family is worth $5 billion but has disavowed him). Bin Laden is a terrorist. He gave sanctuary to Ramzi Ahmed Yousef, mastermind of the 1993 bombing of the New York World Trade Center. More incriminating, in February 1998, bin Laden and other extremists issued a *fatwa* (religious ruling) stating that "to kill the Americans and their allies—civilian and military—is an individual duty for every Muslim who can do it in any country in which it is possible to do it."

Clearly, because Islam strictly forbids Arabs to wage war on or kill each other, the sorts of fundamentalist Islamic terrorist acts discussed above are genocide. Invoking Allah does not change the nature of such acts.

The problem is that practitioners of terrorism are not new, nor do they reside only among fringe extremists. Terrorism is a war tactic designed to kill or violate random victims so as to instill so much fear in the majority of citizens that they will voluntarily act against their best interests and in favor of those of the terrorists. The superficial goals of terrorism are generally political: to overthrow a political regime, to rectify grievances, or to undermine order. All are war goals; all are, in essence, genocidal. Terrorism is *the* strategy of the few to gain tyranny by intimidating and extorting the many. Of all forms of war, terrorism is the least just and most cowardly, because its victims are random, defenseless, and often innocent. The problem is that, as a RAND analysis found, terrorism works. It works so well that professional political terrorist organizations such as that of Abu Nidal (who is accused of attacks on more than nine hundred people in twenty countries) actually train and offer terrorists for hire to customers lacking their own training facilities. The level of international, genocidal terrorism is nightmarish. For example, 1987 saw a record high 666 such attacks, while 1997 saw 304 such acts.

Unfortunately, even our children know that a huge level of terrorism is conducted globally by fringe political and/or religious organizations willing to kill innocent victims in an effort to extort the larger body politic. And our children can see, perhaps better than we desensitized adults do, that the victims these terrorists kill are rarely of the same race, religion, or political stance as the terrorists themselves.

That the desire to commit genocide is encrypted into the male psyche as a means of favorably altering access to critical resources for one's own gene pool is bad enough. But that the tactics of terrorism are now widely considered legitimate by many groups, and that these same terrorists now favor superweapons of mass death—whether they be thermonuclear, biological, or chemical—is far worse. What this should tell us is that security and a civilized world will not come gratis.

• • •

Blanchard knew his eyes were gone. He had tried to close them. He had tried to aim them away from the sun. But his efforts were futile against the tape—exactly as his Japanese torturers had intended.

By late 1944, the United States had broken Japanese power in the Pacific. The U.S. victory at Leyte, followed by the U.S. landing at Luzon in January 1945, worried the U.S. high command that the Japanese would murder all Allied prisoners as defeat loomed closer. (Japan did, for example, execute American fliers well after the United States dropped "Little Boy" and "Fat Man" on Hiroshima and Nagasaki.) To prevent this slaughter of prisoners, the United States trained special Ranger forces to liberate each camp in the Philippines in simultaneous lightning attacks before the Japanese knew for sure that the war was lost.

Step one of this very risky rescue operation was completed by a few special volunteer Rangers. After starving themselves until they resembled prisoners, one special squad brazenly infiltrated Cabanatuan to study and record the layout of the camp. Then they escaped. Armed with a full battle plan for the defense of each camp in the Philippines, the United States began step two. On January 28, a force of 175 Philippine guerrillas and 115 U.S. Rangers marched twenty-five miles behind Japanese lines at night to Cabanatuan. Less than five hundred yards from the camp, they hid in rice paddies during the daylight hours before their attack. Some breathed through straws in the muck. That night, twenty-four hours after their mission had begun, the Rangers split into predetermined groups for specific missions.

Staff Sergeant Richard Moore and two other Rangers crawled seven hundred yards to knock out a Japanese guard tower. They next raked a Japanese barracks with tommy guns. Then they cut a hole in the stockade fence to rescue the prisoners.

Moore found Blanchard lying in a ditch. Blanchard's buddies had hidden him there to prevent his being killed by what they feared was Japanese fire.

In twenty-two minutes, the Rangers and Philippine guerrillas killed all 521 Japanese guards and rescued all 513 Allied prisoners. Just before leaving Cabanatuan, one of the guerrillas treated the freed prisoners to a final gratifying sight: the severed head of the camp commander.

Starved to half his normal weight, paralyzed by beriberi, and blinded by the cruel "sun cure," Blanchard warned Moore that he could not walk—trying to save him might be a waste of time. Moore, however, had decided that Blanchard would damned well be numbered among the four thousand survivors of the twelve thousand Americans who had started on the Bataan Death March three years earlier.

"You've found a mate," Moore told him. Moore hoisted Blanchard, now weighing a mere 107 pounds, on his back and carried him three miles behind Japanese lines to a Philippine oxcart.

After Japan surrendered, the governor of Texas appointed Lloyd Paul Blanchard justice of the peace of Jefferson County. To prepare for his new job, Blanchard studied Braille. Jim Moore went home to Hollywood and became a movie double for film star Robert Mitchum. After Blanchard proposed to the girl next door, Helen Braquet, Moore served as his best man.

• • •

PART THREE

THE ANTIDOTE

We must, however, acknowledge, as it seems to me, that man with all his noble qualities, with sympathy which he feels for the most debased, with benevolence which extends not only to other men but to the humblest living creature, with his god-like intellect which has penetrated into the movements and constitution of the solar system—with all these exalted powers—Man still bears in his bodily frame the indelible stamp of his lowly origin.

— Charles Darwin, 1871

We cannot command nature except by obeying her.

— Sir Francis Bacon

CHAPTER EIGHT

Who, Me?

"Government employees should be treated like street whores," screamed Robert Lezelle Courtney, as he beat thirty-eight-year-old Cynthia Volpe to a pulp. Courtney, a forty-seven-year-old millionaire and low-rent landlord, was enraged because environmental health inspector Volpe had just declared one of his units in Bakersfield, California, uninhabitable due to sewage problems.

As Volpe turned to get into her car, Courtney grabbed her hair, twisted her neck, slammed her against the car, and yanked her to the ground. "You ruin people's lives," Courtney screamed at her as she lay flat on her back trying to ward off his kicks and punches, "and I'm going to ruin yours." Volpe absorbed his blows for five minutes. Finally, knowing she could not win against his 230 pounds, she faked unconsciousness. Courtney left her on the pavement, her nose broken, her eyes swollen shut, her face a bloody pulp.

Two months later, Volpe took Courtney to court, charging him with assault with a deadly weapon and assault with great bodily injury. She was suing for $3 million. Courtney pleaded innocent. The court set him free on bail of $7,500. A spirited defense by Courtney's lawyer charging that Volpe, not Courtney, had started the fight deadlocked the jury. They would have to return to court the next day to resume deliberations. Before court resumed, however, Courtney's guilt or innocence would be a moot point.

At dawn the next morning, a sheriff's dispatcher received a desperate 911 call from Cynthia Volpe. A man was inside her house, she said, shooting a gun. "Please hurry," she pleaded.

Before deputies arrived, Courtney fatally shot Cynthia's husband, Kenneth Volpe, and her mother, Betty Reed. As Cynthia herself tried to crawl under the bed to hide, Courtney shot her four times at point-blank range, killing her. Cynthia's children, Keith, fourteen, and

Andrea, nine, had barricaded themselves in their bedrooms. Courtney either bypassed them by accident, was in a hurry to escape, or just did not give a damn.

The next morning, a convenience store operator in Lamont recognized Courtney at his gas pump and phoned the police. Officers from the Bakersfield police, Highway Patrol, and Kern County Sheriff's Department zeroed in on Courtney's faded green 1973 Lincoln Continental. Thirty-two officers chased Courtney for thirty minutes and thirty miles. Fifteen of the officers fired more than two hundred rounds of ammunition at him. The shots flattened some of the Continental's tires, but Courtney had otherwise "armored" the interior of the car with stacks of newspapers. He was also wearing a bulletproof vest and a military Kevlar helmet. He was armed with three guns. He fired four hundred rounds from one, an illegal, full-auto MAC-9 9 mm.

One police sharpshooter hit Courtney's helmet, knocking it off, but Courtney was unscathed. Driving toward the downtown courthouses, Courtney skidded into the center median and leaped out to open his trunk for more ammunition. More shots by police sent him back to the driver's side door. Surrounded and under fire, Courtney placed a .25-caliber pistol under his chin and shot himself. Meanwhile, officers fired two more fatal shots.

Courtney, it turns out, had a history of almost inconceivable violence. On April 2, 1958, at age thirteen, he had an argument with his nine-year-old brother, Jessie, Jr., and his seven-year-old sister, Bonnie, over a toy. He went to the basement of his Anchorage home for a rifle, then he murdered both of them. Next he shot his brutally disciplining mother to death. Courtney spent three years in Alaska's Bureau of Juvenile Institutions. Later he moved to southern California. His juvenile record was sealed and unavailable to the prosecuting Kern County district attorney, who says that had she known of it, she would have requested that no bail be granted.

• • •

WHY WAS IT, we might ask, that a prosecuting attorney was prevented by law from knowing that a man on trial for brutal violence was already a convicted mass murderer? And why had a convicted murderer been set loose to wander among helpless, trusting citizens? Finally, why does neither of these severe lapses of justice even surprise us? Probably because most of us perceive the criminal justice system in general to have had its own tires shot out a long time ago. Whether we identify the shooters as amoral commercial defense attorneys, plea-bargaining prosecutors, soft-headed liberal judges, or juries from the left side of the bell curve, most of us agree that the criminal justice system has for decades limped when it should have hurdled.

As a consequence, many people today find themselves living in a jungle of men's violence. At last count, the FBI reported that an American became a victim of a violent crime every nineteen seconds. Although none of us is happy about this situation, most of us don't know what we can do to fix it.

Let's review the salient points of men's dark side. First, most men are, by nature, programmed, under specific circumstances, to employ violent solutions to their problems—especially ones that potentially affect their reproductive success, and particularly when they also feel that those problems cannot be solved nonviolently. For different men at different times, these solutions include rape, robbery, murder, warfare, genocide, and terrorism. Second, men, when they decide to use these solutions, do so in the expectation that the outcome to them will be favorable. Most of the rest of us fertilize the jungles of violence that such men create by leaving the job of ensuring justice to someone else.

The Anatomy of Cooperation

TO SEE WHAT I MEAN by this, imagine that we are all herders using a common pasture. Our "commons" is not the owned and controlled sort of old England, but instead a free-for-all pasture available to all of us. We have all the animals grazing on the commons that it can support. It is saturated; our herds are at carrying capacity. One more beast out there will tip the delicate balance of soil and forage and start wrecking the commons by cutting its carrying capacity through overgrazing and erosion. This will hurt all of us. But we each face the same decision daily: do we continue to put only our current number of animals on the commons to graze, or do we add just one more?

The gain to anyone who adds one more is an entire cow or sheep, but the cost to this same individual is only a tiny fraction of the entire deterioration of the commons caused by this beast. Most of that cost is paid by all the rest of us. So what should we do? In his classic article "The Tragedy of the Commons," Garrett Hardin explains the dilemma. The selfish but logical choice in the short run is to add another beast. Best, of course, to do it secretly. We should cheat, but in a small, modest, quiet way. After all, the impact is tiny. The cooperative, enlightened choice is to keep our herd below the size that will damage the commons and hurt every herder. This enlightened choice safeguards the commons forever, and with it our livelihoods.

What do most of us do? We add one more animal. Then another. And another. Until the commons is a desert.

The Earth is our commons, and lest we get bogged down here thinking about farm animals, the commons is our society, the way we interact with each other and the planet: the way we wage war; the way we allow robbers, rapists, and murderers to run free (or not); the way we turn our backs on the future.

We face these decisions every day: do we do what is best for ourselves as individuals, or do what is best for most of us? We know we gain more in the short term by deciding in our own favor. What little each of us shares of the common good is often so diluted that we may not even think of it as real. Furthermore, we do not trust others also to act in the common good. Why should we be the only suckers? This sort of short-term thinking—the seduction of instant self-gratification—tips our scale toward the selfish decision. All too often, we make this decision even if we cut our own throats in the long run.

"Let the authorities deal with it," we may mumble as we refuse to stand up and do something. "That's why we pay taxes, isn't it?"

Meanwhile, people in government look at taxpayers as ignorant beasts out there on the commons, placidly grazing and supremely indifferent to doing their part to solve their own problems. *We* are the problem, politicians and civil servants say, except of course when they are trying to get reelected.

Our primate heritage prejudices us in favor of our own immediate self-interest, and it inclines us to distrust the motives of others. But will we allow these facets of human nature to cripple our efforts to solve the crises of rape, murder, war, and genocide? Are we hopeless slaves to our xenophobia and selfishness? Or were we born with something more?

I think we are born with an antidote within us, a part of our human nature that can free us of the blind selfishness of both sexes and of the violent dark side of the male psyche. This antidote is the human instinct to cooperate in enlightened self-interest.

Cooperation has gotten a bad rap. The cynical saying "Nice guys finish last" is wrong. Rather, smart nice guys finish first. Theoretical biologist Richard Dawkins explains how, via an elegant variation of a classic game, Prisoner's Dilemma. It goes like this. Imagine two players: you and me. We are dealt two cards by the Great Banker in the Sky: one says "cooperate"; the other says "cheat." We play against one another (although, as we will see, the "against" is our decision). We each choose the card we want to play.

There are four possible outcomes. First, we can both choose "cooperate," at which point the Banker will reward each of us with $300. Second, we can each choose "cheat," at which point the Banker will fine each of us $10. Third, I can choose cooperate, and you can choose cheat; the Banker will fine me $100 for being a sucker but reward you with $500 for cheating. Fourth, we can switch our positions in the third scenario. The bottom line in Prisoner's Dilemma is that cooperation pays off, but cheating pays off even more if your partner is predictably honest. If your partner also cheats, however, cheating is a losing strategy for both players every time.

The psychology of this game is infuriating. The only rational decision is to cheat. At worst you lose $10, but you may gain $500. Cooperating is simply too expensive, at the cost of $100 each time you get suckered by a cheater.

In real life, people cheat; some cheat like crazy. And the stakes can be a lot higher. In the original Prisoner's Dilemma, two prisoners in separate cells are each faced with the decision to rat on the other or to remain silent. If both keep quiet, no case can be made against either of them, hence they will be freed. If both squeal on the other, both will stay in prison. But if one squeals and the other is silent, the rat fink goes free for providing state's evidence, while the silent cooperator rots in prison.

Here, too, the decision to cheat is rational—one never knows whether one's partner (or opponent) will cheat. The lesson? *Trust of a partner is impossible to establish reliably in just one game.*

But in real life, trust is possible, because we play the game more than once and, in small communities, we face the same partner enough times to learn whether he cheats or cooperates. And unless we are stupid, we can consistently win by cooperating in game after game. To see whether cooperation really works, Robert Axelrod and W. D. Hamilton pitted against each other sixty-three game strategies of "iterated" (repeated) Prisoner's Dilemma, from nasty to nice, entered by contestants. The strategies battled it out on computer until the losers went broke. *The consistent winner was tit for tat, with no first use of cheating.*

This shocked many strategists who had entered highly sophisticated, hawk-like "cheater" programs. Fourteen of the fifteen winningest strategies, however, were "nice." Dawkins concluded that the "bourgeois" tit-for-tat strategy was the closest thing to an evolutionarily stable strategy. Tit for tat—nice for nice but evil for evil—won because the "always-cooperate" (dove) strategies were such suckers that cheaters crushed them, just like hawks always kill doves. Moreover, variations of tit-for-tat strategies have the added advantage of resisting invasions of new cheating strategies. Nice guys can finish first.

Later, as described entertainingly by Matt Ridley in *The Origins of Virtue: Human Instincts and the Evolution of Cooperation,* more complicated games were won by tit-for-tat relatives that were nicer. These were more sophisticated than the original tit-for-tat winners and more generous and forgiving. All but one of them, however, had flaws. The ultimate winner (as of the mid-1990s) was a player designed by Marcus Frean and named "Firm but fair." "Firm" cooperates with cooperators, returns to cooperating after a mutual defection, quits playing with a cheater, and punishes a sucker by further defection. Paradoxically, it also continues to cooperate nicely after being a sucker in the previous round. As Ridley says, "Firm" greets its opponent with a smile and rules supreme as an evolutionary stable strategy.

Thus cooperation *can* be the best policy. But tit-for-tat strategies will consistently pay off only if your opponent never demands that he or she win more than you. Of course, if your opponent is not demanding but instead is happy to make equal gains, he or she is not an opponent at all; he or she is a partner—as one in a trio of lions bonded by reciprocal altruism. Even so, it is clear that *decisive retaliation must be included in the tit-for-tat arsenal, or cooperation will fail.*

Linnda R. Caporeal and her colleagues devised an experiment to watch how cooperation ticks with real people. Each of nine strangers was given $5. If five or more of them contributed their $5 to a general fund, everyone would get $10 in return. Thus cooperators would end up $10 ahead. Cheaters (noncontributors), however, would end up $15 ahead. If the group failed to produce five contributors, cooperators would end up with zero, while cheaters would keep their original $5.

Some groups did very badly; others cleaned up. A single factor made all the difference: whether or not the strangers had a chance to discuss their strategies first. Those who talked always won, usually ending up with seven or eight contributors, most of whom felt that their own decisions were not critical to the group. In contrast, groups who did not talk in advance won bonuses only 60 percent of the time. In short, 40 percent of groups who did not get a chance to talk would not trust blindly, and because of their distrust, they lost their bonuses.

Granted, blind trust is a lot to ask for. The human psyche instinctively knows that trust is risky unless it also knows that cooperation is likely. Our genes do program us to be selfish. But cooperating with those we have reason to trust *is* acting in self-interest, however enlightened. The lesson? *Cooperation demands communication and/or experience with the other players.* This is exactly what Robert Trivers predicted in his excellent article on reciprocal altruism (see Chapter 6). Our genes program us to cooperate when we know we are dealing with a cooperator.

Also important is knowing that we will face the same players in the future. When the "shadow of the future" is foreseeably long—months or years—note Robert Axelrod and Douglas Dion, people stop cheating and quickly use tit for tat. British and German troops in the trenches during World War I, for example, simply quit firing their rifles at each other, even when within range. They did so because they had to "live" together in those trenches for the foreseeable future and because retaliation for "cheating"—by taking potshots to please an officer—cost them too heavily in lives.

An intriguing additional dimension of the cooperation game emerged in 1994. Philosopher Philip Kitcher and computer scientist John Batali entered a tit-for-tat program that included the tactic to "opt out" by refusing to play against (or with) a known cheater. For a while, it won against all other strategies. When faced with a cheater, the "opt out" move was the winner. Opting out of the "game" of life, however, has serious repercussions in the real world. It is, in itself, a decision not to cooperate. Does this apparent "opt out" option explain why three-quarters of voting-age Americans do not bother to vote?

Congress and the president both do offer monumental examples of cheating. The budget of the United States, notes political scientist Howard E. Shuman, is the nation's number one political and priorities document. Wrapped in gold paper and crowned with sealing wax, it represents the intent of where and how to spend more than one-fifth of America's gross national product. This money is the most immense raw power in the world. But due to many acts of Congress designed to lessen its own culpability, the annual budget is decided by one man, the president. And although Congress always rejects the budget, it stays under presidential control by virtue of veto power. Because most presidents and most Congresses in most years have spent much more than the federal income, the United States' national debt is more than $5 trillion—about ten times more money than is in circulation on the entire planet.

Instead of cooperating with American taxpayers, our politicians, hopelessly addicted to deficit spending and to squandering money to favor the special interests that support them, all too often have chosen to cheat taxpayers at large by favoring those in their districts. But whereas congressmen cheat taxpayers, they cooperate with each other. U.S. presidents look like saints, Shuman notes, in comparison to the pork barrel deals of congressmen at the public's expense. "The rapaciousness with which members propose funds for their pet projects must be seen to be believed," Shuman explains. "One of the unwritten rules is that a senator 'does not press objections to another's pet project unless that project adversely affects vital interests in his own state. To do otherwise would violate the norm of reciprocity.'"

As Congress shows, cooperation is impossible unless punishment is imposed for cheating. Indeed, if the punishment is severe enough, nearly any kind of cooperation—even irrational cooperation—is possible. "Punishment," conclude Robert Boyd and Peter J. Richerson, "allows the evolution of reciprocity (or even nonadaptive behaviors) in sizable groups."

Both research and history tell us that cooperation on a large scale cannot evolve in a big, anonymous group unless cheaters are found and punished. With punishment, however, even a few men with enough power can force everyone to cooperate—or to do nearly anything. But, note Boyd and Richerson, what is gained by punishing must eventually exceed the cost of punishing, or cooperation stops. Hence, perhaps, the failure of communism. Otherwise, punishment works.

One problem is finding someone to do the punishing. Most likely this will be a cooperator. But, by definition, a cooperator has already paid a price simply to cooperate. To enforce cooperation, he now must pay an even bigger price by finding and punishing cheaters—a dangerous role. In reality, few cooperators are willing to punish. This reticence leaves but one strategy that is absolutely guaranteed to create widespread cooperation.

Boyd and Richerson call it the "moralist" strategy. Moralists are all enforcers; they accuse and also punish everyone they see for not cooperating, or for not being in "good standing" morally. *They also punish cooperators who fail to be enforcers.* (This need to be in good standing as an enforcing cooperator creates in us one more compelling "conformist" drive to adopt and wear symbols—the crucifix or Shriner's fez or swastika. We know we must appear to be a "moral" cooperator whether we are or not.) Severe enough moralists can nearly eliminate cheaters (look at the low rates of crime in Saudi Arabia, for example).

What history teaches is that the moralist strategy usually begins as a blessing, then quickly becomes a curse. "Moralism" can be so powerful that it forces insane behaviors: religions using human sacrifice or suicide, political ideologies that abolish individual freedom, socialistic government, suicidal wars.

The cohesiveness of cults, for example, does not depend on the charisma or divine power of their leaders (although these do help) as much as on our instinct to be suckers for moralist psychology—a "shared belief system." Frequently, leaders of weak character abuse moralist strategies for their personal gain. Later, as cooperation dwindles due to their abuse, they resort to punishment on such an intense scale that they must recruit police forces—and secret police forces—to find and punish members who deviate from their "moral" norm of "cooperation." In short, moralists may evolve into terrorists, who resort to severe taxation, theft of human rights, pogroms, inquisitions, torture, and genocidal concentration camps to convince people to "cooperate." Americans saw this in 1993, when they watched the FBI, with Attorney General Janet Reno at the helm, raid the Branch Davidian compound in Waco, Texas, with an M60 tank. This raid ended in the killing of 86 people—24 of them children—and was ultimately conducted for minor Bureau of Alcohol, Tobacco, and Firearms infractions the FBI only suspected.

Eventually, people who are oppressed by moralist police cooperate by banding together in guerrilla action to revolt against moralist rulers—as in the American Revolution. Or, alternatively, if the moralist strategy grows too weak to punish cheaters, people band together as vigilantes to enforce useful cooperation on their own. Within the United States before 1900, for example, at least five hundred full-fledged vigilante groups formed to enforce laws not otherwise being enforced. In so doing, they executed about seven hundred "offenders."

Despite all this analysis, *neither biologists nor psychologists invented anything about cooperation that we did not already know intuitively. That cooperation is the right thing to do is instinctive knowledge in the human psyche—and in the psyches of lions, baboons, and chimps, too.*

Indeed, the most revered philosophers and prophets in history have known that, of all the options, cooperation works best. Lao-Tzu, K'ung-Fu-tzu (Confucius), the Old Testament, Zoroastrians, Jesus Christ, and Muhammad all taught that cooperation and sharing were among the highest and most important rules to follow. Many also taught the moralist message that failing to follow this universal code of reciprocity (the Golden Rule) would result in punishment, either in this life or later. But these prophets did not invent the supreme value of cooperation. Instead, they vastly appreciated the importance of this instinct within us. Charles Darwin concluded, "The social instincts—the prime principle of man's moral constitution—with the aid of active intellectual powers and the effects of habit, naturally lead to the golden rule."

Christian allegory places our guardian angel on one shoulder and the devil on the other. This is not far from the truth. Natural selection did equip each of us with instincts from the dark side to be evil, "pathologically" selfish, and unfair. Yet it also equipped us with the more noble instincts to be cooperative, fair, trusting, and, to a degree, self-sacrificing. These instincts of light versus dark do battle in us daily. Each time this happens, *we* must *decide* which side wins. Often the dark side wins. This happens even in American presidents.

Battle or not, *fairness* is our first yardstick for measuring others. We treasure friends who treat us fairly, especially if they have been tempted to cheat but have resisted. We resent and despise those who treat us unfairly, even when that unfairness is trivial. This ability—or compulsion—to analyze and categorize the other players in our social world as cheaters or cooperators so as to manage our reciprocal altruism relationships was likely just as critical in our nonhuman, social ancestors.

Primatologist Frans de Waal reports that even chimps can rise above the law of the jungle. In captivity, the group unanimously and aggressively rebuffs transgressors who fail to reciprocate fairly. Conversely, colony members share food with cooperators. These chimps, he suggests, live the moral message that "justice" and cooperation require revenge. "[It] is safe to assume," de Waal concludes, "that the actions of our ancestors were guided by gratitude, obligation, retribution, and indignation long before they developed enough language capacity for moral discourse. . . . Morality is as firmly grounded in neurobiology as anything else we do or are." De Waal's observations suggest that the ability to represent to ourselves what goes on in other people's minds—

long considered to be the basis for moral decision making and the parent of both compassion and cruelty—may not be unique to *Homo.* Indeed, our problem now is the opposite: too many current *Homo sapiens* have abandoned their morality and decided not to cooperate.

Why are human emotions so finely tuned to judge fairness and so powerful that they spur us to sever unfair relationships and to burn bridges with "friends" who have cheated us? And why do we so quickly forge relationships of reciprocal altruism with people who do enough of the right things to convince us that they are trustworthy partners?

The functional roots of these emotions are easy to dissect. We live in a world in which cheaters, robbers, rapists, murderers, and warmongers lurk in every human landscape. Not only are their selfish, though natural, strategies "unfair" and uncooperative, but they also cost us cooperative types dearly. So we must be vigilant to avoid being victimized. Life for social primates has been like this for eons. The survivors of this evolutionary gauntlet have bequeathed to us instincts to evaluate the intentions of our fellows, and these instincts thunder at us to watch out for, then condemn and punish, unfairness. All this is simple self-preservation. We cannot help but rejoice when villains are punished, especially when their own misdeeds have trapped them. We detest villains so much that some of us mob together as vigilantes to lynch them. By contrast, we sympathize with innocents who are treated wrongly. Injustice tears at our hearts. We respect and admire and root for underdogs who risk everything in a struggle against adversity for a just cause. And we love it when the underdog finally has his day. Again, we cannot help it; we are programmed too well. Hollywood knows this and earns billions because of it.

Our sense of fairness ranks third in the top priorities of the human psyche, right after the welfare of our children and the loyalty of our spouses. Nothing else is more important to us than identifying cheaters and cooperators because it is human nature to depend on being social to survive and succeed. But being social is exactly how cheaters bleed the unwary. Hence it is also human nature for us to endlessly reevaluate our peers so that we can accurately predict how they will treat us when the chips are down. Not only is this the most critical ability we possess, but our emotions enhance this ability as a powerful compass to steer us away from cheaters and toward cooperators.

Fragile as it may seem, cooperation is powerful. On a large scale, it works wonders. The most basic standard of success in today's political world is freedom from oppression. The Swiss, for example, enjoy the greatest freedom from crime and war—despite assault rifles and other high-powered weapons in most homes—simply because they, as individuals, made the decision to cooperate with one another. Every Swiss, they say, is his or her own police officer.

Our instinct for cooperation is so powerful that even in America's concrete jungles, it can appear like a genie from a magic lamp. For example, after the 1989 San Francisco earthquake that killed one hundred people, injured three thousand, and destroyed $10 billion in property in fifteen seconds, social experts predicted rampant looting. Instead, people of all races and socioeconomic levels leaped into action to rescue victims trapped in crushed vehicles and collapsed buildings. Finding and rescuing the last victim, a fifty-seven-year-old man buried alive in his car for ninety hours, was a triumph for the rescuers akin to NASA's landing on the moon. Meanwhile, arrests for looting dropped to 25 percent of normal.

Biologist Richard D. Alexander concludes that the human psyche evolved so that individuals will evaluate, direct, and use social scenarios to win singly or cooperatively against others for status, resources, and, ultimately, reproductive success. Humans, he adds, have become not only the most moral and cooperative creatures on earth but also the most dangerous hostile force against other humans. *To Alexander, understanding the evolutionary origin of the dark side of human nature is the key to unlocking the chains that bind us to our atavistic and internecine relations with other groups.*

But will we cooperate merely because we understand ourselves better? No. No more than a drunk will quit drinking just because he peers into a mirror and sees an alcoholic. Until he sees *himself* as a drunk, however, nothing will change.

If we do hope truly to see ourselves, we must look at ourselves as natural selection does. To natural selection we each "look" like nothing more than a coadapted and integrated gene complex. Richard Dawkins claims that we are gene machines. We are the vehicles our immortal DNA "drives" selfishly and blindly into the future on its ride into eternity. Meanwhile, mutation and natural selection keep improving this DNA to make its vehicles ("us") compete better against other such vehicles ("them").

Though elegant, this perspective is hateful to those who think of the "soul" or the "spirit" as who we really are. Dawkins is quick to admit that his metaphor of genes "purposefully manipulating" us for the sake of their own replication is only metaphor, but we do already know that some genes are selfish, and the metaphor does instruct us. In fact, in looking evolutionarily at what we were designed to do, Dawkins's metaphor is essential. Even so, our eject button from a fate as genetic robots is clear: "To say that we are evolved to serve the interests of our genes," notes Richard Alexander, "in no way suggests that we are obliged to serve them."

The complexity of DNA and its vehicles is a case of the whole exceeding the sum of its parts. Our intelligence, self-awareness, morality, and culture make us the most amazing and capable beings in the known universe—but not so amazing that we can safely ignore our evolutionary roots in natural selection. These roots are still with us—for evil, as in the lethal and genocidal violence by men, or for good, as in understanding and cooperating to solve the atavistic aggression that is our evolutionary legacy. Our fate lies in our hands.

Both history and science tell us that the road to cooperation and trust will be the roughest road humanity has ever trod—especially now that 6 billion of us are scrambling to monopolize and suck dry Earth's last natural resources. We are lucky that natural selection has dealt us such a high card, the king of intellect. The trick will be to play it well.

Social Violence: A Failure of Cooperation

OUR UNRELENTING PROBLEMS of rape, murder, war, and genocide require even greater levels of effort than rare catastrophes such as devastating earthquakes. But what are we doing about them? Most of us are wishing they would just go away.

Rape in America, for example, is rampant due in part to a lack of cooperation between our system of justice and the victim. Many courts fail to identify, assign, or allow the victim rights. Instead, they bend over backward to protect the accused. Fifty-five percent of suspects arrested for rape are released from custody, often the same day. Lawyer Elizabeth J. Swasey explains:

> In most states, [rape] victims are not notified of crucial events, such as the trial. They're not consulted before their cases are plea bargained away. They're not informed about sentencing. They're not alerted to parole hearings or told of the criminal's release. They can't describe how the crime affected their lives. *Many aren't even allowed to attend the*

trial. The bottom line is, victims have no constitutional rights in the judicial process.

Even so, unless rape victims take the pains to cooperate with all other women by prosecuting their assailants, these men will remain free to rape other women—sometimes even the original victim again. Moreover, a judicial system that is unsympathetic to victims due in part to those victims' own ambivalence, not only gives license to the rapist to rape again, but it also makes that rapist invisible to that judicial system. True, less than one in four women who reported a rape in the mid-1990s took her rapist off the street (for an average of 7.25 years; see Chapter 4), but each rapist in prison prevents some women from becoming victims. Between 1980 and 1991, the 300 percent increase in prisoners in jail correlated with a 30 percent drop in rates of victimization in America overall. Victims who fight back in court are truly cooperating, despite the criminal justice system, and even rape counselors, often being weak partners.

What about the problem of American murder? America is a country where police are *not* required by law to protect anyone (the duty of police is to arrest offenders, not to protect any given potential victim from violent or deadly assault) and also a land where only four felons go to prison for every hundred violent crimes committed.

The only control people exert on murder rates is what they as a group do about it. Most social groups today use lethal retribution to punish murders and to deter new offenders, as most have in the past. Only three in the past—the Cabag (South America), the Thai, and the Dogon (West Africa)—lacked capital punishment. Otherwise, capital punishment for "murder one" was for years every culture's revenge for murder. Students of "natural law" call this *lex talionis,* or "comparable punishment." The most famous definition of *lex talionis* exists in Exodus: "It is life for life, eye for eye, tooth for tooth, hand for hand, foot for foot, burn for burn, wound for wound and lash for lash." Most people think of this formula as the essence of justice.

Any effective strategy to deter murder or any other crime demands at least (as it does for war) that people be ready to extract revenge based on equity—certain and swift—no matter what it costs. But actually doing this is dangerous. Remember that committee of !Kung executioners described in Chapter 5? To sidestep this danger, civilized peoples have handed this risky task over to government. In theory, this is not a bad idea—even if it does cost taxpayers about $100 billion per year. In practice, however, the U.S. judicial system, designed to protect the innocent, is now married to a huge commercial legal industry staffed by 70 percent of all the lawyers on Earth and dedicated, for $200 billion per year in fees, to protect the guilty. Together the two have stolen equity from America's crime victims.

The results? The last published *annual* tally (in the mid-1990s) reported 42,361,840 crimes committed in the United States (an arrest was made for one out of three of these). By 1991, the chance that a twelve-year-old American child would be the victim of a violent felony sometime in his or her lifetime had climbed to 83 percent. Three-quarters of all serious crimes are committed by career felons, a third of whom, when rearrested, had previous charges still pending on some of the 187 crimes per year *each* (on average) commits. Of the 4.3 million people serving criminal sentences in 1994, only 26 percent were in jail.

Am I lawyer bashing? Maybe, but with some reason. Lawyer Ray G. Clark, for example, made the following statement when the jury convicted his client, "Night Stalker" Richard Ramirez, on forty-three felony counts of the most brutal and sadistic rape-murders ever known: "I'm disappointed. I felt that we had raised a reasonable doubt. . . . I could not even condone taking Hitler's life." Clark admitted that he did not know whether Ramirez was guilty or not. "I never asked," he said.

At which side of the table would you seat Clark, with cheaters or cooperators?

More to the point, although a U.S. prison sentence may be inconvenient, it is no longer punishment. U.S. prisons have become "universities of crime," where the students eat good food; stay in lighted rooms (cells); and have good plumbing, roommates, exercise facilities, radio, color TV, conversation with anyone, phone privileges, and nearly every other modern convenience—including conjugal visitation rights and the right to sue the state for discomfort or lack of dessert. (Some even raise a stink for not receiving the latest editions of law texts.) Some prisoners get furloughs to wander freely among their victims. *Meanwhile, they teach one another the latest techniques on how to disarm police.*

Behavioral change expert Anthony Robbins notes that because convicted felons do not reap pain while in prison—and thus fail to link pain to their criminal behavior—their sentences fail to change their behavior once they are released. The proof, Robbins says, is in French prisons, which still have small, dark cells isolated from the outside world and lacking all amenities. In the 1980s, the French spent only about $200 per year for each convicted felon. The recidivism rate for French prisoners was a mere 1 percent. By contrast, Americans spend $19,000 per year on each prisoner to perpetuate a nightmarish 82 percent recidivism rate. Some U.S. prisons have become so pleasant that many former prisoners (in California) have brutally murdered random victims just to get back inside.

In *Crime and Human Nature,* James Q. Wilson and Richard J. Herrnstein conclude:

> Whatever factors contribute to crime—the state of the economy, the competence of the police, the nurturance of the family, the availability of drugs, the quality of the schools—they must all affect the behavior of *individuals* if they are to affect crime. . . . Behavior is determined by its consequences; a person will do that thing the consequences of which are perceived by him or her to be preferable to the consequences of doing something else. . . . The punishments of a legal system are an essential part of the story of why criminal behavior may or may not take place.

As mentioned in Chapter 5, U.S. murder rates have dropped in the 1990s by one-third. This means that more than seven thousand people—a small city—each year are still alive who otherwise would be corpses. Why?

Analysis of murder rates in various regions versus changes, or their lack, in those regions' local economies, police numbers and actions, prevention programs, prison sentencing, and illegal drug use and sales shows differences. Many of these changes, however, are inconclusive, inconsistent, or even paradoxical. Consider first economic changes—or the lack thereof. The murder rate decreased 66 percent [1990–1996] in New York City, but the unemployment rate remained a serious 9 percent, compared to 4.3 percent nationwide. (Note that while robberies normally drop with an improving economy, murders do not.) A more reliable predictor is prevention programs (after-school activities and block watches), which are credited with positive, though modest, effects in reducing murder.

More and smarter policing has helped in some cities. New York and New Orleans, for example, attribute their 49 percent [1993–1996] and 37 percent drops in murders to more diligent, energetic, and honest police officers. Meanwhile, Washington, D.C.'s horrendous murder rate also dropped despite poor policing, and Nashville's rate *increased* 55 percent despite a 16 percent increase in police.

The huge growth of the American prison system also has been credited with curbing murder. By 1998, the United States had 1,500 prisons and 3,000 jails, the largest system in the world—even if also the most pleasant. By mid-1997, these institutions held an astonishing 1,725,842 prisoners, a huge number of them drug offenders. Holding murderers in jail longer has lowered murder rates in most regions, but not all. Salt Lake City, for example, increased its prison population by 19 percent but saw its crime rate rise.

Where does this leave us? All of the above processes acting in concert almost certainly lower rates of murder. Beyond these measures, however, two other external processes are tied firmly to declines in homicides. The first one, mentioned in Chapter 5, is that several hundred thousand citizens now legally carry concealed weapons to protect themselves, and where they do, murders and rapes decline.

The second process is the decline and changes in the illegal crack cocaine trade in the inner cities. In the late 1980s, this trade "kicked off an incendiary chain reaction" of juveniles—both crack users and sellers—carrying and using guns to kill offensively and defensively over the cash needed in the crack market. Unfortunately, they also used these guns to kill during disputes in general. Crack, which required dealers to carry weapons for self-defense against almost everyone else, has now disappeared from street-corner sales and has moved to more clandestine, indoor, and even phone sales. Meanwhile, crack use has dropped in many cities. The decline in drug-related killings is so impressive that journalist Gordon Witkin concludes, "Virtually the entire violent-crime wave of the late 1980s and early 1990s could be blamed on young people with guns."

That's the good news. The bad news is that most murder dynamics among adults, especially in states not allowing concealed weapons, are roughly the same as they have been.

Revisiting for a moment America's astronomical prison population reveals that, on average, 14 people are murdered, 48 women are raped, and 578 people are robbed each day—*all by convicted criminals returned to the streets on probation or early parole.* It is no surprise, then, that only 19 percent of Americans in 1996 expressed great confidence in the criminal justice system. An astounding number, including 90.2 percent of high school seniors and a majority of adults, said that they still live in fear of crime and violence. Eighty-five percent of Americans in general (and 86 percent of crime victims) also feel that guilty offenders are not punished enough. When asked if governments should make greater efforts to rehabilitate or to punish violent criminals, 24 percent of Americans said "rehabilitate" and 67 percent said "punish." Most believe, apparently, that *lex talionis* is appropriate justice for violence.

By 1997, 75 percent of Americans supported the death penalty for murder (a figure nearly double that of thirty years earlier)—even if the other option were life imprisonment with no parole. A nearly unanimous 92 percent of police chiefs and sheriffs supported the death penalty. Sixty percent of Americans agreed to capital punishment even for teenage killers. The most revealing question asked by pollsters was this: "Some experts estimate that one out of a hundred people who have been sentenced to death were actually innocent. If that estimate were right, would you still support the death penalty for a person convicted of murder, or not?" Seventy-four percent said yes, they would still support the death penalty. Twenty percent said no.

A 1997 look at surviving relatives of murder victims offers a deeper insight into society's role in punishing murderers. Some of these survivors had been wounded and left for dead. Others had been elsewhere when the crime occurred. When interviewed, several survivors of different murderers said vehemently that death by state execution alone was *too easy and too small a punishment* for the hideous crime the murderer had perpetrated.

Having a death penalty on the books and actually using it, however, are two different things producing entirely different results in deterring murder. By April 30, 1996, 3,122 inmates sat on death row (98.6 percent of them men), all of them convicted murderers. How long will they sit there before they are executed? Most of them never will be executed. *Only one murderer per month was executed in the United States from 1977 to 1996 (this rose to four per month in 1996). Meanwhile, nearly two thousand murders are committed per month.* Does executing one in a thousand murderers deter violent criminals? Probably not. Instead, it likely sends the message that committing murder is a reasonably good bet.

Other than by cooperating to administer *lex talionis* justice, how else might America curb its murder and other violent crime? Many experts insist that the best prevention is teaching our children to respect all people as individuals with full rights over their own person. This agrees with everything we know about the cultural transmission of values (see Chapter 3). Freda Alder, for example, found that the most common factor leading to low violent crime was "some form of strong social control, outside and apart from the criminal justice system . . . [to] transmit and maintain values. . . . Among those social control systems, there is, above all, the family."

Likewise, in 1996 the primary cause Americans cited (24 percent) for early violence in schools was lack of parental discipline and control. Sadly, by 1991 only half of American children under eighteen years old lived in households with both parents (only 25.6 percent of black children lived with both parents). Clearly, parental education is a serious challenge.

Larger-scale violence can be curbed by the same processes, though not nearly as easily. As discussed in the previous two chapters, the world is marred by small wars and cruel acts of terrorism. Thousands of terrorist acts have been perpetrated. Some offer us unequivocal lessons as to appropriate responses. We'll look at just two.

First, on June 27, 1976, Islamic terrorists of the Popular Front for the Liberation of Palestine hijacked Air France Flight 139 from Athens, refueled it in Libya, and then landed in Entebbe, Uganda. They held 105 Israeli passengers hostage in the airport, with Idi Amin Dada's permission, to be released only if Israel released a long list of convicted Palestinian terrorists from prison.

The Israelis assembled a commando team. They built a mock-up of the Entebbe facilities to devise, refine, practice, and perfect an extremely complex rescue plan. After getting their dry run down to fifty-five minutes, they loaded the commandos, armed jeeps, and infantry carriers into four huge Hercules propjets and flew them 2,500 miles undetected. With them were two Boeing 707s, one as a mobile command center and communication post, the other as a mobile hospital.

Lieutenant Colonel Yonni Netanyahu led Operation Thunderbolt on the ground. At one minute after midnight on July 4, Netanyahu's command fought it out with the Ugandan army and the ten Islamic and German terrorists. After fifty-three minutes, the four Hercules transports lifted off the Ugandan runway. Aboard were all Israeli equipment and personnel and 103 living hostages of the 105 held captive. (One had been killed; another was in a Kampala hospital and would be killed.) The commando force had killed about four dozen of Amin's Ugandan troops and seven of the ten terrorists. They took the other three prisoner. One Israeli commando was killed—Colonel Netanyahu himself.

Second, America's response to Osama bin Laden's 1998 terrorist bombings of the U.S. embassies in Kenya and Tanzania, mentioned in

Chapter 7, was different. America launched dual retaliatory attacks using remote-controlled Tomahawk cruise missiles, each armed with thousand-pound nonnuclear warheads. Seventy of these missiles converged on bin Laden's terrorist training facilities south of Kabul, Afghanistan. Another six Tomahawks pounded a chemical plant allegedly manufacturing precursor chemicals for deadly VX nerve gas in Khartoum, Sudan. Both attacks destroyed their targets, but both missed bin Laden himself.

There exists no single public strategy or policy to create a complete shield or immunity to terrorist attacks. But, as in all forms of war, *the greatest immunity possible relies on the individuals of a social group maintaining a firm will and resolve to employ swift, decisive, and massive retaliation in response to any terrorist incident*. Negotiating with terrorists is simply an invitation to future terrorist attacks.

The Road to Freedom

As we have seen, people trust and cooperate actively only if the stakes are high *and* if they talk first. Encouragingly, people have already cooperated like this on a massive scale, even on a global scale—but on environmental issues, not violence. Perhaps this is because everyone knows that destruction of this planet's ecosystem *will* impact their lives. Meanwhile, many people may be in denial about the true threat of violence. Someone else's wife or daughter usually gets raped, and someone else's son gets murdered or drafted and blown to bits fighting a war. Either way, one major lesson has emerged from environmental litigation that applies equally to litigation against crime: *laws work only when enough people cooperate to enforce them*.

It would be nice if someone else would solve our problems of violence for us, but that is not going to happen. Our freedom to pursue our vision of happiness has a price: *assume responsibility and cooperate*. Although many people believe that they can coast through life without paying this price, their "opting out" will ultimately exact from all of us a far more monstrous price. In reality, the solution to the problem of men's violence—our real-life game of Prisoner's Dilemma—rests with you and me. We cannot wait for someone else to cooperate; it is up to us. We have become the ruling creature on this planet, and we have the power to change not only it but also our own species.

Proof of this already exists. Citizens in many communities have recently taken this lesson to heart and made serious differences by cooperating. They have done this in several ways: by monitoring spontaneous play groups of children, by willingly intervening to prevent acts such as truancy and street-corner "hanging" by teenage peer groups, by confronting men who disturb public space, and by being flexible when levels of public services drop. All of these actions are forms of informal, resident-based social control. By comparing and analyzing neighborhoods in the Chicago area that did or did not do these things, researchers found that "the combined measure of informal social control and cohesion and trust remained a robust predictor of lower rates of violence."

The antidote to men's violence in America is for the vast majority of U.S. citizens to make the individual decision to cooperate as a group to achieve two processes not currently happening. The first process is to teach children, all children, from day one, self-control, self-discipline, and self-responsibility in a world where we ourselves show that offensive violence is wrong. We must make the teaching of fairness, justice, and human values our primary goals. Boys becoming young men must already have been socialized with these deep human values (in a palatable form, more or less as Boy Scouts of America wishes to do) *by their parents.* Second, we must decide to cooperate to make felonious violence—rape, murder, offensive war, genocide, and terrorism—not only "not pay" for the perpetrator but also reap pain. In short, to stop violence, we must decide that our justice is *lex talionis* justice.

Doing all this not only requires that we assume a personal responsibility to cooperate to overcome these threats of violence; it also will eventually require us to make a gigantic leap—on a level never before achieved—away from our instincts of individual and kin group selfishness, xenophobia, and distrust, all of which fuel war and the male violence we face in rape and murder. This leap must propel us to patriotic loyalty within our national community and carry us beyond it toward global cooperation between nations. That this latter goal is not a natural human tendency anyone can realistically expect (outside Earth being invaded by hostile aliens) almost goes without saying. But it is the only way to win against men's violence.

If we fail as individuals to commit to a disciplined effort to attack the dark side of the male psyche by taking action to make rape, murder, offensive war, genocide, and terrorism true capital crimes, this dark side will plague us forever. That we can, as cooperating individuals, make this difference is clear—whether we are talking about Earth First! or Mothers Against Drunk Driving or a hypothetical Citizens

Against Violence. As Anthony Robbins points out, "Changing an organization, a company, a country—or a world—begins with the simple step of changing yourself."

The price of our freedom—and of the antidote to the violence in our genes and in our world—is nonnegotiable. "The prime factor in each man's or woman's success," wrote Theodore Roosevelt, "must normally be that man's or woman's own character . . . above all the qualities of honesty, of courage, and of common sense. Nothing will avail a nation if there is not the right type of character among the average men and women . . . who make up the great bulk of our citizenship."

Can we change our characters toward increased individual responsibility? As the road ahead forks, one direction will lead to self-gratification, the other to self-discipline and cooperation. One decision will lead to violence and destruction, the other to survival, trust, and a better world. Does taking the right road seem impossible? Once upon a time, so did flying.

CHAPTER NOTES

Front Matter

vi "In a world full of hatred": Hamburg, D. A. 1986. New risks of prejudice, ethnocentrism, and violence. *Science* 231:533.

viii chimps in Uganda and ape violence in general: Ghiglieri, M. P. 1988. *East of the Mountains of the Moon: Chimpanzee Society in the African Rain Forest*. New York: Free Press. See also Ghiglieri, M. P. 1984. *The Chimpanzees of Kibale Forest*. New York: Columbia University Press. Ghiglieri, M. P. 1984. The mountain gorilla: last of a vanishing tribe. *Mainstream* 15(3–4):36–40. Ghiglieri, M. P. 1985. The social ecology of chimpanzees. *Scientific American* 252(6):102–113. Ghiglieri, M. P. 1986. River of the red ape. *Mainstream* 17(4):29–33. Ghiglieri, M. P. 1986. A river journey through Gunung Leuser National Park, Sumatra. *Oryx* 20(2):104–110. Ghiglieri, M. P. 1987. Sociobiology of the great apes and the hominid ancestor. *Journal of Human Evolution* 16(4):319-357. Ghiglieri, M. P. 1987. War among the chimps. *Discover* 8(11):66–76. Ghiglieri, M. P. 1989. Hominid sociobiology and hominid social evolution. In P. G. Heltne and L. A. Marquardt (eds.), *Understanding Chimpanzees,* pp. 370–379. Cambridge: Harvard University Press.

Part One: Roots

1 "Sexual selection apparently has acted on man": Darwin, C. 1871. *The Descent of Man and Selection in Relation to Sex*. New York: Modern Library, p. 918.

1 "Compelled by the urge": Small, M. F. 1993. *Female Choices: Sexual Behavior of Female Primates*. Ithaca, N.Y.: Cornell University Press, pp. 10–11.

Chapter One: Born to Be Bad?

5 surgical gender "reassignment" does not work: Diamond, M. 1982. Sexual identity: monozygotic twins reared in discordant sex roles and the BBC follow-up. *Archives of Sexual Behavior* 11:181–186. See also Durden-Smith, J., and D. Desimone. 1983. *Sex and the Brain*. New York: Arbor House. Frieze, I., J. E. Parsons, P. B. Johnson, D. N. Ruble, and G. L. Zellman. 1978. *Women and Sex Roles: A Social Psychological Perspective*. New York: Norton.

5 boys mostly play contest games: Eibl-Eibesfeldt, I. 1989. *Human Ethology*. New York: Aldine de Gruyter, 277, pp. 268–288, 589–602.

5 girls copy feminine behavior: Barkley, R. A., D. G. Ullman, L. Otto, and J. M. Brecht. 1977. The effects of sex-typing and sex appropriateness of modelled behavior on children's imitation. *Child Development* 48:721–725. See also Bandura, A., D. Ross, and S. A. Ross. 1961. Transmission of aggression through imitation of aggressive models. *Journal of Abnormal and Social Psychology* 63:575–582.

5 the Israeli experiment in creating monogenderal roles in the kibbutz led to self-identified maternal role models: Spiro, M. E. 1958. *Children of the Kibbutz*. Cambridge: Harvard University Press. See also Spiro, M. E. 1979. *Gender and Culture: Kibbutz Women Revisited*. Durham, N.C.: Duke University Press. Tiger, L., and J. Shepher. 1975. *Women in the Kibbutz*. New York: Harcourt Brace Jovanovich. Eibl-Eibesfeldt, 1989. pp. 279–283.

5 parents nurture boys and girls differently: Rubin, J. Z., F. J. Provenzano, and Z. Luria. 1979. The eye of the beholder: parents' views on sex of newborns. In J. H. Williams (ed.), *Psychology of Women,* pp. 134–144. New York: Norton.

5 mothers burp, rock, arouse, stimulate: Konner, M. 1982. *The Tangled Wing: Biological Constraints on the Human Spirit*. New York: Holt, Rinehart and Winston, pp. 113–114.

6 "It ain't so much": Shapiro, A. M. 1993. Review of *Biology and Conservation of Monarch Butterflies* by S. B. Malcolm and M. P. Zalucki (eds.). *Science* 260:1983–1984. See also Reiss, A. J., Jr., and J. A. Roth (eds.). 1993. *Understanding and Preventing Violence*. Washington, D.C.: National Academy Press, pp. 38–39.

6 the social sciences at odds with biology: Gross, P. R., and N. Levitt. 1994. *Higher Superstition: The Academic Left and Its Quarrels with Science*. Baltimore: Johns Hopkins University Press.

7 behavior is controlled by genes in animals: Scheller, R. H., and R. Axel. 1984. How genes control an innate behavior. *Scientific American* 250(3):54–62.

7 obsessive-compulsive disorders also originate biologically: Rapoport, J. L. 1989. The biology of obsessions and compulsions. *Scientific American* 260(3):83–89. See also Holden, C. 1993. Hyperactivity linked to genes. *Science* 260:295.

9 inheritance plays a role in human behavior: Plomin, R. 1990. The role of inheritance in behavior. *Science* 248:183–188.

7 IQs of 245 adopted children more closely matched those of their biological parents: Wright, K. 1988. Nature, nurture, and death: a study of adoptees suggests there is no escaping death. *Scientific American* 258(6):86–94. See also Aldhous, P. 1992. The promise and pitfalls of molecular genetics. *Science* 257:164–165.

7 "monozygotic twins reared apart are about as similar as monozygotic twins reared together": Bouchard, T. J., Jr., D. T. Lykken, M. McGue, N. L. Segal, and A. Tellegen. 1990. Sources of human psychological differences: the Minnesota study of twins reared apart. *Science* 250:223–228, p. 223.

7 other behaviors known to be based on genetics: Plomin, R. M., J. Owen, and Peter McGuffin. 1994. The genetic basis of complex human behaviors. *Science* 264:1733–1739. See also Herbert, W. 1997. Politics of biology. *U.S. News & World Report* 122(15):72–80.

7 the tendency to divorce is influenced by one's genes: Holden, C. 1992. Why divorce runs in families. *Science* 258:1734.

7 "nice" personality traits are genetically linked: Holden, C. 1993. Nice guys can thank their genes. *Science* 259:33.

7 our genes do influence our behavior and our personalities: Bower, B. 1991. Same family, different lives. *Science News* 140(23):376–378. See also Wills, C. 1998. *Children of Prometheus: The Accelerating Pace of Human Evolution*. Reading, Mass.: Helix/Perseus Books. Horgan, J. 1993. Eugenics revisited. *Scientific American* 268(6):122–131.

7 "Any analysis of the causes": Konner 1982, p. 80.

8 social behavior is shaped by reproductive strategies, often violent: Ghiglieri, M. P. 1987. Sociobiology of the great apes and the hominid ancestor. *Journal of Human Evolution* 16(4):319–357. See also Goss-Custard, J. D., R. I. M. Dunbar, and F. P. G. Aldrich-Blake. 1972. Survival, mating, and rearing strategies in the evolution of primate social structure. *Folia Primatologica* 17:1–19.

9 "As many more individuals": Darwin, C. 1859 [1964]. *On the Origin of Species by Means of Natural Selection, or the Preservation of Favoured Races in the Struggle for Life*. Cambridge: Harvard University Press, pp. 6, 81.

9 the neo-Darwinian definition: Williams, G. C. 1992. *Natural Selection: Domains, Levels, and Challenges*. New York: Oxford University Press. See also Fisher, R. A. 1930. *The Genetical Theory of Natural Selection*. Oxford: Clarendon.

9 127 species of plants and animals exist in which natural selection has been observed: Endler, J. A. 1986. *Natural Selection in the Wild*. Princeton, N.J.: Princeton University Press.

9 natural selection in the Galápagos: Weiner, J. 1994. *The Beak of the Finch*. New York: Knopf.

9 neo-Darwinian theory predicts that this will happen on every planet in the universe: Dawkins, R. 1991. Darwin triumphant: Darwinism as a universal truth. In M. H. Robinson and L. Tiger (eds.), *Man and Beast Revisited,* pp. 23–39. Washington, D.C.: Smithsonian Institution Press, pp. 24, 29, 37, 38.

9 "For a biologist": Barash, D. P. 1979. *The Whisperings Within*. New York: Harper and Row, p. 9.

9–10 a special form of natural selection called sexual selection: Darwin 1859, pp. 87–90. See also Darwin, C. 1871. *The Descent of Man and Selection in Relation to Sex*. New York: Modern Library. Clutton-Brock, T. H. 1983. Selection in relation to sex. In D. S. Bendall (ed.), *Evolution from Molecules to Men,* pp. 457–481. Cambridge: Cambridge University Press. West-Eberhard, M. J. 1991. Sexual selection and social behavior. In Robinson and Tiger, pp. 159–172.

10 pretty male and macho male sexual selection: Darwin 1871, pp. 570, 818–819, 863, 874–875, 907–908, 916.

10 "any investment by the parent": Trivers, R. L. 1972. Parental investment and sexual selection. In B. Campbell (ed.), *Sexual Selection and the Descent of Man: 1871–1971,* pp. 136–179. Chicago: Aldine.

10 lifetime reproductive success of women: MacCleur, J. W., J. V. Neel, and N. A. Chagnon. 1971. Demographic structure of a primitive population: a simulation. *American Journal of Physical Anthropology* 35:193–207.

10 a man can fertilize a different woman every day: Short, R. V. 1981. Sexual selection in man and the great apes. In C. E. Graham (ed.), *Reproductive Biology of the Great Apes,* pp. 319–341. New York: Academic Press.

10–11 cassowary behavior and biology: Bentruppenbaumer, J. 1996. Personal communication, Mission Beach, Queensland, Australia.

11 in mammals, polyandry does not happen: Crook, J. H., and S. J. Crook. 1988. Tibetan polyandry: problems of adaptation and fitness. In L. Betzig, M. Borgerhoff Mulder, and P. Turke (eds.), *Human Reproductive Behavior: A Darwinian Perspective,* pp. 97–114. New York: Cambridge University Press.

11 male birds of paradise: Beehler, B. M. 1989. The birds of paradise. *Scientific American* 261(6):117–123.

11 A clever test of the "pretty male strategy" and sexual dimorphism was done among long-tailed widow birds in Kenya. Males' tails are ponderously long, and females seem to love them. To test this, ornithologist Malte Andersson caught thirty-six males and subjected them to one of four treatments: he trimmed group 1's normal 20-inch tails to only $5^1/_2$ inches; he superglued extra trimmings onto tails of group 2 to make them a superlong 30 inches; he trimmed group 3's tails but superglued them back to their original length; he did nothing to the control group but let them go free. The results? Males with superglued superlong tails attracted four times more females, who built nests and laid eggs in them, than did the unfortunate males whose tails Andersson had shortened. Birds with normal tails (glued or untouched) had intermediate success: Andersson, M. 1982. Female choice selects for extreme tail length in a widow bird. *Nature* 299:818–820. See also Ridley, M. 1992. Swallows and scorpionflies find symmetry beautiful. *Science* 257:327–328.

11 sexual selection has equipped polygynous male primates with larger body size and canines and a combat psychology: Small, M. F. 1989. Female choice in nonhuman primates. *Yearbook of Physical Anthropology* 32:103–127.

12 body and antler size determine a stag's reproductive success: Clutton-Brock, T. H., F. E. Guiness, and S. D. Albon. 1982. *Red Deer: Behavior and Ecology of Two Sexes.* Chicago: University of Chicago Press. See also Clutton-Brock, T. H. 1985. Reproductive success in red deer. *Scientific American* 252(2):86–92.

12 "Next to being human": Katchadourian, H. 1989. *Fundamentals of Sexuality.* 5th ed. San Francisco: Holt, Rinehart and Winston, p. 281.

12 Evils of the unfair double standard were once supported even by "science." For example, physiologist Paul Mobius wrote a century ago, "All progress is due to man. Therefore woman is like a dead weight on him, she prevents much restlessness and meddlesome inquisitiveness, but she also restrains him from noble actions, for she is unable to distinguish good from evil": Shields, S. 1975. Functionalism, Darwinism, and the psychology of women. *American Psychologist* 30:739–754, p. 745.

12 Gender is due to socialization, while sex is due to chromosomes: Archer, D., and J. Lloyd. 1985. *Sex and Gender.* New York: Cambridge University Press.

13 search, find, and fertilize: Daly, M., and M. Wilson. 1983. *Sex, Evolution, and Behavior.* Belmont, Calif.: Wadsworth, p. 72.

13 why sex?: Small, M. F. 1993. *Female Choices: Sexual Behavior of Female Primates.* Ithaca, N.Y.: Cornell University Press, pp. 15–28.

13 being sexual carries a price tag: Betzig, L. 1988. Mating and parenting in Darwinian perspective. In Betzig, Borgerhoff Mulder, and Turke, pp. 3–20. See also Anderson, A. 1992. The evolution of sexes. *Science* 257:324–326.

13 "Females and males are apples and oranges": Small 1993, p. 97.

13 the basic blueprint of mammals is female: Konner 1982, p. 122.

14 interval 1A2: Roberts, L. 1988. Zeroing in on the sex switch. *Science* 239:22. See also Morell, V. 1994. Rise and fall of the Y chromosome. *Science* 263:171–172.

14 gene SRY: Haqq, C. M., C.-Y. King, E. Ukiyama, S. Falsafi, T. N. Haqq, P. K. Donahoe, and M. A. Weiss. 1994. Molecular basis of mammalian sexual determination: activation of Müllerian inhibiting substance gene expression by SRY. *Science* 266:1494–1500.

14 severe maternal stress during the second trimester also seems to predispose some males to homosexuality: Ellis, L., W. Peckham, M. A. Ames, and D. Burke. 1988. Sexual orientation of human offspring may be altered by severe maternal stress during pregnancy. *Journal of Sex Research* 25(1):152–157.

14 the human brain is, by default, female: Gibbons, 1991. The brain as "sexual organ." *Science* 253:957–959.

14 Pillard suspects that Müllerian inhibiting hormone helps defeminize the brain. C. Burr. 1993. Homosexuality and biology. *Atlantic* 271(3):47–65, pp. 59–60.

14 infant girls with higher levels of testosterone: Konner 1982, pp. 101, 123–124.

14 finally feeling like normal women: Symons, D. 1979. *The Evolution of Human Sexuality.* New York: Oxford University Press, p. 311.

15 low levels of testosterone in utero predispose men to homosexuality: Dorner, G. 1976. *Hormones and Brain Differentiation.* Amsterdam: Elsevier.

15 research shows that male sexual orientation is genetic: Bower, B. 1992. Gene influence tied to sexual orientation. *Science News* 141(1):6. See also C. Burr. 1993. Homosexuality and biology. *Atlantic* 271(3):47–65, p. 64. LeVay, S. 1993. *The Sexual Brain.* Cambridge: MIT Press. Grady, D. 1993. Gay genes. *Discover* 14(1):55.

15 a genetic component to women's homosexuality: Bower, B. 1992. Genetic clues to female homosexuality. *Science News* 142(8):117.

15 interval Xq28 and male homosexuality: Hamer, D. H., S. Hu, V. L. Magnuson, N. Hu, and A. M. L. Pattatucci. 1993. A linkage between DNA markers on the X chromosome and male sexual orientation. *Science* 261:321–327. See also Risch, N., E. Squires-Wheeler, and B. J. B. Keats. 1993. Male sexual orientation and genetic evidence. *Science* 262:2063–2064. Hamer, D. H., S. Hu, V. L. Magnuson, N. Hu, and A. M. L. Pattatucci. 1993. Response. *Science* 262:2065.

15 testosterone reduces fear: Speltz, M. L., and D. A. Bernstein. 1976. Sex differences in fearfulness: verbal report, overt avoidance, and demand characteristics. *Journal of Behavior Therapy and Experimental Psychiatry* 7:177–222.

15 testosterone increases aggression: Reinish, J., and S. Sanders. 1985. A test of sex differences in aggressive response to hypothetical conflict situations. *Journal of Personality and Social Psychology* 50:1045–1049.

15 at adolescence testosterone levels soar: Eibl-Eibesfeldt 1989, pp. 270–279.

15 sporting events in which women and men compete: Cassidy, W. 1990. Women and the NRA. *American Rifleman* 138(6):7.

16 "Attitude counts": Sapolsky, R. M. 1990. Stress in the wild. *Scientific American* 262(1):116–123, pp. 122, 123.

16 three- to five-year-old boys and girls: Girch, L. L., and J. Billman. 1986. Preschool children's food sharing with friends and acquaintances. *Child Development* 57:387–395.

16 by age nine, boys create peer hierarchies: Coie, J. D., K. A. Dodge, R. Terry, and

V. Wright. 1991. The roles of aggression in peer relations: an analysis of aggressive episodes in boys' play groups. *Child Development* 62(4):812–826, pp. 823–825.

16 the brain is structurally and functionally organized: Gazzaniga, M. S. 1989. Organization of the human brain. *Science* 245:947.

16 dendritic connections and terminal branches: Kalil, R. E. 1989. Synapse formation in the developing brain. *Scientific American* 261(6):79. See also Barinaga, M. 1994. Watching the brain remake itself. *Science* 266:1475–1476.

16 women performed better on verbal tests: Feingold, A. 1988. Cognitive gender differences are disappearing. *American Psychologist* 43:95–103.

16 men outscore women in math: Hedges, L. V., and A. Nowell. 1995. Sex differences in mental test scores, variability, and numbers of high scoring individuals. *Science* 269:41–45. See also Holden, C. 1991. Is "gender gap" narrowing? *Science* 253:959–960. Savani, J. 1992. Letter: women in mathematics. *Science* 257:309–310.

16 "Much independent evidence": Gaulin, S. J. C., and H. A. Hoffman. 1988. Evolution and development of sex differences in spatial ability. In Betzig, Borgerhoff Mulder, and Turke, pp. 129–152, pp. 139–139.

17 Changes women's cognitive performance: Staff. 1988. Sex hormones linked to task performance. *Science* 242:1509.

17 women commit most crimes just before their menstrual periods: Wallace, R. A. 1979. *The Genesis Factor.* New York: Morrow, p. 116.

17 the top ten mathematics departments: Selvin, P. 1991. Does the Harrison case reveal sexism in math? *Science* 252:1781–1783.

17 hormones implicated in math ability: Kolata, G. 1983. Math genius may have hormonal basis. *Science* 222:1312.

17 reinforcement of sex roles is not the origin of sex difference: Geary, D. C. In press. A tentative model for representing gender differences in the pattern of cognitive abilities. *American Psychologist*. See also Thorne, B. 1993. *Gender Play: Girls and Boys in School*. New Brunswick, N.J.: Rutgers University Press. Hedges and Nowell 1995, p. 45. Gaulin and Hoffman 1988, p. 144.

17 male and female brains in mammals differ: Gibbons 1991.

17 sex differences include number and size of neurons: Toran-Allerand, C. D. 1984. On the genesis of sexual differentiation of the central nervous system. In G. J. DeVries et al. (eds.), *Sex Differences in the Brain: Structure and Function,* pp. 63–98. New York: Elsevier. See also LeVay 1993.

17 differences in locations of verbal processing: Holloway, M. 1990. Profile: Vive la difference: Doreen Kimura plumbs male and female brains. *Scientific American* 263(4):40, 42.

17 the *Titanic:* Beesley, L. 1960. The loss of the S.S. *Titanic:* its story and its lessons. In J. Winocour (ed.), *The Story of the Titanic as Told by Its Survivors,* pp. 1–109. New York: Dover, pp. 9, 23.

18 American women earn less than men: Krugman, P. R. 1991. Women have made slow progress. *U.S. News & World Report* 111(19):63.

18 "One of the more surprising discoveries": Getty, J. P. 1976. *As I See It*. Englewood Cliffs, N.J.: Prentice Hall, pp. 106-107.

18 "When all the factors of prejudice": Morgan, E. 1972. *The Descent of Woman*. New York: Stein and Day, p. 213.

18 "Whatever women say in public": Harley, W. F., Jr. 1986. *His Needs, Her Needs.* Old Tappan, N.J.: Fleming H. Revell, pp. 116–117, 119.

19 women prefer to marry an economically successful man: Betzig 1988.

19 wealthier men had more wives: Ibid, p. 5.

19 Ache women and men: Hill, K., and H. Kaplan. 1988. Tradeoffs in male and female reproductive strategies among the Ache, parts 1 and 2. In Betzig, Borgerhoff Mulder, and Turke, pp. 277–289, 291–305.

19 Ache hunters with shotguns: Hill, K., and A. M. Hurtado. 1989. Hunter-gatherers of the New World. *American Scientist* 77(5):437–443.

20 preferences in mates in thirty-seven countries: Buss, D. M. 1994. The strategies of human mating. *American Scientist* 82(3):238–249. See also Buss, D. M. 1989. Sex differences in human mate preferences: evolutionary hypotheses tested in 37 cultures. *Behavioral and Brain Sciences* 12:1–49. Schlosser, E. 1997. The business of pornography. *U.S. News & World Report* 122(5):42–50, pp. 47–48.

20 lonely hearts ads: Dunbar, R. 1995. Are you lonesome tonight? *New Scientist* no. 1964:26–31, p. 30.

20–21 mate preferences in America: Townsend, J. M. 1989. Sex differences in process and criteria. Paper presented at the first annual meeting of the Human Behavior and Evolution Society, Northwestern University, Evanston, Ill., August 25–27.

21 Men rated the composite "average" face in the top five: Holden, C. 1990. Ordinary is beautiful. *Science* 248:306. See also Dunbar 1995, p. 29.

21 women named intelligence and a sense of humor as top traits: Gallup, G. G., Jr., and S. D. Suarez. 1983. Unpublished survey. State University of New York at Albany. (Cited in Kelly, K., and D. Byrne. 1992. *Exploring Human Sexuality.* Englewood Cliffs, N.J.: Prentice Hall, p. 10.)

22 "It turns out": Buss 1994, p. 249.

22 adolescent children of professional parents: Tanner, J. M. 1988. Human growth and constitution. In G. A. Harrison, J. M. Tanner, D. R. Pilbeam, and P. T. Baker, *Human Biology,* pp. 337–435. New York: Oxford University Press.

22 American wives of wealthier men: Essock-Vitale, S. M., and M. T. McGuire. 1988. What 70 million years hath wrought: sexual histories and reproductive success of a random sample of American women. In Betzig, Borgerhoff Mulder, and Turke, pp. 221–235.

22 "A successful man": Winokur, J. (ed.). 1991. *A Curmudgeon's Garden of Love.* New York: Plume, p. 176.

22 "His needs are not hers": Harley 1986, pp. 10, 176–177. See also Hite, S. 1986. *Women and Love: A Cultural Revolution in Progress.* New York: Knopf, p. 469.

23 women use language to seek confirmation: Tannen, D. 1990. *You Just Don't Understand: Women and Men in Conversation.* New York: Ballantine.

23 laughter: Provine, R. R. 1996. Laughter. *American Scientist* 84(1):38–45, p. 42.

23 American Ph.D.'s in engineering: 1990. *Statistical Abstract of the United States,* table 1004, p. 591.

23 National Academy of Sciences: Flam, F. 1992. What should it take to join science's most exclusive club? *Science* 256:960–961.

23 68 percent of men teaching science: Healy, B. 1992. Women in science: from panes to ceilings. *Science* 255:1333.

23 women in top executive jobs: Saltzman, A. 1991. Trouble at the top. *U.S. News & World Report* 110(23):40–48.

24 "training women for nonachievement": Frieze, Parsons, Johnson, Ruble, and Zellman 1978, pp. 246, 372.

24 toddlers of working mothers: Gottfried, A. E., and A. W. Gottfried. 1988. *Maternal Employment and Children's Development.* New York: Plenum.

24 women's lack of "success": Greer, G. 1971. *The Female Eunuch.* New York: McGraw-Hill, pp. 2, 19, 222.

24 According to Engels, "The liberation of the woman first requires the reintroduction of the entire feminine sex into public industry and . . . this also necessitates the dissolution of the single family as society's economic unit. . . . The care and upbringing of children becomes a public matter": Engels, F. 1884 [1964]. *The Origin of the Family, Private Property, and the State.* New York: International Publishers, pp. 213–214.

25 "These women [scientists]": Konner 1982, p. 107.

25 only one job in five: Zuckerman, M. B. 1995. The nanny state of the nation. *U.S. News & World Report*. 118(13):72.

25 age-old problems among female social primates: Altmann, J. 1980. *Baboon Mothers and Infants.* Cambridge: Harvard University Press.

25 divorce in half of all marriages: Fisher, H. 1991. Monogamy, adultery, and divorce in cross-species perspective. In Robinson and Tiger, pp. 95–126. See also Betzig, L. Cited in Buss 1994, p. 249.

26 people in all known cultures: Daly and Wilson 1983, pp. 262–263. See also Database. 1995. Feminine power. *U.S. News & World Report* 118(6):15.

26 baby and child care by women: Konner 1982, p. 110.

26 emotional distress depresses secretion of growth hormone: Tanner 1988, p. 388.

27 "Communally reared children": Rossi, A. S. 1977. A biosocial perspective on parenting. *Daedelus* 106:1–31, p. 24. See also Limbaugh, R. 1993. *See, I Told You So.* New York: Simon and Schuster, pp. 108–109.

27 "What's distinctive": Mead, M. 1949 [1971]. *Male and Female.* New York: Dell, pp. 193, 194, 195–196, 229.

27–28 Mead's Samoans: Freeman, D. 1983. *Margaret Mead and Samoa: The Making and Unmaking of an Anthropological Myth.* Cambridge: Harvard University Press, pp. 93, 104, 106, 107, 157–173, 220–222, 227, 243–249, 258–259. See also Mead, M. 1928 [1973]. *Coming of Age in Samoa: A Psychological Study of Primitive Youth for Western Civilization.* New York: Morrow, pp. 52, 54–56, 58–60, 84, 86, 110–111, 112, 116–117, 118, 123.

29 all hunter-gatherers divide labor by sex: Bettinger, R. L. 1991. *Hunter-Gatherers: Archaeological and Evolutionary Theory.* New York: Plenum. See also Winterhalder, B., and E. A. Smith (eds.). 1981. *Hunter-Gatherer Foraging Strategies: Ethnographic and Archeological Analyses.* Chicago: University of Chicago Press. Lee, R. B. 1968. What hunters do for a living, or, how to make out on scarce resources. In R. B. Lee and I. DeVore (eds.), *Man the Hunter,* pp. 30–48. Chicago: Aldine. Silberbauer, G. 1981. Hunter/gatherers of the central Kalahari. In R. S. O. Harding and G. Teleki (eds.), *Omnivorous Primates: Gathering and Hunting in Human Evolution,* pp. 455–498. New York: Columbia University Press.

29 "If women didn't exist": Winokur 1991, p. 202.

29 "hunting, fighting": Symons 1979, p. 163.

30 great apes share need and drive to be programmed: Latinen, K. 1989. Demography of chimpanzees in captivity. In P. G. Heltne and L. A. Marquardt (eds.), *Understanding Chimpanzees,* pp. 354–359. Cambridge: Harvard University Press.

30 gender perceptions by parents play a big role in nurturing: Pillard, R. C., and J. D. Weinrich. 1988. The periodic table model of the gender transpositions, part I: a theory based on masculinization and defeminization of the brain. *Journal of Sex Research* 23(4):425–454.

30 a gene for hyperaggression in men: Morell, V. 1993. Evidence found for a possible "aggression gene." *Science* 260:1722–1723. See also Cohen, P. 1995. Sex, violence, and the single gene. *New Scientist* no. 2005:17.

30 "Men are more violent than women": Konner 1982, p. 109.

30 statistics on homicide: Daly, M., and M. Wilson. 1988. *Homicide.* New York: Aldine de Gruyter, p. 149.

Chapter Two: The Puppet Masters

31 Uganda: Ghiglieri, M. P. 1988. *East of the Mountains of the Moon: Chimpanzee Society in the African Rain Forest.* New York: Free Press.

33 brain statistics: Galan, M. 1993. *Mind and Matter: Journey Through the Mind and Body.* New York: Time-Life, p. 10.

33–35 how the old mammalian brain is organized: Fincher, J. 1984. *The Brain: Mystery of Matter and Mind*. New York: Torstar.

34 the amygdala allows us to read people's emotions: Bower, B. 1994. Brain faces up to fear, social signs. *Science News* 146(25):406.

34 the hypothalamus is ruled by more than thirty hormones: Restak, R. M. 1991. The brain as a supercomputer. In M. H. Robinson and L. Tiger (eds.), *Man and Beast Revisited,* pp. 225–230. Washington, D.C.: Smithsonian Press, p. 229.

34 the hypothalamic pleasure center: Galan 1993, pp. 67, 73.

34 the sexually dimorphic nucleus of the hypothalamus: Gibbons, A. 1991. The brain as "sexual organ." *Science* 253:957–959.

34 the INAH-3 nucleus: Barinaga, M. 1991. Is homosexuality biological? *Science* 253:956–957.

35 serotonin, suicide, and violence: Holden, C. 1992. A new discipline probes suicide's multiple causes. *Science* 256:1761–1762.

36 anger in the female septum: Douglas, K. 1996. Cherchez la difference. *New Scientist* Supplement no. 2027:14–16.

36 jealousy for men and women: Buss, D. M. 1994. The strategies of human mating. *American Scientist* 82(3):238–249, p. 247. See also Buss, D. *The Evolution of Desire: Strategies of Human Mating.* New York: Basic Books. Buss, D. M., R. Larsen, D. Westen, and J. Semmelroth. 1992. Sex differences in jealousy. *Psychological Science* 3:251–255. Symons, D. 1979. *The Evolution of Human Sexuality.* New York: Oxford University Press, p. 27. Daly, M., and M. Wilson. 1988. *Homicide.* New York: Aldine de Gruyter, p. 183.

37 fear is generated in the amygdala: Barinaga, M. 1992. How scary things get that way. *Science* 258:887–888. See also Galan 1993, p. 65.

38 women, testosterone, and sex: Persky, H. 1987. *Psychoendocrinology of Human Sexual Behavior.* New York: Praeger.

38 porn in the United States: Schlosser, E. 1997. The business of pornography. *U.S. News & World Report* 122(5):42–50.

39 the pheromones present in a man's armpit sweat: Cutler, W. B., G. Preti, A. Krieger, G. R. Huggins, C. R. Garcia, and H. J. Lawley. 1986. Human axillary secretions influence women's menstrual cycles: the role of donor extract from men. *Hormones and Behavior* 20:463–473.

39 sex every day: Small, M. F. 1995. *What's Love Got to Do with It?* New York: Anchor Books, pp. 15–24, 81.

39 in India, men even kill wives: Brodie, F. M. 1967. *The Devil Drives.* New York: Norton, p. 64.

39–40 the custom of droit du seigneur: Tiger, L. 1992. *The Pursuit of Pleasure*. New York: Little, Brown.

40 "Much more than women": Symons 1979, p. 27.

40 Wilt "the Stilt" Chamberlain: Chamberlain, W. 1991. *A View from Above.* New York: Villard, p. 251.

40 only 15 percent of wives: Schrof, J. M., and B. Wagner. 1994. Sex in America. *U.S. News & World Report* 117(15):74–82, p. 77.

40 "there is no necessary relationship": Rindfuss, R. B., S. P. Morgan, and G. Swicegood. 1988. *First Births in America.* Berkeley: University of California Press, p. 37.

40 kipsigis men in Kenya: Borgerhoff Mulder, M. 1988. Kipsigis bridewealth payments. In L. Betzig, M. Borgerhoff Mulder, and P. Turke (eds.), *Human Reproductive Behavior: A Darwinian Perspective,* pp. 65–82. New York: Cambridge University Press. See also Borgerhoff Mulder, M. 1988. Reproductive success in three Kipsigis cohorts. In Clutton-Brock, T. H. (ed.), *Reproductive Success: Studies of Individual Variation in Contrasting Breeding Systems,* pp. 419–435. Chicago: University of Chicago Press.

40 men continue to choose polygyny: Hern, W. M. 1989. Polygyny, fertility, and reproductive success among the Shipibo. Paper presented at the first annual meeting of the Human Behavior and Evolution Society, Northwestern University, Evanston, Ill., August 25–27.

40–41 "One of the strangest studies": Harley, W. F., Jr. 1986. *His Needs, Her Needs.* Old Tappan, N.J.: Fleming H. Revell, pp. 41–42, 39.

41 "Because women": Fisher, H. E. 1992. *Anatomy of Love.* New York: Norton, p. 69.

41 "It is a woman's business": Winokur, J. (ed.), 1991. *A Curmudgeon's Garden of Love.* New York: Plume, p. 174.

41 "Gay men and straight men": Burr, C. 1993. Homosexuality and biology. *Atlantic* 271(3):47–65, p. 62.

41 "The double standard implies": Faunce, P., and S. Phipps-Yonas. 1979. Women's liberation and human sexual relations. In J. H. Williams (ed.), *Psychology of Women,* pp. 228–240. New York: Norton, p. 231.

41–42 "the woman they marry": Hite, S. 1986. *Women and Love: A Cultural Revolution in Progress.* New York: Knopf, p. 190.

42 the sexual double standard is natural: Irons, W. 1988. Reproductive behavior in humans. In Betzig, Borgerhoff Mulder, and Turke, pp. 307–314, p. 313.

42 "In about two-thirds": Buss 1994, pp. 246–247.

42 relatives of the groom: Strassmann, B. 1989. Menstrual huts and menstrual synchrony among the Dogon. Paper presented at the first annual meeting of the Human Behavior and Evolution Society, Northwestern University, Evanston, Ill., August 25–27.

42 Middle-class Los Angeles women: Essock-Vitale, S. M., and M. T. McGuire. 1988. What 70 million years hath wrought: sexual histories and reproductive success of a random sample of American women. In Betzig, Borgerhoff Mulder, and Turke, pp. 221–235.

42 "77 percent of single women": Hite 1986, pp. 250, 395.

43 only 15 percent of married women have ever had an affair: Schrof and Wagner 1994, p. xx.

43 65 percent of 116 societies: Wilder, R. 1982. Are sexual standards inherited? *Science Digest* 90(7):69.

43 nearly half of 104 societies: Betzig, L. L. 1986. *Despotism and Differential Reproduction: A Darwinian View of History.* New York: Aldine.

43 nonhuman primate liars: Whiten, A., and R. W. Byrne. 1988. Tactical deception in primates. *Behavior and Brain Sciences* 11:233–273. See also Chevalier-Skolnikoff, S. 1986. An exploration of the ontogeny of deception in human beings and nonhuman primates. In R. W. Mitchell and N. S. Thompson (eds.), *Deception: Perspectives on Human and Nonhuman Deceit,* pp. 205–220. Albany: State University of New York Press. Miles, H. L. 1986. How can I tell a lie? Apes, language, and the problem of deception. In Mitchell and Thompson, pp. 245–266. De Waal, F. 1986. Deception in the natural communication of chimpanzees. In Mitchell and Thompson, pp. 221–244. De Waal, F. B. M. 1984. *Chimpanzee Politics: Power and Sex Among Apes.* New York: Harper Colophon.

43 rates of lying: Kiernan, V. 1995. Lies, damned lies, and here are the statistics. *New Scientist* no. 1992:5.

44 toddlers start lying: Vasek, M. E. 1986. Lying as a skill: the development of deception in children. In Mitchell and Thompson, pp. 271–292.

44 lying is so common: Alexander, R. D. 1987. *The Biology of Moral Systems.* New York: Aldine de Gruyter.

44 "Deception is an adaptive behavior": Bower, B. 1989. Deceptive successes in young children. *Science News* 136(22):343.

44–45 what happens when people are threatened: Tooby, J., and L. Cosmides. 1989. The logic of threat. Paper presented at the first annual meeting of the Human

Behavior and Evolution Society, Northwestern University, Evanston, Ill., August 25–27.

45 54 percent of men think about sex: Schrof and Wagner 1994, p. 78.

45 94 percent of sex scenes on soap operas: Abelson, P. H. 1992. A major generation gap. *Science* 256:945.

45 87 percent of intercourse during prime time: Impoco, J., and M. Guttman. 1995. Hollywood: right face. *U.S. News & World Report* 118(19):66–72, p. 72.

45 a mere glimpse of a woman's genitalia: Symons 1979, p. 27.

45 The double standard in aging is severe. Take, for example, Susan Sontag's observations:

> Suppose, when both husband and wife are already in their late forties or early fifties, they divorce. The husband has an excellent chance of getting married again, probably to a younger woman. His ex-wife finds it difficult to remarry. Attracting a second husband younger than herself is improbable; even to find someone her own age she has to be lucky, and she will probably have to settle for a man considerably older than herself, in his sixties or seventies. Women become sexually ineligible much earlier than men do. A man, even an ugly man, can remain eligible well into old age. He is an acceptable mate for a young attractive woman. Women, even good-looking women, become ineligible (except as partners of very old men) at a much younger age.

Sontag, S. 1979. The double standard in aging. In J. H. Williams (ed.), *Psychology of Women,* pp. 462–478. New York: Norton.

45 $1.7 billion cosmetic surgery industry: Staff. 1994. Database. *U.S. News & World Report* 117(15):15.

45 "in preliterate societies": Symons 1979, p. 27.

46 to test this idea: Betzig 1986, p. 34.

46 the king of Dahomey had first pick: Ibid., pp. 70–82.

46 Moulay Ismail the Bloodthirsty: McWhirter, N., and R. McWhirter. 1975. *The Guiness Book of World Records.* New York: Sterling.

46 men seek women above all else: Betzig 1986, p. 97.

46 "Power is an aphrodisiac": Dean, J. 1976. *Blind Ambition.* New York: Simon and Schuster.

46 women are often mutilated: Diamond, J. 1992. *The Third Chimpanzee: The Evolution and Future of the Human Animal.* New York: HarperPerennial, p. 96. See also Daly, M., and M. Wilson. 1983. *Sex, Evolution, and Behavior.* Belmont, Calif.: Wadsworth, pp. 296–297. Hosken, F. P. 1979. *The Hosken Report: Genital and Sexual Mutilation of Females.* 2nd rev. ed. Lexington, Mass.: Women's International Network News. Brodie 1967, p. 110.

46 Cahokia chiefs: Fiedel, S. J. 1987. *Prehistory of the Americas.* New York: Cambridge University Press, pp. 251, 333.

46 "Absolute power corrupts absolutely": Acton, J. E. E. D. 1948. *Essays on Freedom and Power.* Boston: Beacon Press.

47 oxytocin and love: Brownlee, S. 1997. Can't do without love. *U.S. News & World Report* 122(6)58–60.

47 love is spurred by a flood of phenylethylamine: Liebowitz, M. R. 1983. *The Chemistry of Love.* Boston: Little, Brown.

49 fatness and fertility in hunter-gatherers: Hill, K., and H. Kaplan. 1988. Tradeoffs in male and female reproductive strategies among the Ache, part 2. In Betzig, Borgerhoff Mulder, and Turke, pp. 291–305.

49 weight loss can cause permanent infertility: Frisch, R. 1988. Fatness and fertility. *Scientific American* 258(3):88–95.

49 30 percent of women remained sterile: Marx, J. 1988. Sexual responses are—almost—all in the brain. *Science* 241:903–904.

50 suicide rate of American teenage boys: Maguire, K., and A. L. Pastore (eds.). 1997. *Sourcebook of Criminal Justice Statistics, 1996*. Washington, D.C.: U.S. Department of Justice, p. 343.

50 emotions are vital biological compasses: Concar, D. 1996. You're wrong, Mr. Spock. *New Scientist* Supplement no. 2027:4–7, p. 5.

51 "[The] continued pretense": Konner, M. 1982. *The Tangled Wing: Biological Constraints on the Human Spirit*. New York: Holt, Rinehart and Winston, p. 207.

51 examples of writings by sociologists condemning biology: Connell, B. 1989. Masculinity, violence, and war. In M. S. Kimmel and M. A. Messner (eds.), *Men's Lives,* pp. 194–200. New York: Macmillan. See also Leakey, R. 1981. *The Making of Mankind*. New York: Dutton, p. 21.

51 "If the system": Hite 1986, p. 764.

52 "Intelligence can be understood": Sternberg, R. J. 1985. Human intelligence: the model is the message. *Science* 230:1111–1118, p. 1117.

Chapter Three: What Manner of Creature?

54 "Is it not unreasonable": Williams, G. C. 1966. *Adaptation and Natural Selection: A Critique of Some Current Evolutionary Thought*. Princeton, N.J.: Princeton University Press, p. 16.

54–56 finding Lucy: Johanson, D., and M. Edey. 1981. *Lucy: The Beginnings of Humankind*. New York: Simon and Schuster, pp. 16–17.

56 a twenty-five-year-old female: Ibid. See also Johanson, D. C. 1996. Face to face with Lucy's family. *National Geographic* 189(3):96–117, p. 114.

56 Lucy was "even better designed": Lovejoy, O. 1988. Evolution of human walking. *Scientific American* 259(5):116–125, p. 122.

56 Lucy was an imperfect biped: Jungers, W. L. 1988. Relative joint size and hominid locomotor adaptations with implications for the evolution of hominid bipedalism. *Journal of Human Evolution* 17(1/2):247–265. See also Susman, R. L., J. T. Stern, and W. L. Jungers. 1985. Locomotor adaptations in the Hadar hominids. In E. Delson (ed.), *Ancestors: The Hard Evidence,* pp. 184–192. New York: Alan R. Liss. Gibbons, A. 1994. Walkers and swingers. *Science* 264:350.

56 Site 333 yielded Afar's "First Family": Johanson, D. 1976. Ethiopia yields "First Family" of early man. *National Geographic* 150(6):790–811.

56 extrapolations on the "First Family": Johanson, D. C., and K. O'Farrell. 1990. *Journey from the Dawn: Life with the World's First Family*. New York: Villard.

56 Dean Falk calculated: Simons, E. L. 1989. Human origins. *Science* 245:1343–1350.

57 twenty years passed before an *afarensis* skull turned up: Shreeve, J. 1994. "Lucy," crucial early human ancestor, finally gets a head. *Science* 264:34–35.

57 controversially, they named Lucy *Australopithecus afarensis:* Johanson, D., and T. White. 1979. A systematic assessment of early African Hominidae. *Science* 203:321–330.

57 Sexual dimorphism: Short, R. V. 1981. Sexual selection in man and the great apes. In C. E. Graham (ed.), *Reproductive Biology of the Great Apes,* pp. 319–341. New York: Academic Press. See also Pilbeam, D. 1986. Distinguished lecture: hominid evolution and hominid origins. *American Anthropologist* 88:295–312.

57 *A. afarensis* was an impossible ancestor: Falk, D. 1986. Letters: hominid evolution. *Science* 234:11.

58 *Australopithecus anamensis:* Leakey, M. G., C. S. Feibel, I. MacDougall, and A. Walker. 1995. New four-million-year-old hominid species from Kanapoi and Allia Bay, Kenya. *Nature* 376:565–571. See also Leakey, M., and A. Walker. 1997. Early hominid fossils from Africa. *Scientific American* 276(6):74–79.

58 *Australopithecus ramidus:* White, T. D., G. Suwa, and B. Asfaw. 1995. *Australopithecus ramidus,* a new species of early hominid from Aramis, Ethiopia. *Nature* 371:306–312. See also Fischman, J. 1994. Putting our oldest ancestors in their place. *Science* 265:2011–2012. Lewin, R. 1995. Bones of contention. *New Scientist* 2002:14–15.

58 *Homo habilis:* Johanson, D., and J. Shreeve. 1989. *Lucy's Child: The Discovery of a Human Ancestor.* New York: Morrow.

58 In 1923, Raymond Dart discovered the first *Australopithecus africanus,* long thought to be an ancestor of Leakey's *Homo habilis,* but this became impossible due to the contemporaneity of both: Dart, R. A. 1925. *Australopithecus africanus:* the man-ape of South Africa. *Nature* 115:195–199. See also Brain, C. K. (ed.). 1993. *Swartkrans: A Cave's Chronicle of Early Man.* Pretoria, South Africa: Transvaal Museum. Lewin, R. 1987. *Bones of Contention.* New York: Simon and Schuster, p. 52. Susman, R. 1994. Fossil evidence for early hominid tool use. *Science* 265:1570–1573.

58–59 Confusingly, because so many of these fossils were contemporary from 1.8 million years ago but were once thought to be sequential in lineage, paleoanthropologists have been forced to scrap the idea of *Australopithecus africanus, A. aethiopithecus, A. boisei, A. robustus,* or any similar species being the ancestor of *Homo habilis* and *H. erectus.* They have even eliminated *habilis* as an ancestor to *erectus:* Leakey, R. E. F., and A. C. Walker. 1976. *Australopithecus, Homo erectus,* and the single species hypothesis. *Nature* 261:572–574. See also Falk, D. 1983. Cerebral cortices of East African early hominids. *Science* 221:1072–1074. Walker, A., R. E. Leakey, J. M. Harris, and F. H. Brown. 1986. 2.5 Myr *Australopithecus boisei* from west of Lake Turkana, Kenya. *Nature* 322:517–522. Conroy, G. C., M. W. Vannier, and P. T. Tobias. 1990. Endocranial features of *Australopithecus africanus* revealed by 2- and 3-D computerized tomography. *Science* 247:838–841.

59 99 percent of species: Raup, D. M. 1986. Biological extinction in earth history. *Science* 231:1528–1533.

59 average life spans of mammal species: Simpson, G. G. 1983. *Fossils and the History of Life.* New York: Scientific American, p. 135.

59 sixteen mass extinctions: Kerr, R. A. 1992. Origins and extinctions: paleontology in Chicago. *Science* 257:486–487. See also Kerr, R. A. 1992. Another impact extinction? *Science* 256:1280.

59 Terminal Cretaceous Event: Alvarez, W. 1997. *T. rex and the Crater of Doom.* Princeton, N.J.: Princeton University Press. See also Alverez, W., P. Claeys, and S. W. Walker. 1995. Emplacement of Cretaceous-Tertiary Boundary shocked quartz from Chicxulub Crater. *Science* 269:930–935. Sheehan, P. M., D. E. Fastovsky, R. G. Hoffmann, C. B. Berghaus, and D. L. Gabriel. 1991. Sudden extinction of the dinosaurs: latest Cretaceous, Upper Great Plains, U.S.A. *Science* 254:835–839.

59 advanced dinosaurs: Bakker, R. T. 1986. *Dinosaur Heresies.* New York: Morrow. See also Horner, J. 1988. *Digging Dinosaurs.* New York: Workman. Lessem, D. 1992. *Dinosaurs Rediscovered: New Findings Which Are Revolutionizing Dinosaur Science.* New York: Touchstone.

59 "Had the dinosaurs survived": Stanley, S. M. 1987. *Extinction.* New York: Scientific American, p. 8.

59 NASA is trying to monitor: Matthews, R. 1992. A rocky watch for earthbound asteroids. *Science* 255:1204–1205.

60 *erectus* emerged in Africa: Darwin, C. 1871. *The Descent of Man and Selection in Relation to Sex.* New York: Modern Library, p. 520.

60 The earliest evidence of *Homo* on earth comes from Africa. Andrew Hill found a fragment of a *Homo* jaw 2.4 million years old sitting unrecognized for 25 years in the National Museum of Kenya: Gibbons, A. 1992. Human ancestor found—in museum. *Science* 255:1071.

60 the crown jewel *Homo erectus:* Walker, A., and R. Leakey. (eds.). 1993. *The Nariokotome* Homo erectus *Skeleton.* Cambridge: Harvard University Press. See also Leakey, R., and A. Walker. 1985. *Homo erectus* unearthed. *National Geographic* 168(5):624–629. Leakey, R., and R. Lewin. 1992. *Origins Reconsidered.* New York: Doubleday. Morrell, V. 1995. *Ancestral Passions: The Leakey Family and* *the Quest for Humankind's Beginnings.* New York: Simon and Schuster.

60 a ten-year-old brain: Tanner, J. M. 1988. Human growth and constitution. In G. A. Harrison, J. M. Tanner, D. R. Pilbeam, and P. T. Baker, *Human Biology,* 337–435. New York: Oxford University Press, p. 350.

60 bending and flexing of the basicranium: Bower, B. 1989. Talk of ages. *Science News* 136(2):24–26.

60 language is what made humans human: O'Neill, L., M. Murphy, and R. B. Gallagher. 1994. What are we? Where did we come from? Where are we going? *Science* 263:181–183. See also Darwin 1871, p. 912.

60 the human propensity to use language is genetic: Pinker, S. 1991. Rules of language. *Science* 253:530–535. See also Allman, W. F. 1991. The clues in idle chatter. *U.S. News & World Report* 111(8):61–62.

60 infants distinguish between phonetics: Kuhl, P. K., K. A. Williams, F. Lacerda, K. N. Stevens, and B. Lindblom. 1992. Linguistic experience alters phonetic perception in infants by six months of age. *Science* 255:606–608.

60 children can learn three or four languages at once: Galan, M. 1993. *Mind and Matter: Journey Through the Mind and Body.* New York: Time-Life, p. 102.

60 a child learns up to ten new words daily: Miller, G. A., and P. M. Gildea. 1987. How children learn words. *Scientific American* 257(3):94–99. See also Locke, J. L. 1993. *The Child's Path to Spoken Language.* Cambridge: Harvard University Press.

61 "The development of human speech": Wilson, E. O. 1975. *Sociobiology: The New Synthesis.* Cambridge, Mass.: Belknap Press, p. 556.

61 language depends on the left brain: Corina, D. P., J. Vaid, and U. Bellugi. 1992. The linguistic basis of left hemisphere specialization. *Science* 255:1258–1260.

61 thirty-eight species of nonhuman primates hunt vertebrate prey: Butynski, T. M. 1982. Vertebrate predation by primates: a review of hunting patterns and prey. *Journal of Human Evolution* 11:421–430.

61 meat is so important: Hamilton, W. J., III, and C. D. Busse. 1978. Primate carnivory and its significance to human diets. *BioScience* 28(12):761–766.

62 definition of culture: Boyd, R., and P. J. Richerson. 1985. *Culture and the Evolutionary Process.* Chicago: University of Chicago Press, p. 33.

62 most belief systems: Cavalli-Sforza, L. L., M. W. Fledman, K. H. Chen, and S. M. Dornbusch. 1982. Theory and observation in cultural transmission. *Science* 218:19–27.

62 an octopus can learn by observing others: Fiorito, G., and P. Scotto. 1992. Observational learning in *Octopus vulgaris. Science* 256:545–547.

63 the greater the advantage of an idea: Boyd and Richerson. 1985.

63 people more often conform by imitating: Boyd, R., and P. J. Richerson. 1985. Ibid.

63–64 "Once the !Kung": Yellen, J. E. 1990. The transformation of the Kalahari !Kung. *Scientific American* 262(4):96–105.

64 culture backfires: Darwin 1871, p. 501.

65 *erectus* and fire: Balter, M. 1995. Did *Homo erectus* tame fire first? *Science* 268:1570. See also Benditt, J. 1989. Cold water on the fire: a recent survey casts doubt on evidence for early use of fire. *Scientific American* 260(5):21–22. Wu, R. and S. Lin. 1983. Peking man. *Scientific American* 248(6):86–94. Theunissen, B. 1988. *Eugene Dubois and the Ape Man from Java: The History of the First "Missing Link" and Its Discoverer.* Norwell, Mass.: Kluwer. Leakey and Lewin 1992, p. 53.

65 the spread of *erectus:* Sartono, S. 1975. Implications from *Pithecanthropus VIII.* In R. H. Tuttle (ed.), *Morphology and Paleoecology,* pp. 327–360. Chicago: Aldine. See also Swisher, C. C., III, G. H. Curtis, T. Jacob, A. G. Getty, A. Suprijo, and Widiasmoro. 1994. Age of the earliest known hominids in Java, Indonesia. *Science* 263:1118–1121. De Vos, J. 1994. Dating hominid sites in Indonesia. *Science* 266:1726. Swisher, C. C., III. 1994. Response. *Science* 266:1727.

65 *erectus* in eastern Europe: Davis, R. S., V. A. Ranov, and A. E. Dodonov. 1980. Early man in Soviet Central Asia. *Scientific American* 243(6):130–137.

65 differences between advanced African *erectus* and the Javanese and Peking specimens: Larick, R., and R. L. Ciochom. 1996. The African emergence and early Asian dispersals of the genus *Homo. American Scientist* 84(6):538–551.

65 Goodman pioneered: Goodman, M., and R. Tashian (eds.). 1976. *Molecular Anthropology.* New York: Plenum.

65 scientists calibrated a molecular clock of evolution: Tanner, N. M. 1987. The chimpanzee model revisited and the gathering hypothesis. In W. G. Kinzey (ed.), *The Evolution of Human Behavior: Primate Models,* pp. 3–27. New York: State University of New York Press.

65 unraveling the secret of DNA: Watson, J. 1968. *The Double Helix.* New York: Atheneum.

65 reading the genetic code: Dickerson, R. E. 1983. The DNA helix and how it is read. *Scientific American* 249(6):94–111.

65–66 complex information: Marx, J. 1992. Homeobox genes go evolutionary. *Science* 255:399–341. See also Genome Issue. 1993. *Science* 262:20–49.

66 "The full information contained therein": Wilson, E. O. 1988. *Biodiversity.* Washington, D.C.: National Academy Press, p. 7. See also Guyer, R. L., and D. E. Koshland, Jr. 1989. The Molecule of the Year. *Science* 246:1543–1546. Mullis, K. B. 1990. The unusual origin of the polymerase chain reaction. *Scientific American* 262(4):56–65.

66 comparing the DNA of one bird species with another: Sibley, C. G., and J. E. Ahlquist. 1986. Reconstructing bird phylogeny by comparing DNAs. *Scientific American* 254(2):82–92.

66 humans diverged 6.6 million years ago: Sibley, C. G., and J. E. Ahlquist. 1984. The phylogeny of the hominid primates, as indicated by DNA-DNA hydridization. *Journal of Molecular Evolution* 20:2–15.

66 newest dates for divergence from Sibley and Ahlquist: Lewin, R. 1988. Molecular clocks turn a quarter of a century. *Science* 239:561–563.

66 attacks on Sibley and Ahlquist: Lewin, R. 1988. Conflict over DNA clock results. *Science* 241:1598–1600, p. 1598.

66 "I know my molecules had ancestors": Lewin, R. 1988. *In the Age of Mankind.* Washington, D.C.: Smithsonian Institution Press, p. 49.

66 general agreement with Sibley and Ahlquist: Menon, S. 1997. Enigmatic apes. *Discover* 18(8):26–27. See also Norell, M. A., and M. J. Novacek. 1992. The fossil record and evolution: comparing cladistic and paleontological evidence for vertebrate history. *Science* 255:1690–1693. Gibbons, A. 1992. Sorting the hominoid bone pile. *Science* 256:176–177. Simons, E. L. 1990. Discovery of the oldest known anthropoidean skull from the Paleogene of Egypt. *Science* 247:1567–1569. See also Begun, D. R. 1992. Miocene fossil hominids and the chimp-human clade. *Science* 257:1929–1933. Walker, A., and M. Teaford. 1989. The hunt for Proconsul. *Scientific American* 260(1):76–82.

66–67 Sibley and Ahlquist data replicated by Jeffrey Powell: Lewin, R. 1988. DNA clock conflict continues. *Science* 241:1756–1759, p. 1759.

67 "It is now clear": Pilbeam, D. 1984. The descent of hominoids and hominids. *Scientific American* 250(3):84–96.

67 *erectus,* droughts, and population dispersal: DeMenocal, P. B. 1995. Pio-Pleistocene African climate. *Science* 270:53–59. See also Larick and Ciochom 1996.

67 Our "one ancestor" is the consensus of most—but not all—paleoanthropologists. A few—Milford Wolpoff, for example—still propose a very unlikely model of human evolution, the multiple or "regional" model, which says modern races of humans evolved in parallel in multiple populations, one of which was in Africa. But evolutionary theory and paleontological data support a single-species model for human origin: Wolpoff, M. H., et al. 1988. Letter: modern human origins. *Science* 241:772. Stringer, C. B., and P. Andrews. 1988. Genetic and fossil evidence for the origin of modern humans. *Science* 239:1263–1268. See also Stringer, C. B., and P. Andrews. 1988. Letter: modern human origins. *Science* 241:773–774. Stringer, C. B. 1990. The emergence of modern humans. *Scientific American* 263(6):98–104. Singer, R., and J. Wymer. 1982. *The Middle Stone Age at Klasies River Mouth in South Africa*. Chicago: University of Chicago Press.

67 best-selling *Newsweek* article: Lewin, R. 1989. Species questions in modern human origins. *Science* 243:1666–1667.

67 "Eve's [genes]": Tierney, J., L. Wright, and K. Springen. 1988. The search for Adam and Eve. *Newsweek* January 11: 46–52, p. 46.

68 "none of Eve's descendants interbred": Bower, B. 1992. *Erectus* unhinged. *Science News* 141(25):408–411.

68 extensive repeats of work on mitochondrial Eve: Horai, M., et al. 1995. Recent African origin of modern humans revealed by complete sequences of human mitochondrial DNA. *Proceedings of the National Academy of Sciences, USA* 92:532–536. See also Hasegawa, M., and S. Horai. 1991. Time of the deepest root for polymorphism in human mitochondrial DNA. *Journal of Molecular Evolution* 32:37–42. Ruvolo, M., et al. 1993. Mitochondrial COH sequences and modern human origins. *Molecular Biology and Evolution* 10:1115–1135. L. Vigilant, M. Stoneking, H. Harpending, K. Hawkes, and A. C. Wilson. 1991. African populations and the evolution of human mitochondrial DNA. *Science* 253:1503–1507. Cavalli-Sforza, L. L., and F. Cavalli-Sforza. 1995. *The Great Human Diasporas: The History of Diversity and Evolution*. New York: Addison-Wesley, pp. 62–73.

68 nuclear DNA confirmed an out-of-Africa *sapiens:* Stringer, C. B., and R. McKie. 1996. *African Exodus: The Origins of Modern Humanity*. London: Jonathan Cape, p. 122. See also Stringer and Andrews 1988, Genetic and fossil evidence. Cavalli-Sforza, L. L., P. Menozzi, and A. Piazza. 1993. Demic expansions and human evolution. *Science* 259:639–646. Gibbons, A. 1994. African origins theory goes nuclear. *Science* 264:350–351. Dorit, R. L., H. Akashi, and W. Gilbert. 1995. Absence of polymorphism at the ZFY locus on the human Y chromosome. *Science* 268:1183–1185. Yellen, J. E., A. S. Brooks, E. Cornelissen, M. J. Mehlman, and K. Stewart. 1995. A Middle Stone Age worked bone industry from Katanda, Upper Semliki Valley, Zaire. *Science* 268:553–556.

68 archaeology dates modern humans in the Middle East: Valladas, H., et al. 1988. Thermoluminescence dating of Mousterian "Proto-Cro-Magnon" remains from Israel and the origin of modern man. *Nature* 331:614.

68 early humans in Australia: Murray, P. 1984. Extinctions down under: a beastiary of extinct Australian Late Pleistocene monotremes and marsupials. In Martin, P. S., and R.G. Klein, (eds.), *Quarternary Extinctions: A Prehistoric Revolution*, pp. 600–628. Tucson: University of Arizona Press.

68 early humans in Tasmania: Cosgrove, R. 1989. Thirty thousand years of human colonization in Tasmania: new Pleistocene dates. *Science* 243:1706–1708. See also Turner, C. G., III. 1989. Teeth and prehistory in Asia. *Scientific American* 260(2):88–96. Cavalli-Sforza, L. L. 1991. Genes, peoples, and languages. *Scientific American* 265(5):104–110.

68 Neandertal differences from modern man: Trinkaus, E., and W. W. Howells. 1979. The Neandertals. *Scientific American* 241(6):118–123.

68 the Neandertal face: Rak, Y. 1986. The Neandertal: a new look at an old face. *Journal of Human Evolution* 15(3):151–164.

68 Neandertals ate almost nothing but meat: Dorozynski, A., and A. Anderson. 1991. Collagen: a new probe into prehistoric diet. *Science* 254:520–521.

69 Neandertals buried their dead with tools, flowers, or food: Trinkaus, E. 1983. *The Shanidar Neanderthals*. New York: Academic Press.

69 Neandertal bones: Gore, R. 1996. Neanderthals. *National Geographic* 189(1):2–35, p. 26.

69 Neandertals may have killed one another in combat: Trinkaus, E., and M. R. Zimmerman. 1982. Trauma among the Shanidar Neanderthals. *American Journal of Physical Anthropology* 57:61–76.

69 some scientists insist that Neandertals were our ancestors: Jelinek, A. J. 1982. The Tabun Cave and Paleolithic man in the Levant. *Science* 216:1369–1375. See also Smith, F. H. 1984. Fossil hominids from the upper Pleistocene of central Europe and the origin of modern Europeans. In F. H. Smith and F. Spencer (eds.), *The Origins of Modern Europeans,* pp. 187–209. New York: Alan R. Liss. Shreeve, J. 1995. *The Neandertal Enigma: Solving the Mystery of Modern Human Origins.* New York: Morrow. Stringer and McKie 1996. Bower, B. 1997. Neandertals make big splash in gene pool. *Science News* 152(3):37.

69 technology of European modern *sapiens:* Nelson, H., and R. Jurmain. 1994. *Introduction to Physical Anthropology.* 5th ed. Thousand Oaks, Calif.: West.

69 Neandertal hunting: Gore 1996, p. 26.

69 mammoth-bone dwellings: Gladkih, M. I., N. L. Kornietz, and O. Soffer. 1984. Mammoth-bone dwellings on the Russian plain. *Scientific American* 251(5):164–175. See also Hedges, R. E. M., and J. A. J. Gowlett. 1986. Radiocarbon dating by accelerator mass spectrometry. *Scientific American* 254(1):100–107.

69 "A small demographic advantage": Lewin 1988, *In the Age of Mankind,* p. 133.

69 the last Neandertals were in Spain: Gore 1996, p. 34.

70 humanity's closest living relatives: Tanner 1987.

70 98.4 percent of human and chimp DNA: Gibbons, A. 1990. Our chimp cousins get that much closer. *Science* 250:374. See also Sibley and Ahlquist 1984.

70 chimps and humans are "sibling species": Lewin, R. 1988. Family relationships are a biological conundrum. *Science* 242:671.

70–71 social behaviors of chimps, bonobos, and humans: Ghiglieri, M. P. 1987. Sociobiology of the great apes and the hominid ancestor. *Journal of Human Evolution* 16(4):319–357.

71 179 hunter-gatherer societies: Ember, C. R. 1978. Myths about hunter-gatherers. *Ethnology* 17:439–448. See also Berté, N. 1988. K'ekchi' horticultural labor exchange: productive and reproductive implications. In L. Betzig, M. Borgerhoff Mulder, and P. Turke (eds.), *Human Reproductive Behavior: A Darwinian perspective,* pp. 83–96. New York: Cambridge University Press.

71 women worldwide cooperate less: Hrdy, S. B. 1981. *The Woman That Never Evolved.* Cambridge: Harvard University Press, p. 130.

71 rise and decline of feminism: Staff. 1992. The year that was. *U.S. News & World Report* 111(27):100.

71 Eskimos on dangerous hunts: Morgan, C. J. 1979. Eskimo hunting groups, social kinship, and the possibility of kin selection in humans. *Ethology and Sociobiology* 1:83–86.

71 the problems monkeys face in going it alone is predators: Hamilton, W. D. 1971. Geometry for the selfish herd. *Journal of Theoretical Biology* 31:295–311. See also Anderson, C. M. 1986. Predation and primate evolution. *Primates* 27:15–39. Van Schaik, C. P. 1983. Why are diurnal primates living in groups? *Behaviour* 87:120–144. Van Schaik, C. P., and J. A. R. A. M. Van Hooff. 1983. On the ultimate causes of primate social systems. *Behaviour* 85:91–117. Cheney, D. L., and R. W. Wrangham. 1986. Predation. In B. B. Smuts, D. L. Cheney, R. M. Seyfarth, R. W. Wrangham, and T. T. Struhsaker (eds.), *Primate Societies,* pp. 227–239. Chicago: University of Chicago Press. Rijksen, H. D. 1978. *A Field Study on Sumatran Orangutans (*Pongo pygmaeus abelii *Lesson 1827): Ecology, Behaviour, and Conservation.* Wageningen, Netherlands: H. Veenman and Zonen, B. V., p. 135. Crook, J. H. 1970. The socio-ecology of primates. In J. H. Crook (ed.), *Social Behavior of Birds and Mammals,* pp. 103–166. New York: Academic Press. Crook, J. H., and J. S. Gartlan. 1966. Evolution of primate societies. *Nature* 210:1201–1203. Eisenberg, J. F., N. A. Muchenhirn, and R. Rudran. 1972. The relation between ecology and social structure. *Science* 176:863–874. Clutton-Brock, T. H. 1974. Primate social organization and ecology. *Nature* 250:539–542. Clutton-Brock, T. H., and P. Harvey. 1977. Primate ecology and social organization. *Journal of the Zoological Society London* 183:1–39.

71–72 availability of food: Janson, C. H., and M. L. Goldsmith. 1995. Predicting group size in primates: foraging costs and predation risks. *Behavioral Ecology* 6(3):326–336.

72 human division of labor: Smith, A. 1776. [1986]. *Wealth of Nations.* New York: Penguin.

72 fissioning and fusion: Hayden, B. 1981. Subsistence and ecological adaptations of modern hunter-gatherers. In R. S. O. Harding and G. Teleki (eds.), *Omnivorous Primates: Gathering and Hunting in Human Evolution,* pp. 344–421. New York: Columbia University Press. See also Harako, R. 1981. The cultural ecology of hunting behavior among the Mbuti pygmies in the Ituri Forest, Zaire. In Harding and Teleki, pp. 499–555. Silberbauer, G. 1981. Hunter/gatherers of the central Kalahari. In Harding and Teleki, pp. 455–498. Woodburn, J. 1968. An introduction to Hadza ecology. In R. B. Lee and I. DeVore (eds.), *Man the Hunter,* pp. 49–55. Chicago: Aldine. Schultz, J. W. 1907 [1981]. *My Life as an Indian.* New York: Fawcett Columbine.

72 chimps sacrifice feeding to socialize: Bauer, H. R. 1980. Chimpanzee society and social dominance in evolutionary perspective. In F. F. Strayer and D. Freedman (eds.), *Dominance Relations: Ethological Perspectives on Human Conflict,* pp. 97–119. New York: Garland.

72 no known human society has come close to being promiscuous: Van den Berghe, P. L. 1979. *Human Family Systems: An Evolutionary View.* New York: Elsevier, p. 52.

72 83.5 percent of societies worldwide: Murdock, G. P. 1967. Ethnographic atlas: a summary. *Ethnology* 6:109–236.

72 853 societies: Caldwell, J. C., and P. Caldwell. 1990. High fertility in Sub-Saharan Africa. *Scientific American* 262(5):118–125, p. 120. See also Low, B. S. 1988. Pathogen stress and polygyny in humans. In Betzig, Borgerhoff Mulder, and Turke 1988, pp. 115–127.

72 Kipsigis polygyny: Borgerhoff Mulder, M. 1988. Reproductive success in three Kipsigis cohorts. In T. H. Clutton-Brock (ed.), *Reproductive Success: Studies of Individual Variation in Contrasting Breeding Systems,* Chicago: University of Chicago Press. pp. 419–435.

72 polyandry among brothers who own land: Crook, J. H., and S. J. Crook. 1988. Tibetan polyandry: problems of adaptation and fitness. In Betzig, Borgerhoff Mulder, and Turke, pp. 97–114.

73 Mormon men: Faux, S. F. 1981. A sociobiological perspective of the doctrinal development of Mormon polygyny. Sunstone Theological Symposium. (Cited in Daly, M., and M. Wilson, 1983. *Sex, Evolution, and Behavior.* Belmont, Calif.: Wadsworth.)

73 "Nothing in male sexuality": Symons, D. 1979. *The Evolution of Human Sexuality.* New York: Oxford University Press, p. 292.

73 in all cultures, men support their children: Van den Berghe 1979, p. 48. See also Hewlett, B. S. 1988. Sexual selection and parental investment among Aka pygmies. In Betzig, Borgerhoff Mulder, and Turke, pp. 263–276.

73 Miocene global warming: Allman, W. F. 1992. Climate and the rise of man. *U.S. News & World Report* 112(22):60–67. See also Kingston, J. D., B. D. Marino, and A. Hill. 1994. Isotopic evidence for Neogene hominid paleoenvironments in the Kenya Rift Valley. *Science* 264:955–959.

73–74 "bulldozer" herbivores: Kortlandt, A. 1984. Vegetation research and the "bulldozer" herbivores of tropical Africa. In A. C. Chadwick and C. L. Sutton (eds.), *Tropical Rain-Forest: The Leeds Symposium,* pp. 205–226. Special publication of the Leeds Philosophical Society.

74 apes living on the forest-savanna fringe: Kortlandt, A. 1983. Marginal habitats of chimpanzees. *Journal of Human Evolution* 12:231–278. See also Sinclair, A. R. E., M. D. Leakey, and M. Norton-Griffiths. 1986. Migration and hominid bipedalism. *Nature* 324:305. Lewin, R. 1987. Four legs bad, two legs good. *Science* 235:969–971.

74 "We looked at the data": Rodman, P. S., and H. M. McHenry. 1980. Bioenergetics and the origin of hominid bipedalism. *American Journal of Physical Anthropology* 52:103–106. See also Leakey, R., and R. Lewin. 1992. *Origins Reconsidered: In Search of What Makes Us Human*. London: Little, Brown, p. 90.

74 "If you're an ape": Rodman and McHenry 1980.

74 a 60 percent reduction in heat: Stringer and McKie 1996, p. 27.

74 "the primary adaptation": Leakey and Lewin 1992, p. 91. See also Rodman and McHenry 1980.

74–75 volcanic ash at Laetoli: Johanson and Edey 1981, p. 245.

75 during an elephant-dung-throwing fight: Lewin, 1987, *Bones of Contention,* p. 278.

76 "The Laetoli footprints": Johanson and Shreeve 1989, p. 188.

76 Mary Leakey agrees: Leakey, M. D. 1979. Footprints in the ashes of time. *National Geographic* 155(4):446–457.

76 "Laetoli National Forest": Johanson and Shreeve 1989, p. 191. See also Holden, C. 1993. Hominid sites ravaged by *Homo sapiens*. *Science* 262:34.

76 the stride lengths between prints: Hay, R. L., and M. D. Leakey. 1982. The fossil footprints of Laetoli. *Scientific American* 249(2):50–57.

76 "Except in relatively": Lewin 1987, Four legs bad, p. 971.

77 *Australopithecus robustus* and culture: Grine, F. E. (ed.). 1989. *Evolutionary History of the "Robust" Australopithecines*. New York: Aldine de Gruyter.

77 greater demands due to complicated ecology: Pagel, M. D., and P. H. Harvey. 1989. Taxonomic differences in the scaling of brain on body weight in mammals. *Science* 244:1589–1592. See also Harvey, P. H., and J. R. Krebs. 1990. Comparing brains. *Science* 249:140–146.

77 "extractive foraging played no significant role in the evolution of primate intelligence": King, B. 1986. Extractive foraging and the evolution of primate intelligence. *Human Evolution* 1(4):361–372, p. 361.

77 intelligence evolved under the pressure to outwit: Cheney, D., R. Seyfarth, and B. Smuts. 1986. Social relationships and social cognition in nonhuman primates. *Science* 234:1361–1366, p. 1364.

77 prime cause of the quick expansion of the human brain: Trivers, R. L. 1971. The evolution of reciprocal altruism. *Quarterly Review of Biology* 46:35–57. See also Trivers, R. 1991. Deceit and self-deception: the relationship between communication and consciousness. In M. H. Robinson and L. Tiger (eds.), *Man and Beast Revisited,* pp. 175–191. Washington, D.C.: Smithsonian Institution Press, p. 176.

77 "It is incontestable": Small, M. 1990. Political animal social intelligence and the growth of the primate brain. *The Sciences* March/April:36–42, p. 42.

77 "only significant hostile force": Alexander, R. D. 1987. *The Biology of Moral Systems.* New York: Aldine de Gruyter, p. 261.

77–78 "may well have benefitted": Fox, R. 1991. Aggression then and now. In Robinson and Tiger, pp. 81–93.

78 human nature: West-Eberhard, M. J. 1991. Sexual selection and social behavior. In Robinson and Tiger, pp. 159–172, p. 171.

78 how our earliest human ancestors behaved: Ghiglieri 1987.

78 *Homo erectus* behavior developed early: Foley, R. A., and P. C. Lee. 1989. Finite social space, evolutionary pathways, and reconstructing hominid behavior. *Science* 243:901–906, p. 905.

PART TWO: VIOLENCE

79 "For my own part": Darwin, C. 1871. *The Descent of Man and Selection in Relation to Sex*. New York: Modern Library, pp. 919–920.

79 "It is a law of nature": Manchester, W. 1978. *American Caesar: Douglas MacArthur, 1880–1964*. New York: Little, Brown, p. 9.

Chapter Four: Rape

82–83 The assault on Kay: Kline, D. 1989. Personal communication.

83 the tendency of TV to report bizarre violence as news: Bok, S. 1998. *Mayhem*. Reading, Mass.: Addison-Wesley, p. 63.

83 rapes in 1996: Federal Bureau of Investigation (FBI). 1997. *Crime in the United States, 1996*. Uniform Crime Reports, Release date, Sunday, September 28, 1997. Washington, D.C.: U.S. Department of Justice, pp. 23–25.

83 reports of rape have increased: Weiner, N. A., and M. E. Wolfgang. 1990. The extent and character of violent crime in America, 1969 to 1982. In N. A. Weiner, M. A. Zahn, and R. Sagi (eds.), *Violence: Patterns, Causes, Public Policy,* pp. 24–37, p. 27. New York: Harcourt Brace Jovanovich. See also Staff. 1992. UCR shows sixth straight crime jump; BJS reports rape up by 59 percent. *Law Enforcement News* 28(358):1, 4.

83 71 rape victims per 100,000 women: FBI 1997, p. 24.

83 rape accounts for 1 in every 19 reported violent crimes: Ibid., p. 8.

83 reported rates of rape vary: Crowell, N. A., and A. W. Burgess (eds.). 1996. *Understanding Violence Against Women*. Washington, D.C.: National Academy Press, pp. 1, 23.

83 a survey of middle-class Los Angeles women: Essock-Vitale, S. M., and M. T. McGuire. 1988. What 70 million years hath wrought: sexual histories and reproductive success of a random sample of American women. In L. Betzig, M. Borgerhoff Mulder, and P. Turke (eds.), *Human Reproductive Behavior: A Darwinian Perspective,* pp. 221–235. New York: Cambridge University Press.

83 between five and twenty unreported rapes: Kilpatrick, D. G., H. S. Resnick, B. E. Saunders, and C. L. Best. 1994. Survey research on violence against women: results from National Women's Survey. Paper presented at the 46th annual meeting of the American Society of Criminology, Miami, November 11. National Crime Victims Research and Treatment Center, Medical University of South Carolina.

83–84 men convicted of rape serve 7.25 years in prison: Maguire, K., and A. L. Pastore (eds.). 1997. *Sourcebook of Criminal Justice Statistics, 1996*. Washington, D.C.: U.S. Department of Justice, Bureau of Justice Statistics, pp. 398, 470, 476. See also Zawitz, M. W. (ed.). 1988. *Report to the Nation on Crime and Justice*. Washington, D.C.: U.S. Department of Justice, Bureau of Justice Statistics, pp. 97, 100.

84 a woman is eight times more likely to be raped in America than in Europe: Staff. 1992. Unsafe sex. *U.S. News & World Report* 111(1):100–102.

84 rape victims are typically young: Schlesinger, S. R. 1985. *The Crime of Rape*. Washington, D.C.: U.S. Department of Justice, Bureau of Statistics Bulletin, pp. 1, 2. See also Maguire and Pastore 1997, pp. 232–233.

84 two-thirds of victims were raped by strangers: Schlesinger 1985, p. 2. See also Thornhill, R., and N. W. Thornhill. The evolutionary psychology of human rape. Manuscript.

84 88 percent of rapists were solitary: Reiss, A. J., Jr., and J. A. Roth (eds.). 1993. *Understanding and Preventing Violence*. Washington, D.C.: National Academy Press, p. 75.

84–85 typical rapists are young: FBI 1997, pp. 226–230. See also Zawitz 1988, p. 43. Reaves, B. 1990. *Felony Defendants in Large Urban Counties,* 1988. Washington, D.C.: Bureau of Justice Statistics Executive Summary, p. 5.

85 only a third of rapists had been using alcohol or drugs: Zawitz 1988, p. 51.

85 uneducated, unemployed, or underemployed: Wright, J. D., and P. H. Rossi. 1986. *Armed and Considered Dangerous: A Survey of Felons and Their Firearms*. New York: Aldine de Gruyter, p. 48.

85 blacks accounted for 42 percent of rapes: FBI 1997, p. 232.

85 rapists fit: Wolfgang, M. E., and F. Ferracuti. 1967. *The Subculture of Violence*. London: Tavistock.

85 "Rapists are usually": Brownmiller, S. 1975. *Against Our Will*. New York: Simon and Schuster, p. 208.

85 two-thirds of rapists have prior arrest records: Reaves, B. 1990. *Felony Defendants in Large Urban Counties, 1988*. Washington, D.C.: U.S. Department of Justice, Bureau of Justice Statistics Executive Summary.

85 repeat rapists: Dillingham, S. D. 1991. *Violent Crime in the United States*. Washington, D.C.: U.S. Department of Justice, Bureau of Justice Statistics, p. 15.

85 about 14.8 percent of rapists use weapons: Maguire and Pastore 1997, p. 215. See also Dillingham 1991, p. 10. Schlesinger 1985, p. 4.

85 one-quarter of rapists use threats: Schlesinger 1985, p. 5.

85 the use of a weapon increases success: Ibid., p. 4. See also Zawitz 1988, p. 20.

85–86 study of convicted felons: Wright and Rossi 1986, p. 76.

86 "the typical rapist's preferred victim is the 'American dream ideal'": Geis, G. 1977. Forcible rape: an introduction. In D. Chappell, R. Geis, and G. Geis (eds.), *Forcible Rape: The Crime, the Victim, and the Offender*, pp. 1–44, p. 27. New York: Columbia University Press.

86 American men blame the victim for her rape: Larsen, K. S., and E. Long. 1988. Attitudes towards rape. *Journal of Sex Research* 24:299–304.

86 men in India and elsewhere also blame women: Kanekar, S., and M. B. Kolsawalla. 1980. Responsibility of a rape victim in relation to her respectability, attractiveness, and provocativeness. *Journal of Social Psychology* 36:156–179.

87 women are to blame because they already know that men will rape: Beneke, R. E. 1982. *Men on Rape*. New York: St. Martin's Press.

87 by doing so she forfeits: Gordon, M. T., and S. Riger. 1988. *The Female Fear*. New York: Free Press.

87 41 percent of urban women: Knowlton, R. W. 1989. Rape victimizes all women. *Utne Reader* November/December:42.

87 more men than women believe that at least some women want to be raped: Malamuth, N. M., S. Haber, and S. Feshbach. 1980. Testing hypotheses regarding rape: exposure to sexual violence, sex differences, and the "normality" of rapists. *Journal of Research in Personality* 14:121-137. See also Hite, S. 1986. *Women and Love: A Cultural Revolution in Progress*. New York: Knopf, p. 187. The following testimony documented by Shere Hite, seems to support this belief: "When I was raped, during the middle I thought, 'This is just like some fantasy.' This messed me up, because it was like a fantasy and I came. But all I felt was fear and disgust for the creep. It was so horrible, yet I had this nagging feeling that I'd asked for it. It was like all men say—you wanted it—you liked it. But even though I had an orgasm, I didn't like it and I didn't want it."

87 the most common sexual fantasies of women: Pelletier, L. A., and E. S. Herold. 1988. The relationship of age, sex guilt, and sexual experience with female sexual fantasies. *Journal of Sex Research* 24:250–256.

87–88 "Not only do women": Bond, S. B., and D. L. Mosher. 1986. Guided imagery of rape: fantasy, reality, and the willing victim myth. *Journal of Sex Research* 22(2):162–183, p. 177.

88 "Women's responses to being pressured": Christopher, F. S. 1988. An initial investigation into a continuum of premarital sexual pressure. *Journal of Sex Research* 25(2):255–266, p. 263.

89 "an overall sadism effect": Heilbrun, A. B., Jr., and D. T. Seif. 1988. Erotic value of female distress in sexually explicit photographs. *Journal of Sex Research* 24(1):47–57.

89 sexual sadists are most likely to rape: Heilbrun, A. B., Jr., and M. P. Loftus. 1986. The role of sadism and peer pressure in the sexual aggression of male college students. *Journal of Sex Research* 22(3):320–332.

89 the macho male image as a cause of rape: Segel-Evans, K. 1987. Rape prevention and masculinity. In F. Abbot (ed.), *New Men, New Minds: Breaking the Male Tradition,* pp. 117–121. Freedom, Calif.: Crossing Press.

89 the most pervasive rape myth: Palmer, C. T. 1988. Twelve reasons why rape is not sexually motivated: a skeptical examination. *Journal of Sex Research* 25(4):512–530.

89–90 "Indeed, one of the earliest": Brownmiller 1975, pp. 14–15.

90 "female fear": Ibid., p. 16.

90 "The opportunity for free play": Wright, Q. 1942. *A Study of War.* Chicago: University of Chicago Press, p. 135. See also Elshtain, J. B. 1997. Women and war. In C. Townshend (ed.), *The Oxford Illustrated History of Modern War,* pp. 264–277. New York: Oxford University Press.

90 the fall of Rome: Ferrill, A. 1986. *The Fall of the Roman Empire: The Military Explanation.* New York: Thames and Hudson, p. 117.

90 Nazi invaders in Russia: Brownmiller 1975, pp. 64–72.

90 mass military rapes in the 1990s: Crossette, B. 1998. Violation: an old scourge of war becomes its latest crime. *New York Times,* June 12:4-1, 4-6.

91 "Rape of Nanking": Brownmiller 1975, pp. 57–61.

91–92 "There's a thing": Hackworth, D. H., Colonel, and J. Sherman. 1989. *About Face: The Odyssey of an American Warrior.* New York: Simon and Schuster, p. 132.

92 "It is becoming": Crossette 1998, p. 4-1.

93 "It is now generally accepted": Palmer 1988, p. 513.

93 "for sex and that's all": Ibid., p. 518.

93 "efforts to negotiate": Groth, N. 1979. *Men Who Rape.* New York: Plenum, p. 28. See also Groth, N. 1990. Rape: behavioral aspects. In Weiner, Zahn, and Sagi, pp. 72–79.

93 "Men who commit rape": Brownmiller 1975, p. 266.

93 for every three women violently assaulted: Maguire and Pastore 1997, pp. 210–211.

93 Brownmiller's ideas on why men rape are mere speculation unsubstantiated by scientific data from the real world; evolutionary biologists disagree: Symons, D. 1979. *The Evolution of Human Sexuality.* New York: Oxford University Press, pp. 278–279.

93 rape as punishment "does not prove": Ibid., p. 279. See also Hoebel, E. A. 1961. *The Cheyenne Indians of the Great Plains.* New York: Holt, Rinehart and Winston, pp. 94–95. See also Paige, K. E., and J. M. Paige. 1981. *The Politics of Reproductive Ritual.* Berkeley: University of California Press, p. 31.

94 Patricia was raped in Central Park: Staff. 1989. A Clockwork Orange in Central Park. *U.S. News & World Report* 106(18):10.

94 one victim among 3,400: Holtzman, E. 1989. Rape—the silence is criminal. *New York Times,* May 5:A19.

94–95 Mayor Ed Koch asked: Gibbs, N. 1989. Wilding in the night. *Time* 133(19):20–21.

95 spent at least $15 on a woman: Holtzman 1989.

95 an orangutan's diet: Janzen, D. E. 1979. How to be a fig. *Annual Review of Ecology and Systematics* 10:13–53. See also Ghiglieri, M. P. 1986. River of the red ape.

Mainstream 17(4):29–33. Ghiglieri, M. P. 1986. A river journey through Gunung Leuser National Park. *Oryx* 20(2):104–110.

95 mothers and grown daughters?: Galdikas, B. M. F. 1984. Adult female sociality among wild orangutans at Tanjung Puting Reserve. In M. F. Small (ed.), *Female Primates: Studies by Female Primatologists,* pp. 217–235. New York: Alan R. Liss.

96 average distance covered per day by orangutans: MacKinnon, J. 1971. The orangutan in Sabah today. *Oryx* 11:141–191. See also MacKinnon, J. 1973. Orangutans in Sumatra. *Oryx* 12:234–242. Rijksen, H. D. 1974. Orangutan conservation and rehabilitation in Sumatra. *Biological Conservation* 6:20–25.

96 solitary adult orangutans: Van Schaik, C. P., and J. A. R. A. M. Van Hooff. 1996. Toward an understanding of the orangutan's social system. In McGrew, W. C., L. F. Marchant, and T. Nishida (eds.), *Great Ape Societies,* pp. 3–15, New York: Cambridge University Press. See also Rodman, P. S. 1988. Diversity and consistency in ecology and behavior. In J. H. Schwartz (ed.), *Orangutan Biology,* pp. 31–51. New York: Oxford University Press.

96 ape mothers average 4, 5, and 7 years of investment per offspring among gorillas, chimps, and orangutans, respectively: Harcourt, A. H., K. Stewart, and D. Fossey. 1981. Gorilla reproduction in the wild. In C. E. Graham (ed.), *Reproductive Biology of the Great Apes,* pp. 265–279. New York: Academic Press. See also Galdikas, B. M. F. 1981. Orangutan reproduction in the wild. In Graham, pp. 281–300. Goodall, J. 1983. Population dynamics during a 15 year period in one community of free-living chimpanzees in the Gombe National Park, Tanzania. *Zeitschrift fur Tierpsychologie* 61:1–60.

96 males can impregnate another female every day: Short, R. V. 1981. Sexual selection in man and the great apes. In Graham, pp. 319–341.

96 adult male orangutans use two opposite mating strategies: Mitani, J. C. 1985. Sexual selection and adult male orangutan calls. *Animal Behaviour* 33:272–283.

96 territorial males: Rodman, P. S. 1973. Population composition and adaptive organization among orangutans of the Kutai Reserve. In R. P. Michael and J. H. Crook (eds.), *Ecology and Behaviour of Primates,* pp. 171–209. London: Academic Press. See also Schurmann, C. L. 1982. Mating behavior of wild orangutans. In L. E. M. deBoer (ed.), *The Orangutan: Its Conservation and Biology,* pp. 269–284. The Hague: Dr. W. Junk.

96 males advertise with long calls: Mitani 1985, Sexual selection. See also Galdikas 1981. Rodman, P. S., and J. C. Mitani. 1986. Orangutans: sexual dimorphism in a solitary species. In B. B. Smuts, D. L. Cheney, R. M. Seyfarth, R. W. Wrangham, and T. T. Struhsaker (eds.), *Primate Societies,* pp. 146–154. Chicago: University of Chicago Press. Rijksen, R. D. 1978. *A Field Study on Sumatran Orangutans (*Pongo pygmaeus abelii *Lesson 1827): Ecology, Behaviour, and Conservation.* Wageningen, Netherlands: H. Veenman and Zonen, B. V. Mitani, J. C. 1985. Mating behavior of adult male orangutans in the Kutai Reserve, Indonesia. *Animal Behaviour* 33:392–402.

96 The second male strategy: MacKinnon, J. 1974. The behaviour and ecology of wild orangutans, *Pongo pygmaeus. Animal Behaviour* 22:3–74. See also Galdikas, B. M. F. 1979. Orangutan adaptation at Tanjung Puting Reserve: mating and ecology. In D. A. Hamburg and E. R. McCown (eds.), *The Great Apes,* pp. 195–233. Menlo Park, Calif.: Benjamin/Cummings.

96 adult females are scarce: MacKinnon 1971. See also MacKinnon 1973. Rijksen 1974. Rijksen, H. D. 1982. How to save the mysterious "man of the forest." In deBoer, pp. 313–341.

96 subadult males rape females: Galdikas 1981. See also MacKinnon 1974. Rijksen 1978. Schurmann 1982.

96–97 Gundul rapes a woman: Galdikas, B. M. F. 1995. *Reflections of Eden: My Years with the Orangutans of Borneo.* New York: Little, Brown, p. 294.

97 scorpion flies rape: Thornhill, R. 1980. Rape in *Panorpa* scorpionflies and a general rape hypothesis. *Animal Behaviour* 28:52–59.

97 mallard ducks rape: Barash, D. P. 1977. Sociobiology of rape in mallards *(Anes platyrrynchos)*: responses of the mated male. *Science* 197:788–789.

97 several fish species, snow geese, and mountain bluebirds rape: Thornhill, R., and N. W. Thornhill. 1987. Human rape: the strengths of the evolutionary perspective. In C. Crawford, M. Smith, and D. Krebs (eds.), *Sociobiology and Psychology: Ideas, Issues, and Applications,* pp. 269–291. Hillsdale, N.J.: Lawrence Erlbaum.

97 chimpanzees rape: Tutin, C. E. G., and W. McGinnis. 1981. Chimpanzee reproduction in the wild. In Graham pp. 239–264. See also Goodall, J. 1986. *The Chimpanzees of Gombe.* Cambridge: Harvard University Press, p. 481.

97 human rape seems to be aimed at copulating: Palmer 1988, p. 515.

97 castrated rapists have far lower repeat rate: Berlin, F. S., and C. F. Meinecke. 1981. Treatment of sex offenders with anti-androgenic medication: conceptualization, review of treatment modalities, and preliminary findings. *American Journal of Psychiatry* 138:601–607. See also Palmer 1988, p. 520. Abel, G. G. 1983. The relationship between treatment for sex offenders and the court. In S. N. Verdun-Jones and A. A. Keltner (eds.), *Sexual Aggression and the Law,* pp. 15–26. Vancouver: Simon Fraser University. Kelley, K., and D. Byrne. 1992. *Exploring Human Sexuality.* Englewood Cliffs, N.J.: Prentice Hall, pp. 402–403.

98 instrumental force: Palmer 1988, p. 521.

98 one study found that only 11 percent: Kaufman, A. 1984. Rape of men in the community. In I. R. Stuart and J. G. Greer (eds.), *Victims of Sexual Aggression: Treatment of Children, Women, and Men,* pp. 156–179. New York: Van Nostrand Reinhold.

98 "an extremely small proportion": Schlesinger 1985, pp. 4, 5.

98 88 percent of rapists used only instrumental force: Smithyman, S. D. 1978. The undetected rapist. Ph.D. dissertation, Claremont Graduate School, Ann Arbor, Mich.

98 "All I wanted": Symons 1979, p. 246.

98–99 "Most individuals using self-protection": Schlesinger 1985, p. 4. See also Crowell and Burgess 1996, pp. 98–99. Maguire and Pastore 1997, p. 217.

99 Orlando, Florida, police: Kates, D. 1989. Should you own a gun for protection? *U.S. News & World Report* 106(18):28.

99 the primary concern of women confronted by a rapist: Bart, P., and P. O'Brien. 1985. *Stopping Rape: Successful Survival Strategies.* New York: Pergamon Press.

99–100 The "Hillside Strangler," Ken Bianchi, was a schizophrenic psychotic. In the 1970s, he murdered three girls in Rochester, New York, as the "Double Alphabet" murderer/rapist, then, as the "Hillside Strangler," he raped at least ten more victims, ages twelve to twenty-eight, and left them in the hills of Los Angeles, strangled, nude, and posed sexually. Bianchi was arrested in 1979 for rape-murdering two other women in Bellingham, Washington. His likely total was seventeen victims: Schwarz, T. 1981. *The Hillside Strangler: A Murderer's Mind.* New York: Signet, p. 178. See also Dustin Hurlbut. 1997. Personal communication.

100 such rapists are crazy: Thornhill and Thornhill 1987, p. 285.

100 the "Night Stalker" terrorized Los Angeles: Chen, E. 1989. Ramirez guilty on all Night Stalker charges. *Los Angeles Times,* September 21:1, 28, 29, 30. See also Harris, S. 1989. Night Stalker's attacks leave lingering legacy of suffering. *Los Angeles Times,* September 21:26. Feldman, P. 1989. Killer's chair is empty in solemn courtroom. *Los Angeles Times,* September 21:26. Sahagun, L. 1989. Scene of capture unaltered by touch of fame. *Los Angeles Times,* September 21:27. Chen, E. 1989. Nightmares continue for some jurors. *Los Angeles Times,* October 5:1, 34–35.

100 up to 80 percent of convicted rapists: Glueck, B. 1956. Final report: research project for study and treatment of persons convicted of crimes involving sexual aberrations. Unpublished report submitted to governor of New York. See also McCaldon, R. J. 1967. Rape. *Canadian Journal of Corrections* 9(1):37–59.

101 "Given the extremely high rate": Symons 1979, p. 283.

101 "Rape must be understood": Weiner, Zahn, and Sagi 1990, p. 45.

101 twenty-six large American cities: Ellis, L., and C. Beattie. 1983. The feminist explanation for rape: an empirical test. *Journal of Sex Research* 19(1):74–93.

102 "a violent sexual crime": Kelley and Byrne 1992, p. 396.

102 "First, to claim that coitus": Katchadourian, H. 1989. *Fundamentals of Human Sexuality.* 5th ed. San Francisco: Holt, Rinehart and Winston, p. 545.

102 "Although all the women": Bart, P. B., and K. L. Scheppe. 1980. There ought to be a law: women's definitions and legal definitions of sexual assault. Paper presented at the meeting of the American Sociological Association, New York, August.

102 "It was just an ape": Galdikas 1995, p. 294.

103 "A captured woman is raped by all": Chagnon, N. A. 1983. *Yanomamo: The Fierce People,* 3d ed. New York: Holt, Rinehart and Winston, p. 176.

103 "because women incite": Symons 1979, p. 284.

104 men are willing to copulate with unwilling women: Landolt, M. A., M. L. Lalumière, and V. L. Quinsey. 1995. Sex differences and intra-sex variations in human mating tactics: an evolutionary approach. *Ethology and Sociobiology* 16:23. See also Clark, R. D., and E. Hatfield. 1989. Gender differences in receptivity to sexual offers. *Journal of Psychology and Human Sexuality* 2:39–55. Thornhill, R., and N. W. Thornhill. 1989. The evolutionary psychology of human rape. Paper presented at the first annual meeting of the Human Behavior and Evolution Society, Northwestern University, Evanston, Ill., August 25–27.

104 "Most men use": Thornhill and Thornhill 1989, The evolutionary psychology of human rape.

104–105 primate mating takes place in an arena of coercion and violence: Small, M. F. 1993. *Female Choices: Sexual Behavior of Female Primates.* Ithaca, N.Y.: Cornell University Press, p. 105.

105 out of fear of men, women recommend harsher sentences for men who rape than for women who commit identical rapes: Smith, R., C. Pine, and M. Hawley. 1988. Social cognitions about male victims of female sexual assault. *Journal of Sex Research* 24:101–112.

105 married women and rape: Alexander, R., and K. Noonan. 1979. Concealment of ovulation, parental care, and human social evolution. In N. A. Chagnon and W. Irons (eds.), *Evolutionary Biology and Human Social Behavior: An Anthropological Perspective,* pp. 436–453. North Scituate, Mass.: Duxbury.

105 married women are the most traumatized by rape: Thornhill, N. W., and R. Thornhill. 1989. Rape victim psychological pain. Paper presented at the first annual meeting of the Human Behavior and Evolution Society, Northwestern University, Evanston, Ill., August 25–27.

105 responses/reactions to being raped: Kelley and Byrne 1992, pp. 378, 405–406.

105 married women are the most likely to be raped at gunpoint: Thornhill and Thornhill 1989, Rape victim psychological pain.

105 women raped by force are less traumatized psychologically: Ibid. See also Thornhill and Thornhill. The evolutionary psychology of human rape. Manuscript in preparation.

105 the most common reaction by men to rape: Miller, W., and A. M. Williams. 1984. Marital and sexual dysfunction following rape: identification and treatment. In Stuart and Greer, pp. 197–210.

106 twenty-five thousand children produced by West Pakistani rapists: Brownmiller 1975, p. 84.

106 Men have a "psychological adaptation to rape": Thornhill and Thornhill 1989, The evolutionary psychology of human rape. Manuscript in preparation.

Chapter Five: Murder

110–112 the gunfight between Donald Hawley and Ray Martinez: Martinez, R. 1991. Personal communication.

112 O. J. Simpson's trial and media coverage: Gleick, E. 1995. Did he or didn't he? *Time* (Australia) February 6:54–59.

112 "Mystery and detective fiction": Winks, R. W. 1982. *Modus Operandi: An Excursion into Detective Fiction.* Boston: Godine, p. 4.

112 the typical mass murderer, including Charles Whitman: Levin, J., and J. A. Fox. 1990. Mass murder: America's growing menace. In N. A. Weiner, M. A. Zahn, and R. Sagi (eds.), *Violence: Patterns, Causes, Public Policy,* pp. 65–69. New York: Harcourt Brace Jovanovich.

112 juvenile mass murders in schools: Witkin, G., M. Tharp, J. M. Schrof, T. Toch, and C. Scattarella. 1998. Again: in Springfield, a familiar school scene: bloody kids, grieving parents, a teen accused of murder. *U.S. News & World Report* 124(21):16–21. See also Blank, J., J. Vest, and S. Parker. 1998. The children of Jonesboro. *U.S. News & World Report* 124(13):16–22.

113 Patrick Purdy: Pickles, A. 1989. Stockton: where the blame belongs. *New Gun Week,* February 10:1, 5. See also Staff. 1989. Should "assault weapons" be banned? *New Dimensions* 3(12):12–15.

113 Joseph T. Wesbacker: Barry, B., and S. Durbin. 1989. Gunman kills 7 "looking for bosses." *Flagstaff (Arizona) Sun* September 15:13. See also Staff. 1989. What causes crime? *New Gun Week* 24(1185):2.

113 George Hennard: Annin, P. 1991. "You could see the hate": a bloodbath in a small Texas town. *Newsweek* October 28:35.

113 Martin Bryant: O'Neill, A.-M. 1996. Death in the ruins. *Time* (Australia) May 13:20–25.

113 thirteen recent mass killers were on psychiatric drugs, one predisposing condition to mass murder given little attention by professionals: Lesmeister, B. 1990. Should psychiatrists be held accountable for mass murders their patients commit? *Blue Press* 1(1):1, 10. See also Lesmeister, B. 1991. Psychiatrists should be held accountable for mass murderers not gun dealers. *American Firearms Industry* January:6, 85, 87–89.

113 the modus operandi of serial murderers, including Ed Gein: Levin and Fox. 1990, p. 65.

113 Donald Leroy Evans: Staff. 1991. Rapist who says he killed 60 faces 1 murder count. *Arizona Republic* August 16:A5.

113 Jeffrey Dahmer: Schwartz, A. W. 1992. *The Man Who Could Not Kill Enough: The Secret Murders of Milwaukee's Jeffrey Dahmer.* New York: Carol. See also Gelman, D., P. Rogers, K. Springen, F. Chideya, and M. Miller. 1991. The secrets of apt. 213. *Newsweek* August 5:40–42.

113 19,645 murders in 1996: Federal Bureau of Investigation (FBI). 1997. *Crime in the United States, 1996.* Washington, D.C.: U.S. Department of Justice, p. 13.

113 every tribe and nation has murderers: Daly, M., and M. Wilson. 1988. *Homicide.* New York: Aldine de Gruyter, p. 14.

114 cycles of U.S. homicide: Reiss, A. J., Jr., and J. A. Roth (eds.). 1993. *Understanding and Preventing Violence.* Washington, D.C.: National Academy Press, p. 50.

114 murder rates in large cities versus small: FBI 1997, pp. 193–194.

114 1991 saw the highest rate of violent crimes: FBI. 1992. Uniform Crime Report, 1991, p. 11. See also Weiner, N. A., and M. E. Wolfgang. 1990. The extent and character of violent crime in America. In Weiner, Zahn, and Sagi, pp. 24–37.

114 thirteen- to twenty-four-year-old black males: FBI 1997, p. 16.

114 1 chance in 133: Langan, P., and C. Innes. 1990. The risk of violent crime. In Weiner, Zahn, and Sagi, pp. 37–42.

115 males are 75 to 80 percent of murder victims: FBI 1997, p. 16.

115 racial patterns of homicide victims: O'Carrol, P. W., and J. A. Mercy. 1990. Patterns and recent trends in black homicide. In Weiner, Zahn, and Sagi, pp. 55–59. See also Zawitz, M. W. (ed.). 1988. *Report to the Nation on Crime and Justice*. Washington, D.C.: U.S. Department of Justice, Bureau of Justice Statistics, p. 28. Church, G. 1989. The other arms race. *Time* 133(6):20–26.

115 murder in North America: Daly and Wilson 1988, *Homicide*. See also Wilbanks, W. 1984. *Murder in Miami*. Lanham, Md.: University Press of America. Wolfgang, M. E. 1958. *Patterns in Criminal Homicide*. Philadelphia: University of Pennsylvania Press.

115 10,350 victims: FBI 1997, p. 19.

115 homicide is most likely to be committed by young males: Ibid., pp. 14, 16. See also Wolfgang, M. E., and M. A. Zahn. 1990. Homicide: behavioral aspects. In Weiner, Zahn, and Sagi, pp. 50–55. Reiss and Roth 1993, p. 72. Blumstein, A., and J. Cohen. 1987. Characterizing criminal careers. *Science* 237:985–991.

115 52 to 56 percent of murderers are black: FBI 1997, pp. 14–16.

115 "on a typical day": Loury, G. 1996. The impossible dilemma. *New Republic*, January 1:21–25.

115 The average age of men who murder: FBI 1997, p. 16. See also Zawitz 1988, p. 43.

115 61 percent of murderers arrested: Reaves, B. (ed.). 1990. *Felony Defendants in Large Urban Counties, 1988*. Washington D.C.: Bureau of Justice Statistics Executive Summary, p. 6.

115 young convicted murderers on parole: Zawitz 1988, pp. 111, 45.

115–118 "What criminal career": Holden, C. 1986. Growing focus on criminal careers. *Science* 233:1377–1378.

118 70 percent of all violent crime: Corbin, R. K. 1992. The president's column. *American Rifleman* 140(6):50. See also Metesa, T. K. 1997. Attacking gangs not civil rights. *American Rifleman* 145(12):42–43.

118 sociologists and the causes for murder: Eibl-Eibesfeldt, I. 1989. *Human Ethology*. New York: Aldine de Gruyter, p. ix.

118 sociological approaches to understanding murder are flawed: Turner, S. P., and J. H. Turner. 1990. *The Impossible Science: An Institutional Analysis of American Sociology*. Newbury Park, Calif.: Sage.

119 Melvin Konner notes how biology undermines the philosophical structure of the behavioral and social sciences so deeply that if biology were acknowledged, both fields would collapse. The philosophical structures of both, he explains, rest on two pillars, or premises, which "are matters of faith rather than knowledge, poetry rather than science. . . . Each is beautiful, but each is wrong." The first pillar is the metaphor that society is a giant organism whose individual cells are people. Selfish people are considered pathological but also correctable. This "organism" metaphor is wrong "because it requires that society itself be a plausible unit of natural selection, which . . . it has never yet been shown to be." The second premise "is merely a simple article of faith, which I like to call, rather unkindly, the 'tinker theory' of human behavior and experience." This tinker theory paints humans as "basically good and decent and healthy and warm and cooperative and intelligent, but something has gone a bit wrong somewhere." The corresponding assumption here is that the situation can be fixed by "tinkering" with the educational system or the political ideology or by electing new leaders, writing new laws, or printing less money: Konner, M. 1982. *The Tangled Wing: Biological Constraints on the Human Spirit*. New York: Holt, Rinehart and Winston, pp. 413–414.

119 "Available research": Reiss and Roth 1993, p. 50.

119 U.S. versus Canadian cities: Kopel, D. B. 1992. *The Samurai, the Mountie, and the Cowboy: Should America Adopt the Gun Controls of Other Democracies?* Buffalo: Prometheus, p. 159.

120 virtually no Swiss commits murder with rifle or pistol: Kopel, D., and S. D'Andrilli. 1990. The Swiss and their guns. *American Rifleman* 138(2):38–39, 74–81.

120 Swiss murder rate: Halbrook, S. P. 1993. Swiss Schuetzenfest. *American Rifleman* 141(5):46–47, 75–76. See also LaPierre, W. 1994. *Guns, Crime, and Freedom*. Washington, D.C.: Regnery Press, p. 171.

120 gun availability does not correlate with murder rate: Kleck, G. 1992. *Point Blank: Guns and Violence in America*. New York: Aldine de Gruyter, pp. 13, 203.

120 54 percent of U.S. murderers: FBI 1997, pp. 18, 20.

120 more people were murdered with baseball bats than guns: Staff. 1989. Traces add up to crime guns in Cox news service "study." *American Rifleman* 137(10):72. See also Cross, M. 1996. Taking arms against the knife. *New Scientist* no. 2002:30–33.

120 best existing study: Wright, J. D., and P. H. Rossi. 1986. *Armed and Considered Dangerous: A Survey of Felons and Their Firearms*. New York: Aldine de Gruyter, p. 68.

120 Wright and Rossi favored stricter gun control: Wright, J. D., P. H. Rossi, and K. Daly. 1983. *Under the Gun: Weapons, Crime, and Violence in America*. New York: Aldine De Gruyter, p. 309.

120 5 percent of all Americans carry guns: Kleck 1992, p. 117. See also Wright, Rossi, and Daly 1983, p. 14.

120–121 Wright and Rossi found: Wright and Rossi 1986, pp. 14, 167.

121–122 Lott surveyed data on guns and murder: Lott, J. R., Jr. 1998. *More Guns, Less Crime*. Chicago: University of Chicago Press, pp. 1, 4, 11, 12, 13.

122 Americans use guns 783,000 times: Kleck 1992, pp. 106–107.

122 randomly polled Americans: Maguire, K., and A. L. Pastore. 1996. *Sourcebook of Criminal Justice Statistics, 1996*. Washington, D.C.: U.S. Department of Justice, pp. 190, 197.

122 to protect themselves, citizens killed 1,382 violent offenders: FBI 1997, p. 22.

122 police were 5.5 times more likely to shoot: Silver, C. R., and D. B. Kates, Jr. n.d. Self-defense, handgun ownership, and the independence of women in a violent, sexist society. In *Restricting Handguns*, pp. 154–155. (Cited in Kopel 1992, pp. 380, 398.)

122–123 "violent crimes": Lott 1998, pp. 19–20, 47, 58, 114, 118, 165.

124 men who scored low on intelligence tests: Wilson, J. Q., and R. J. Herrnstein. 1985. *Crime and Human Nature*. New York: Simon and Schuster, p. 66.

124 "Research consistently places": Bower, B. 1995. Criminal intellects: researchers look at why lawbreakers often brandish low IQs. *Science News* 147(15):232–233, 239.

124 crime in identical versus fraternal twins: Eysenck, H. 1995. Crime and punishment: a review of *Mind to Crime: The Controversial Link Between Mind and Criminal Behaviour* by Anne Moir and David Jessel *New Scientist* no. 2000:51.

124 67 percent of U.S. murders: FBI 1997, pp. 22.

124 38 percent of murderers: Maguire and Pastore 1997, pp. 306, 471.

124 97 percent of convicted murderers: Ibid., pp. 472–476.

125 only a weak correlation between students in poorer neighborhoods: Mayer, S., and C. Jencks. 1989. Growing up in poor neighborhoods: how much does it matter? *Science* 243:1441–1445.

125–126 studies of poverty and homicide: Reiss and Roth 1993, p. 132.

126 "there is little empirical evidence": Widom, C. S. 1989. The cycle of violence. *Science* 244:160–166.

126 20 percent of abused children: Widom, C. S. 1989. *The Intergenerational Transmission of Violence*. Occasional Papers, no. 4. New York: Harry Frank

Guggenheim Foundation, p. 40.

126 Courtwright followed homicide rates: Courtwright, D. T. 1996. *Violent Land: Single Men and Social Disorder from the Frontier to the Inner City.* Cambridge: Harvard University Press.

127 98 percent of households: Bok, S. 1998. *Mayhem.* Reading, Mass.: Addison-Wesley, pp. 51, 58.

127 CDC eliminated: Lamson, S. R. 1993. TV violence: does it cause real life mayhem? *American Rifleman* 141(7):32–33, 89.

127 the average American eighteen-year-old: Abelson, P. H. 1992. A major generation gap. *Science* 256:945. See also Hamburg, D. A. 1992. *Today's Children: Creating a Future for a Generation in Crisis.* New York: Times Books/Random House. Hechinger, F. M. 1992. *Fateful Choices: Healthy Youth for the 21st Century.* New York: Hill and Wang.

127 "the amount of television violence": Lefkowitz, M. M., L. D. Eron, L. O. Walder, and L. R. Huesmann. 1972. Television violence and child aggression: a follow-up study. In G. A. Comstock and E. A. Rubinstein (eds.), *Television and Social Behavior,* vol. 3, pp. 35–135. Washington, D.C.: U.S. Government Printing Office.

127 a 1986 study: Heath, L., C. Kruttschnitt, and D. Ward. 1986. Television and violent criminal behavior: beyond the Bobo Doll. *Violence and Victims.* 1:177–190.

127 Canadian homicide rate: Kopel 1992, p. 413.

127 U.S. prisoners: Ibid.

127 Juvenile mass murderers: Witkin, Tharp, Schrof, Toch, and Scattarella 1998. See also Blank, Vest, and Parker, 1998.

127 "it is estimated": Lamson 1993, p. 33.

127–128 "We are doing": Grossman, D. 1995. *On Killing: The Psychological Cost of Learning to Kill in War and Society.* New York: Little, Brown, pp. 299–330.

128 92 percent of U.S adults: Maguire and Pastore 1996, pp. 222, 223.

128 young people under eighteen: Ibid., p. 377.

128–129 "A young boy": Bok 1998, p. 51.

129 one hundred violent acts per hour: Lamson 1993, p. 89. See also Rainie, H., B. Streisand, M. Guttmann, and G. Witkin. 1993. Warning shots at TV. *U.S. News & World Report* 115(2):48–50. Wolfgang and Zahn 1990, p. 53.

129 Darwin's theory of evolution by natural selection is the basis by which to examine murder: Daly and Wilson 1988, *Homicide,* p. 2.

129 gorillas subsist on common foliage: Schaller, G. B. 1963. *The Mountain Gorilla.* Chicago: University of Chicago Press. See also Fossey, D. 1984. *Gorillas in the Mist.* Boston: Houghton Mifflin. Goodall, A. G. 1977. Feeding and ranging behavior of a mountain gorilla group *(Gorilla gorilla berengei)* in the Tshibinda-Kahusi region (Zaire). In T. H. Clutton-Brock (ed.), *Primate Ecology,* pp. 450–479. New York: Academic Press.

130 the average gorilla group contains eight members: Harcourt, A. H. 1989. The lowland gorillas of Nigeria. Lecture to the California Academy of Sciences, January 26. See also Watts, D. P. 1996. Comparative socioecology of gorillas. In McGrew, W. C., L. F. Marchant, and T. Nishida (eds.), *Great Ape Societies,* pp. 16–28. New York: Cambridge University Press.

130 Silverback harem masters retain females via combat: Harcourt, A. H. 1981. Intermale competition and the reproductive behavior of the great apes. In C. E. Graham (ed.), *Reproductive Biology of the Great Apes,* pp. 301–318. New York: Academic Press.

130 79 percent of encounters: Harcourt, A. H., K. J. Stewart, and D. Fossey. 1981. Gorilla reproduction in the wild. In Graham, pp. 265–279.

130 adult male gorillas are 237 percent the weight of females: Jungers, W. L., and R. L. Susman. 1984. Body size and skeletal allometry in African apes. In R. L. Susman (ed.), *The Pygmy Chimpanzee Evolutionary Biology and Behavior,* pp. 131–177. New York: Plenum.

130 some silverbacks kill infants sired by rival males: Fossey 1984.

130 a female leaves to avoid inbreeding: Gibbons, A. 1993. The risks of inbreeding. *Science* 259:1252.

130 gorilla females fall into stable dominance hierarchies: Harcourt, A. H. 1979. Social relationships between adult female mountain gorillas. *Animal Behaviour* 27:251–264. See also Watts, D. P. 1985. Relations between group size and composition and feeding competition in mountain gorilla groups. *Animal Behaviour* 33:72–85.

130 earlier wives rarely kill a latecomer's infant: Fossey 1984, p. 277.

130 a few females defend their silverback: Watts, D. P. 1989. The mountain gorillas of Rwanda. Lecture to the California Academy of Sciences, January 11.

130 most females do not stay with the first male who recruited them: Harcourt, A. H. 1979. Social relationships between adult male and adult female mountain gorillas in the wild. *Animal Behaviour* 27:325–342.

130–131 Jennie and Nunkie: Ghiglieri, M. P. 1984. The mountain gorilla: last of a vanishing tribe. *Mainstream* 15(3–4):36–40.

131 Simba: Watts 1989, The mountain gorillas.

131 one in seven gorilla infants: Watts, D. P. 1989. Infanticide in mountain gorillas: new cases and a reconsideration of evidence. *Ethology* 81:1–18.

131 Hrdy reckoned that natural selection apparently had produced murderous monkeys. Primatologist Yukimaru Sugiyama had seen the same thing a decade earlier, but his lower-profile conclusions had not raised the same fury among anthropologists who insisted that infanticide was due only to crowding: Hrdy, S. B. 1977. Infanticide as a primate reproductive strategy. *American Scientist* 65:40–49. See also Hrdy, S. B. 1979. Infanticide among animals: a review, classification, and examination of the implications for the reproductive strategies of females. *Ethology and Sociobiology* 1(1):13–40.

131 previous observations of langur infanticides: Sugiyama, Y. 1965. On the social change of Hanuman langurs *(Presbytus entellus)* in their natural conditions. *Primates* 6:381–417. See also Sugiyama, Y. 1967. Social organization of Hanuman langurs. In S. Altmann (ed.), *Social Communication Among Primates,* pp. 221–236. Chicago: University of Chicago Press.

131 infanticide is too "perverted": Darwin, C. 1871. *The Descent of Man and Selection in Relation to Sex*. New York: Modern Library, p. 430.

132 infanticide by chacma baboons: Busse, C., and W. J. Hamilton, III. 1981. Infant carrying by adult male chacma baboons. *Science* 212:1281–1283. See also Crockett, C. M., and R. Sekulic. 1984. Infanticide in red howler monkeys. In G. Hausfater and S. B. Hrdy (eds.), *Infanticide: Comparative and Evolutionary Perspectives,* pp. 173–192. New York: Aldine. Collins, D. A., C. D. Busse, and J. Goodall. 1984. Infanticide in two populations of savanna baboons. In Hausfater and Hrdy, pp. 193–216. Fossey, D. 1984. Infanticide in mountain gorillas *(Gorilla gorilla berengei)* with comparative notes on chimpanzees. In Hausfater and Hrdy, pp. 217–235. Suzuki, A. 1971. Carnivority and cannibalism observed in forest-living chimpanzees. *Journal of the Anthropological Society of Nippon* 74:30–48. Bygott, J. D. 1972. Cannibalism among wild chimpanzees. *Nature* 238:410–411. Goodall, J. V. L. 1977. Infant killing and cannibalism in free-living chimpanzees. *Folia Primatologica* 28:259–282. Butynski, T. M. 1982. Harem male replacement and infanticide in the blue monkey *(Cercopithecus mitis Stuhlmann)* in the Kibale Forest, Uganda. *American Journal of Primatology* 3:1–22. T. T. Struhsaker. 1977. Infanticide and social organization in the redtail monkey *(Cercopithecus ascanius schmidtii)* in the Kibale Forest, Uganda. *Zietschrift fur Tierpsychologie* 45:75–84. Leland, L., T. T. Struhsaker, and T. M. Butynski. 1984. Infanticide by adult males in three primate species of Kibale Forest, Uganda: a test of hypotheses. In Hausfater and Hrdy, pp. 151–172.

132 many species of carnivores commit infanticide: Caro, T. M. 1994. *Cheetahs of the Serengeti Plains: Group Living in an Asocial Species.* Chicago: University of Chicago Press, pp. 88–89.

132 males kill infants of other males, not their own: Whitten, P. L. 1986. Infants and adult males. In B. B. Smuts, D. L. Cheney, R. A. Seyfarth, R. W. Wrangham, and T. T. Struhsaker (eds.), *Primate Societies,* pp. 243–357. Chicago: University of Chicago Press, p. 354.

132 Beethoven sired at least nineteen offspring: Fossey 1984, *Gorillas in the Mist.* Boston: Houghton Mifflin.

132 40 percent of gorilla harems: Weber, A. W., and A. Vedder. 1983. Population dynamics of the Virunga gorillas, 1959–1978. *Biological Conservation* 26:341–366.

132 the silverback leader usually bred exclusively: Harcourt, A. H. 1984. Gorilla. In D. Macdonald (ed.), *All the World's Animals: Primates,* pp. 136–143. New York: Torstar.

132 Zizz recruited seven more females: Watts 1989.

133–134 infanticide among Ayoreo women: Bugos, P. E., and L. M. McCarthy. 1984. Ayoreo infanticide: a case study. In Hausfater and Hrdy, pp. 503–520.

134 the Ache of Paraguay kill infants: Hill, K., and H. Kaplan. 1988. Tradeoffs in male and female reproductive strategies among the Ache, part 2. In L. Betzig, M. Borgerhoff Mulder, and P. Turke (eds.), *Human Reproductive Behaviour: A Darwinian Perspective,* pp. 291–305. New York: Cambridge University Press.

134 infanticide was common everywhere: Scrimshaw, S. C. M. 1984. Infanticide in human populations. In Hausfater and Hrdy, pp. 439–462.

134 in Germany: Voland, E. 1988. Differential infant and child mortality in evolutionary perspective: data from late 17th to 19th century Ostfiesland (Germany). In Betzig, Borgerhoff Mulder, and Turke, pp. 253–261.

134 women in Canada: Daly and Wilson 1988, *Homicide,* pp. 73–83.

135 teenagers killed their newborns: Pinker, S. 1997. Why they kill their newborns. *New York Times,* November 2:6–52.

135 Susan Smith drowns her kids: Rainie, H. 1994. A mother, two kids, and the unthinkable. *U.S. News & World Report* 117(19):16–17.

135 three hundred infanticidal mothers: Pinker 1997.

136 Eskimo fathers abandon: Bower, B. 1994. Female infanticide: northern exposure. *Science News* 146(22):358.

136 in Burma, India, Bangladesh: Scrimshaw 1984.

136 "The anthropological evidence": Johansson, S. R. 1984. Deferred infanticide: excess female mortality during childhood. In Hausfater and Hrdy, pp. 463–485.

136 8,000 fetuses: Heise, L. 1989. The global war against women. *Utne Reader* November/December:40–45.

136 "the most widely used method": Harris, M. 1977. *Cannibals and Kings: The Origins of Cultures.* New York: Random House.

136 the Tapirapé: Wagley, C. 1977. *Welcome of Tears: The Tapirapé Indians of Central Brazil.* New York: Oxford University Press.

136–137 the reasons for infanticide in Human Relations Area Files: Daly and Wilson 1988, *Homicide,* pp. 46–48. See also Daly, M., and M. Wilson. 1989. Evolutionary psychology and family homicide. *Science* 242:519–524, p. 521.

137 the dynamics of child abuse: Gelles, R. J. 1989. Child abuse and violence in single parent families: a test of parent-absent and economic deprivation hypothesis. Paper presented at the annual meetings of the American Sociological Association, San Francisco, August 11.

137 an infant under two years old: Daly and Wilson 1988, *Homicide,* pp. 88–89.

138 "A Wehrmacht memorandum": Dickemann, M. 1984. Concepts and classification in the study of human infanticide: sectional introduction and some cautionary notes. In Hausfater and Hrdy, pp. 427–437.

138 victims are usually abusive fathers: Daly and Wilson 1988, *Homicide,* pp. 98–102.

138 "altercation of relatively trivial origin": Ibid., p. 125.

139 53 percent of murders: Maguire and Pastore 1997, p. 334. See also FBI 1997, p. 21. Daly and Wilson 1988, *Homicide,* p. 175. Zawitz 1988, p. 4.

139 "criminal homicide": Luckenbill, D. F. 1990. Criminal homicide as a situated transaction. In Weiner, Zahn, and Sagi, pp. 59–65, p. 64.

140 Gebusi murder over control of women: Knauft, B. M. 1985. *Good Company and Violence: Sorcery and Social Control in a Lowland New Guinea Society.* Berkeley: University of California Press. Bower, B. 1988. Murder in good company. *Science News* 133(6):90–91.

140–141 "It is not in their nature to fight": Thomas, E. M. 1959. *The Harmless People.* New York: Knopf, pp. 21–24.

141 "We shoot poisoned arrows": Lee, R. B. 1984. *The Dobe !Kung.* New York: Holt, Rinehart and Winston, p. 95.

141 "Among the !Kung": Konner 1982. *The Tangled Wing: Biological Constraints on the Human Spirit.* New York: Holt, Rinehart and Winston, p. 349.

141–142 the Yanomamo: Chagnon, N. A. 1988. Life histories, blood revenge, and warfare in a tribal population. *Science* 239:985–992. See also Chagnon, N. A. 1983. *Yanomamo: The Fierce People.* 3d ed. New York: Holt, Rinehart and Winston, p. 171.

142 "The wife was then given back": Chagnon 1983, p. 173.

142 the Tiwi: Hart, C. W. M., and A. Pilling. 1961. *The Tiwi of Northern Australia.* New York: Holt, Rinehart and Winston.

142–143 "A stranger in the camp": Balikci, A. 1970. *The Netsilik Eskimo.* Garden City, N.Y.: Natural History Press, pp. 182, 184.

143 overview of hunter-gatherer murderers: Coon, C. S. 1971. *The Hunting Peoples.* New York: Atlantic–Little, Brown.

143 affronts to self-esteem: Feshbach, S. 1989. The bases and development of individual aggression. In J. Groebel and R. A. Hinde (eds.), *Aggression and War: Their Biological and Social Bases,* pp. 78–90. New York: Cambridge University Press.

144 the legacy of a Yanomamo headman: Chagnon 1983, pp. 136–137.

144 on average, *unokais:* Chagnon 1988, p. 989.

144 Chagnon's data were attacked: Ferguson, R. B. 1989. Do Yanomamo killers have more kids? *American Ethnologist* 16(3):564–565. See also Albert, B. 1989. Yanomami "violence": inclusive fitness or ethnographer's representation? *Current Anthropologist* 30(5):637–640. See also Chagnon, N. A. 1989. Response to Ferguson. *American Ethnologist* 16(3):565–569. Chagnon, N. A. In press. On Yanomamo violence: reply to Albert. *Current Anthropologist.*

144 90 percent of French prisoners: Halliburton, R. 1929. *New Worlds to Conquer.* Garden City, N.Y.: Garden City Publishing, p. 284.

145 quarter million gang members: Curry, G. D., R. A. Ball, and R. J. Fox. 1994. *Gang Crime and Law Enforcement Recordkeeping.* Washington, D.C.: U.S. Department of Justice, National Institute of Justice Research in Brief, p. 1.

145 1,157 in 1995: Maguire and Pastore 1997, p. 334.

145 "Women are always attracted to power": Winokur, J. (ed.). 1991. *A Curmudgeon's Garden of Love.* New York: Plume, p. 206.

145 "He [Henry] had something": Pilaggi, N. 1985. *Wiseguy.* New York: Pocket Books, pp. 73–74.

146 "The difference [in homicide]": Daly and Wilson 1988, *Homicide,* p. 146.

146 in Los Angeles: Martinez, A. 1990. There goes ol' Ernie, just Rolexing his way to danger in LA. *Reno Gazette-Journal,* September 30:11C.

146 90 percent of robbery arrests: FBI 1997, p. 29.

146 80 percent of Canadian victims: Daly and Wilson 1988, *Homicide,* 178–179.

146 73 percent of all U.S. women: Jones, S. 1994. Women of the NRA. *American Rifleman* 142(6):26.

147 each person decides daily whether or not to steal: Cohen, L. E., and R. Machalek. 1988. A general theory of expropriative crime: an evolutionary ecological approach. *Journal of American Sociology* 94(3):465–501.

147 theft against companies: Thompson, T., D. Hage, and R. F. Black. 1992. Crime and the bottom line. *U.S. News & World Report* 112(14):55–58.

148 $15.3 billion: Federal Bureau of Investigation (FBI). 1994. *Crime in the United States, 1993* Washington, D.C.: U.S. Department of Justice, p. 6.

148 2,301 robbery victims: Ibid., p. 21.

148 interview of Ken: Blinder, M. 1985. *Lovers, Killers, Husbands, and Wives.* New York: St. Martin's Press, pp. 167–168.

149 only 1 woman in 168 people: Bryson, J. B. 1976. The natures of sexual jealousy: an exploratory study. Paper presented to the American Psychological Association, Washington, D.C., September. (Cited in Daly and Wilson 1988, *Homicide,* p. 183.)

149 men admitted their primary worries: Teismann, M. 1975. Jealous conflict: a study of verbal interaction and labeling of jealousy among dating couples involved in jealousy improvisations. Ph.D. dissertation, University of Connecticut. (Cited in Daly and Wilson 1988, *Homicide,* p. 183.)

149 "But still": Pilaggi 1985, p. 151.

149 jealousy ranks high: Daly and Wilson 1988, *Homicide,* p. 186.

149–150 the fate of the mutineers: Nicholson, R. S. 1965. *The Pitcairners.* Sydney: Angus and Robertson.

149–150 "The mutiny": Brown, D. E., and D. Hotra. 1988. Are prescriptively monogamous societies effectively monogamous? In Betzig, Borgerhoff Mulder, and Turke, pp. 153–159.

150 "Women need men": Small, M. F. 1993. *Female Choices: Sexual Behavior of Female Primates.* Ithaca, N.Y.: Cornell University Press, p. 189.

150 "men are much more likely": Daly and Wilson 1988, *Homicide,* p. 192.

151 the parasitic breeding habits of the European cuckoo: Davies, N. B., and M. Brooke. 1990. Coevolution of the cuckoo and its hosts. *Scientific American* 264(1):92–98. See also Aldhous, P. 1995. Cosa Nostra cuckoos rule by terror. *New Scientist,* no. 2006:21.

151 "many American juries": Daly and Wilson 1988, *Homicide,* p. 195.

151 one out of ten husbands were cuckolds: Diamond, J. 1992. *The Third Chimpanzee: The Evolution and Future of the Human Animal.* New York: Harper Perennial, pp. 85–87.

152 Canadian women murdered their spouses: Daly and Wilson 1988. *Homicide,* p. 200.

152 wife beating is common worldwide: Heise 1989. See also MacFarquhar, E. 1994. The war against women. *U.S. News & World Report* 116(12)42–48. Leo, J. 1994. Is it a war against women? *U.S. News & World Report* 117(2):22.

152 violence in American marriage is as commonly initiated by wives: Dutton, D. G. 1988. *The Domestic Assault of Women.* Boston: Allyn and Bacon, pp. 17, 19.

152 violence occurs in 18 percent of lesbian couples: Hite, S. 1986. *Women and Love: A Cultural Revolution in Progress.* New York: Knopf, p. 595.

152 wife beaters are very unlike generally assaultive men: Dutton 1988, p. 82.

152–153 "Natural selection has favored": Smuts, B. B. 1985. *Sex and Friendship in Baboons.* New York: Aldine de Gruyter, p. 105. See also Strum, S. C. 1981. Processes and products of change: baboon predatory behavior at Gilgil. In R. S. O. Harding and G. Teleki (eds.), *Omnivorous Primates: Gathering and Hunting in*

Human Evolution, pp. 255–302. New York: Columbia University Press. Strum, S. C., and W. Mitchell. 1987. Baboon models and muddles. In W. G. Kinzey (ed.), *The Evolution of Human Behavior: Primate Models,* pp. 87–104. New York: State University of New York Press. Hausfater, G., J. Altmann, and S. Altmann. 1982. Long-term consistency of dominance relations among female baboons *(Papio cynocephalus). Science* 217:752–755. Bercovitch, F. B. 1986. Male rank and reproductive activity in savanna baboons. *International Journal of Primatology* 7(6):533–550. Bercovitch, F. B. 1987. Reproductive success in male savanna baboons. *Behavioral Ecology and Sociobiology* 21:163–172.

153 "At least 2 million women": Crowell, N. A., and A. W. Burgess (eds.). 1996. *Understanding Violence Against Women.* Washington, D.C.: National Academy Press, pp. 1, 107.

153 15,537 reports of domestic battery: Staff. 1990. Haphazard homicides at home. *Science News* 137(7):110.

153 "the husband's proprietary concern": Daly, M., and M. Wilson. 1988. Evolutionary social psychology and family homicide. *Science* 242:519–524.

153 "The estranged wife": Daly and Wilson 1988, *Homicide,* p. 219.

153 "Even if I did do this": Staff/Washington Whispers. 1998. *U.S. News & World Report,* January 19:18.

153 societies worldwide: Daly, M., M. Wilson, and S. J. Weghorst. 1982. Male sexual jealousy. *Ethology and Sociobiology* 3:11–27.

154 30 percent of women murder victims: FBI 1997, p. 17.

154 Canadian women are murdered: Crowell and Burgess 1996, pp. 27, 39, 55, 102.

154 homicide is coded into the human male psyche: George, S. A. 1988. Letter. *Science* 243:462. See also Harcourt, A. H. 1988. Letter. *Science* 243:462–463. Daly, M., and M. Wilson. 1988. Letter. *Science* 243:463–464.

Chapter Six: War

156 "If they [the chimpanzees]": Goodall, J. 1986. *The Chimpanzees of Gombe.* Cambridge: Harvard University Press, p. 530.

156–160 William J. Klingenberg's experiences in Vietnam: Klingenberg, W. J. 1968–1998. Personal communications.

157 Vietnam received four times the bomb tonnage used in World War II but the Ho Chi Minh Trail remained open: Arnett, P., I. McLeod, and M. MacLear. 1980. *Vietnam: The Ten Thousand Day War.* Vol. 3: *Days of Decision.* Documentary film, Embassy Home Entertainment, Los Angeles.

157 U.S. casualties caused by friendly fire: Hackworth, D. H., and J. Sherman. 1989. *About Face: The Odyssey of an American Warrior.* New York: Simon and Schuster, p. 594.

157 bombing strikes on Ho Chi Minh Trail: Sheehan, N. 1988. *A Bright Shining Lie: John Paul Vann and America in Vietnam.* New York: Random House, pp. 678–679.

157 the Ho Chi Minh Trail was never severed: Arnett, P., I. McLeod, and M. MacLear. 1980. *Vietnam: The Ten Thousand Day War.* Vol. 5: *The Trail.* Documentary film, Embassy Home Entertainment, Los Angeles.

158 the LZ X-Ray and LZ Albany battles: Moore, H. G., and J. L. Galloway. 1992. *We Were Soldiers Once . . . and Young.* New York: Random House.

160 "War is merely": von Clausewitz, C. 1832 [1976]. *On War.* Princeton, N.J.: Princeton University Press, p. 87. See also Adams, M. C. C. 1990. *The Great Adventure.* Bloomington: Indiana University Press.

160 "What is war anyway": Hackworth and Sherman 1989, p. 137.

161 the offensive goals of war: Durham, W. H. 1976. Resources, competition, and human aggression, part 1: a review of primitive war. *Quarterly Review of Biology* 51(3):385–415.

161 "War is ancient": Rhodes, R. 1986. *The Making of the Atomic Bomb.* New York: Simon and Schuster, pp. 780–781.

161 many biblical wars were exaggerated, but most did happen: Asimov, I. 1981. *Asimov's Guide to the Bible: Two Volumes in One: The Old and New Testaments.* New York: Avenel.

161 Australian rock art: Szalay, A. 1995. Rock art warriors: world's earliest paintings of people at war. *Geo Australia* 17(4):40–52.

161 "The earliest actual image": O'Connell, R. L. 1989. *Of Arms and Men.* New York: Oxford University Press, p. 26. See also Bower, B. 1995. Seeds of warfare precede agriculture. *Science News* 147(1):4.

162 weapons evolved: Brodie, B., and F. Brodie. 1973. *From Crossbow to H-Bomb.* Bloomington: Indiana University Press.

162 "the most dangerous weapon": Tiger, L., and R. Fox. 1971. *The Imperial Animal.* New York: Holt, Rinehart and Winston.

162 dozens of genocidal wars fatten our history books: Diamond, J. 1992. *The Third Chimpanzee: The Evolution and Future of the Human Animal.* New York: HarperPerennial, pp. 284–290.

162 the British slaughtered the Tasmanians: Hughes, R. 1987. *The Fatal Shore.* New York: Knopf, p. 120. See also Diamond 1992, pp. 278–283.

162 the Dutch virtually annihilated the San Bushmen: Lee, R. B. 1984. *The Dobe !Kung.* New York: Holt, Rinehart and Winston, p. 9.

162 German colonists in Namibia: Ibid., pp. 120–121.

162 biological warfare on American Indians: Dobyns, H. F. 1968. *Native American Historical Demography: A Critical Bibliography.* Bloomington: University of Illinois Press. See also Sempowski, M., L. P. Saunders, and J. P. Bradley. 1990. Letter: New World epidemics. *Science* 247:788. Black, F. L. 1992. Why did they die? *Science* 258:1739–1740. Zu Wied, M. 1839 [1976]. *People of the First Man: Life Among the Plains Indians in the Final Days of Glory.* New York: Dutton.

162 blankets contaminated with smallpox: O'Connell 1989, p. 171.

163 "The only good Indian": Brown, D. 1970. *Bury My Heart at Wounded Knee.* New York: Holt, Rinehart and Winston. See also Utley, R. M. 1988. Indian–United States military situation, 1848–1891. In W. E. Washburn (ed.), *Handbook of North American Indians.* Vol. 4: *History of Indian-White Relations,* pp. 163–184. Washington, D.C.: Smithsonian Institution Press.

163 "Probably at least 10 per cent": Wright, Q. 1942 [1965]. *A Study of War.* Chicago: University of Chicago Press, p. 246.

163 wars are ubiquitous: Bower, B. 1991. Gauging the winds of war: anthropologists seek the roots of human conflict. *Science News* 139(6):88–89, 91. See also Appenzeller, T. 1994. Clashing Maya superpowers emerge from a new analysis. *Science* 66:733–734.

163 We do not know: Ferguson, R. B., and L. E. Farragher. 1988. *The Anthropology of War: A Bibliography.* Occasional Papers, no. 1, New York: Harry Frank Guggenheim Foundation, pp. i–vi. See also Nance, J. 1975. *The Gentle Tassaday: A Stone Age People in the Philippine Rain Forest.* New York: Harcourt Brace Jovanovich. MacLeish, K. 1972. The Tassadays: Stone Age cavemen on Mindanao. *National Geographic* 142(2):218–249. Marshall, E. 1989. Anthropologists debate Tassaday hoax evidence. *Science* 247:22–23. Nance, J. 1990. Letter: the Tassaday debate. *Science* 247:790.

163 the Maoris: Cook, Captain J. 1949. *The Voyages of Captain James Cook Round the World,* edited by C. Lloyd. London: Cresset Press, pp. 36–57.

163–164 "If a village": Leahy, M. J., and M. Crain. 1937. *The Land That Time Forgot.* New York: Funk and Wagnalls, p. 156.

164 90 percent of hunter-gatherer societies waged war: Ember, C. R. 1978. Myths about hunter-gatherers. *Ethology* 17:439–448.

164 92 percent of primitive societies engaged in warfare: Otterbein, K. F. 1985. *The Evolution of War.* New Haven, Conn.: Human Relations Area Files Press.

164 prehistoric Indians also fought wars: Fiedel, S. J. 1987. *Prehistory of the Americas.* New York: Cambridge University Press, pp. 134, 209, 219, 248, 256.

164 the "harmless" !Kung and !Ko Bushmen: Eibl-Eibesfeldt, I. 1989. *Human Ethology.* New York: Aldine de Gruyter, pp. 328–330.

164 "No tribes have been adequately described": Wright 1942 [1965], p. 38.

164 war is the biggest (most expensive) business of the governments on earth: Renner, M. 1989. Enhancing global security. In L. Starke (ed.), *State of the World, 1989,* pp. 132–153, p. 133. New York: Norton.

164 "If you wish for peace": Liddell Hart, B. H. 1967. *Strategy.* New York: Praeger, p. 372.

164 humans are innately aggressive due to big-game hunting: Ardrey, R. 1961. *African Genesis.* New York: Atheneum. See also Shipman, P. 1985. The ancestor that wasn't. *The Sciences* 25(2):43–48. Shipman, P. 1986. Scavenging or hunting in early hominids: theoretical framework and tests. *American Anthropologist* 88:27–43. Shipman, P. 1987. An age-old question: why did the human lineage survive? *Discover* 8(4):60–64. Shipman, P. 1988. What does it take to be a meat eater? *Discover* 9(9):38–44.

164–165 aggression is a "drive": Lorenz, K. 1966. *On Aggression.* New York: Harcourt, Brace and World. See also Goldstein, A. 1989. Beliefs about human aggression. In J. Groebel and R. A. Hinde (eds.), *Aggression and War: Their Biological and Social Bases,* pp. 10–24. New York: Cambridge University Press.

165 The idea that war is good for the species due to population control is the most mistaken of all war myths. Men do not wage war and risk their lives to control their own population (so as to avert starvation, for instance). Biologists call such ideas "group selection" arguments. Although they still pop up in natural history books, TV documentaries, and even scientific writing, they are impossible because natural selection cannot perpetuate any behavior in a sexual species that leads to genetic self-sacrifice for the one who performs it. Indeed, natural selection does not act on a species at all; it only acts on genes that affect the reproductive success of individuals. Furthermore, to limit a population, war should kill women most, not men. This is because just one surviving man can impregnate fifty women to produce as many children as fifty men can, while each woman killed absolutely prevents births. World War II, for example, killed millions of men in Europe but did not cut western European population growth much, if at all. Interestingly, when Napoleon Chagnon told some Yanomamo Indians that many anthropologists thought that tribes like them waged war to protect their protein base and thus to prevent overpopulation—not to capture women and breed more—they laughed and said, "Even though we enjoy eating meat, we like women a whole lot more!" More to the point, even if war could control a population at some ecologically safe level, this population then would be a sitting duck for warriors from other tribes who could muster larger armies, wipe them out, and usurp their territory. It would be inevitable. War is too easy to invent. Indeed, men launch wars even when it makes almost everyone else worse off. To rescue the myth of war as population control Marvin Harris says warrior cultures kill their baby girls most to raise more men to make their tribes better fighting societies. But the only reliable data on infanticide reveals that warrior cultures are the least likely of all people to kill their baby girls: Divale, W. T., and M. Harris. 1976. Population, warfare, and the male supremacist complex. *American Anthropologist* 78:521–538. See also Bates, D. G., and S. H. Lees. 1979. The myth of population regulation. In Chagnon, N. A., and W. Irons (eds.). 1979. *Evolutionary Biology and Human Social Behavior: An Anthropological Perspective,* pp. 273–289. North Scituate, Mass.: Duxbury Press, pp. 283–286. Dickemann, M. 1979. Female infanticide, reproductive strategies, and social stratification: a preliminary model. In Chagnon and Irons, pp. 321–367. Dickemann, M. 1984. Concepts and classification in the study of human infanticide: sectional introduction and some cautionary notes. In G. Hausfater and S. B.

Hrdy (eds.). *Infanticide: Comparative and Evolutionary Perspectives,* pp. 427–437. New York: Aldine. Scrimshaw, S. C. M. 1984. Infanticide in human populations. In Hausfater and Hrdy, pp. 439–462. Chagnon, N. A. 1979. Is reproductive success equal in egalitarian societies? In Chagnon and Irons, pp. 374–401. For the most elaborate—and earnest—examples of "scientific" "good for the species" arguments, see Wynne-Edwards, V. C. 1962. *Animal Dispersion in Relation to Social Behavior.* New York: Hafner.

165 good-for-the-species hypotheses and/or group selection are impossible in evolution: Williams, G. C. 1966. *Adaptation and Natural Selection: A Critique of Some Current Evolutionary Thought.* Princeton, N.J.: Princeton University Press.

166 Muslim villagers gave dead chimp a proper burial: Goodall, J. 1989. Area status report. In P. G. Heltne and L. A. Marquardt (eds.), *Understanding Chimpanzees,* pp. 360–361. Cambridge: Harvard University Press.

166 "we have profoundly underestimated the chimpanzee's intellectual operations": Rumbaugh, D. M. 1989. Current and future research on chimpanzee intellect. In Heltne and Marquardt, pp. 296–310.

166 chimps are intelligent: Menzel, E. M., Jr. 1989. Are animals intelligent? In Heltne and Marquardt, pp. 210–219.

166 chimps can perform basic math: Rumbaugh, D. M. 1970. Learning skills of anthropoids. In L. A. Roseblum (ed.), *Primate Behavior,* pp. 1–70. London: Academic Press.

166 chimps hunt cooperatively: Goodall 1986.

166 chimps use medicinal plants: Wrangham, R. W., and J. Goodall. 1989. Chimpanzee use of medicinal leaves. In Heltne and Marquardt, pp. 22–37.

166 chimps manufacture and use tools: Goodall 1986.

166 chimps manufacture and use tools of stone: Kortlandt, A. 1989. Use of stone tools by wild-living chimpanzees. In Heltne and Marquardt, pp. 146–147. See also Kortlandt, A. 1986. Use of stone tools by wild-living chimpanzees and the earliest hominids. *Journal of Human Evolution* 15:77–132.

166 wild chimps use at least thirty-four different calls: Boehm, C. 1989. Methods for isolating chimpanzee vocal communication. In Heltne and Marquardt, pp. 38–59. See also Goodall 1986.

166 chimps are self-aware: Goodall 1986, pp. 34–36, 589.

166 chimps have learned hundreds of signs in American Sign Language: Fouts, R., and S. T. Mills. 1997. *Next of Kin: What Chimpanzees Have Taught Me About Who We Are.* New York: Morrow. See also Gardner, R. A., and B. T. Gardner. 1989. Cross-fostered chimpanzees, Part 1: modulation of meaning. In Heltne and Marquardt, pp. 220–233.

166 chimps learn other languages: Savage-Rumbaugh, S., M. A. Remski, W. D. Hopkins, and R. A. Sevcik. 1989. Symbol acquisition and use by *Pan troglodytes, Pan paniscus, Homo sapiens.* In Heltne and Marquardt, pp. 266–295.

166 chimps communicate to self-advantage: Gill, T. V. 1978. Conversing with Lana. In D. J. Chivers and J. Herbert (eds.), *Recent Advances in Primatology,* pp. 861–866. San Francisco: Academic Press. See also Premack, D., and A. J. Premack.1983. *The Mind of an Ape.* New York: Norton.

166 chimps learn sign language from chimps: Gardner, B. T., and R. A. Gardner. 1989. Cross-fostered chimpanzees, part 2: modulation of meaning. In Heltne and Marquardt, pp. 234–241. See also Fouts, D. H. 1989. Signing interactions between mother and infant chimpanzees. In Heltne and Marquardt, pp. 242–251.

166 Chimps have friendships: Goodall 1986.

166–167 "Old Man" to the rescue: Goodall, J. 1989. Foreword. In Heltne and Marquardt, pp. xi–xvi.

167 monkey rescues keeper: Darwin, C. 1871. *The Descent of Man and Selection in Relation to Sex.* New York: Modern Library, p. 476.

167 chimps learning, breaking, and enforcing rules: de Waal, F. 1996. *Good Natured: The Origins of Right and Wrong in Humans and Other Animals.* Cambridge: Harvard University Press.

167 at least 25 successful transfers were seen by female chimps: Wolf, K., and S. T. Schulman. 1984. Male response to "stranger" females as a function of female reproductive value among chimpanzees. *American Naturalist* 123:163–174. See also Pusey, A., J. Williams, and J. Goodall. 1997. The influence of dominance rank on the reproductive success of female chimpanzees. *Science* 277: 828–831. Pusey, A. E. 1980. Inbreeding avoidance in chimpanzees. *Animal Behaviour* 28:543–552. Bittles, A. H., W. M. Mason, J. Greene, and N. A. Rao. 1991. Reproductive behavior and health in consanguineous marriages. *Science* 252:789–794.

167 males are related as half brothers: Morin, P. A., J. J. Moore, R. Chakraborty, L. Jin, J. Goodall, and D. S. Woodruff. 1994. Kin selection, social structure, gene flow, and the evolution of chimpanzees. *Science* 265:1193–1201.

167 bonobos are one of the rare species in which females emigrate, but males remain: Kano, T. 1982. The social group of pygmy chimpanzees *(Pan paniscus)* of Wamba. *Primates* 23:171–188.

167 other species with female exogamy include gorillas and hamadryas baboons: Stewart, K., and A. H. Harcourt. 1986. Gorillas: variation in female relationships. In B. B. Smuts, D. L. Cheney, R. M. Seyfarth, R. W. Wrangham, and T. T. Struhsaker (eds.), *Primate Societies,* pp. 155–164. Chicago: University of Chicago Press. See also Pusey, A. E., and C. Packer. 1986. Dispersal and philopatry. In Smuts, Cheney, Seyfarth, Wrangham, and Struhsaker, pp. 250–266. Kummer, H. 1968. *Social Organization of Hamadryas Baboons.* Chicago: University of Chicago Press.

168–169 workers and soldiers: Robinson, G. E., R. E. Page, Jr., C. Strambi, and A. Strambi. 1989. Hormonal and genetic control of behavioral integration in honey bee colonies. *Science* 246:109–122, 223. See also Hamilton, W. D. 1964. The genetical theory of social behavior, parts 1 and 2. *Journal of Theoretical Biology* 31:1–52. Hamilton, W. D. 1972. Altruism and related phenomena, mainly in social insects. *Annual Review of Ecology and Systematics* 3:193–232. Wright, S. 1922. Coefficients of inbreeding and relationship. *American Naturalist* 56:330–338.

169 many creatures have an inbred talent: Hepper, P. G. (ed.). 1991. *Kin Recognition.* New York: Cambridge University Press.

169 white-fronted bee-eaters: Emlen, S. T., and P. H. Wrege. 1988. The role of kinship in helping decisions among white-fronted bee-eaters. *Behavioral Ecology and Sociobiology* 23(5):305–315.

170 !Kung San share food with relatives: Konner, M. 1982. *The Tangled Wing: Biological Constraints on the Human Spirit.* New York: Holt, Rinehart and Winston, p. 9.

170 "Me against my brother": Barash, D. 1979. *The Whisperings Within.* New York: Harper and Row, p. 140.

170 Hamilton's two papers "are among the most important contributions": Dawkins, R. 1989. *The Selfish Gene.* New ed. New York: Oxford University Press, p. 90.

170 "Analysis of data": Goodall 1986, p. 473.

170 many chimp "safaris" would be categorized as rapes: Smuts, B. B., and R. W. Smuts. 1993. Male aggression and sexual coercion of females in nonhuman primates and other animals: evidence and theoretical implications. *Advances in the Study of Behavior* 22:1–63.

170 135 copulations before conception: Hasegawa, T., and M. Hiraiwa-Hasegawa. 1990. Sperm competition and mating behavior. In T. Nishida (ed.), *The Chimpanzees of the Mahale Mountains: Sexual and Life History Strategies,* pp. 115–132. Tokyo: University of Tokyo Press.

171 Males at Gombe: Wrangham, R. W., and B. B. Smuts. 1980. Sex differences in the behavioural ecology of chimpanzees in the Gombe National Park, Tanzania. *Journal of Reproduction and Fertility* Supplement, no. 28:13–31. See also Goodall 1986, p. 211.

171 "When the chimpanzees": Boehm 1989, p. 38. See also Boehm, C. 1992. Segmentary "warfare" and the management of conflict: comparison of East African chimpanzees and patrilineal-patrilocal humans. In A. H. Harcourt and F. B. M. de Waal (eds.), *Coalitions and Alliances in Humans and Other Animals,* pp. 137–173. Oxford: Oxford University Press.

171 males at Kibale stay on the move: Ghiglieri 1984. See also Ghiglieri, M. P. 1988. *East of the Mountains of the Moon: Chimpanzee Society in the African Rain Forest.* New York: Macmillan/Free Press.

171 males and females differ in activity patterns: Ghiglieri 1984. See also McGrew, W. C. 1979. Evolutionary implications of sex differences in chimpanzee predation and tool use. In D. A. Hamburg and E. R. McCown (eds.), *The Great Apes,* pp. 440–463. Menlo Park, Calif.: Benjamin/Cummings. Nishida, T. 1989. Local traditions and cultural transmission. In Smuts, Cheney, Seyfarth, Wrangham, and Struhsaker, pp. 462–474. Hiraiwa-Hasagawa, M. 1989. Sex differences in the behavioral development of chimpanzees at Mahale. In Heltne and Marquardt, pp. 104–116. Boesch, C., and H. Boesch. 1984. Possible causes of sex differences in the use of natural hammers by chimpanzees. *Journal of Human Evolution* 13:415–440. Reynolds, V., and F. Reynolds. 1965. Chimpanzees in the Budongo Forest. In I. DeVore (ed.), *Primate Behavior: Field Studies of Monkeys and Apes,* pp. 468–524. New York: Holt, Rinehart and Winston.

171 females preferred other females: Ghiglieri, M. P. 1984. *The Chimpanzees of Kibale Forest.* New York: Columbia University Press. See also Goodall 1986.

171 solidarity among males: Ghiglieri, M. P. 1987. Sociobiology of the great apes and the hominid ancestor. *Journal of Human Evolution* 16(4): 319–357.

171 fusion-fission sociality among bonobos: Badrian, A., and N. Badrian. 1984. Social organization of *Pan paniscus* in the Lomako Forest, Zaire. In R. L. Susman (ed.), *The Pygmy Chimpanzee: Evolutionary Biology and Behavior,* pp. 325–346. New York: Plenum.

171–172 chimp pant-hoots allow them to identify one another: Marler, P., and L. Hobbet. 1975. Individuality in a long-range vocalization of wild chimpanzees. *Zeitschrift fur Tierpsychologie* 38:97–109.

172 chimp behaviors at reunions: Goodall 1986. See also Bauer, H. R. 1974. Behavioral changes about the time of reunion in parties of chimpanzees in the Gombe Stream National Park. In *5th International Congress in Primatology, Nagoya,* pp. 295–299. Basel: Karger.

172 chimp diets: Wrangham, R. W. 1975. The behavioural ecology of chimpanzees in Gombe National Park, Tanzania. Ph.D. dissertation, University of Cambridge. See also Goodall 1986. Nishida, T., and S. Uehara. 1983. Natural diet of chimpanzees *(Pan troglodytes schweinfurthii):* long term record from the Mahale Mountains, Tanzania. *African Studies Monographs* 3:109–130. Ghiglieri 1984. Ghiglieri, M. P. 1985. The social ecology of chimpanzees. *Scientific American* 252(6):102–113.

172 fruit rarely exists in adequate abundance: Ghiglieri 1984, pp. 69, 70–71.

172 weights of wild versus captive chimps: Goodall, J. 1983. Population dynamics during a fifteen-year period in one community of free-living chimpanzees in the Gombe National Park, Tanzania. *Zeitschrift fur Tierpsychologie* 61:1–60.

172 females lose in competition: Wrangham and Smuts 1980, p. 28. See also Goodall 1986, p. 245.

172 sizes and structures of chimp communities: Nishida, T., M. Hiraiwa-Hasegawa, T. Hasegawa, and Y. Takahata. 1985. Group extinction and female transfer in wild chimpanzees in the Mahale National Park, Tanzania. *Zeitschrift fur Tierpsychologie* 67:284–301. See also Itani, J. 1980. Social structure of African great apes. *Journal of Reproduction and Fertility* Supplement, no. 28:33–41. Ghiglieri 1987.

172 males routinely gave food calls at huge fruit trees: Ghiglieri 1984. See also Wrangham 1975.

172 chimps will travel in larger parties when food constraints are lifted: Wrangham,

R. W. 1974. Artificial feeding of chimpanzees and baboons in their natural habitat. *Animal Behaviour* 22:83–93.

172 Kasakela males killed five Kahama males: Goodall, J., A. Bandoro, E. Bergman, C. Busse, H. Matama, E. Mpongo, A. Pierce, and D. Riss. 1979. Intercommunity interactions in the chimpanzee population of the Gombe National Park. In Hamburg and McCown, pp. 13–54. See also Goodall, J. 1979. Life and death at Gombe. *National Geographic* 155(5):592–621. Goodall 1986.

173 "If they had had firearms": Goodall 1986, p. 530.

173 M-Group versus K-Group: Nishida, Hiraiwa-Hasegawa, Hasegawa, and Takahata 1985.

174 Kanyawara males were killed in territorial combat: Wrangham, R., and D. Peterson. 1996. *Demonic Males: Apes and the Origins of Human Violence.* Boston: Houghton Mifflin, pp. 20–21.

174 bonobos are violently territorial against other communities: Kano, T., and M. Mulavwa. 1984. Feeding ecology of the pygmy chimpanzee *(Pan paniscus)* of Wamba. In Susman, pp. 233–274. See also Kano, T. 1980. Social behavior of wild pygmy chimpanzees *(Pan paniscus)* of Wamba. *Primates* 23:171–188. Nishida, T., and M. Hiraiwa-Hasegawa. 1986. Chimpanzees and bonobos: cooperative relationships among males. In Smuts, Cheney, Seyfarth, Wrangham, and Struhsaker, pp. 165–177.

175 group aggression entered the arms race of sexual selection: West-Eberhard, M. J. 1991. Sexual selection and social behavior. In M. H. Robinson and L. Tiger (eds.), *Man and Beast Revisited,* pp. 159–172. Washington, D.C.: Smithsonian Institution.

175 gang violence and homicide: Block, C. R., and R. Block. 1993. *Street Gang Crime in Chicago.* Washington, D.C.: National Institute of Justice Research in Brief. See also Federal Bureau of Investigation (FBI). 1997. *Crime in the United States, 1996.* Washington, D.C.: U.S. Department of Justice, p. 21.

175 Machiavelli, for example, advised, "Indeed, there is no surer way of keeping possession than by devastation": Machiavelli, N. 1514 [1961]. *The Prince.* New York: Penguin, pp. 48, 87.

175 male chimps are only 123 percent the weight of females: Jungers, W. L., and R. L. Susman. 1984. Body size and skeletal allometry in African apes. In Susman, pp. 131–178.

176 resident male chimps kill alien males, infants (mostly male) of nonresident females, and old females: Bygott, J. D. 1972. Cannibalism among wild chimpanzees. *Nature* 238:410–411. See also Suzuki, A. 1971. Carnivority and cannibalism observed in forest-living chimpanzees. *Journal of the Anthropological Society of Nippon* 74:30–48. Kawanaka, K. 1981. Infanticide and cannibalism with special reference to the newly observed case in the Mahale Mountains. *African Studies Monographs* 1:69–99. Nishida, T., and K. Kawanaka. 1985. Within group cannibalism by adult male chimpanzees. *Primates* 26:274–285. Nishida, T., and M. Hiraiwa-Hasegawa. 1985. Responses to mother-son pair in the wild chimpanzee: a case report. *Primates* 26:1–13. Wolf and Schulman 1984. Takasaki, H. 1985. Female life history and mating patterns among the M Group chimpanzees of the Mahale Mountains. *Primates* 26:121–129. Pusey, A. E. 1979. Intercommunity transfer of chimpanzees in Gombe National Park. In Hamburg and McCown, pp. 465–480. Tutin, C. E. G. 1979. Mating patterns and reproductive strategies in a community of wild chimpanzees *(Pan troglodytes schweinfurthii). Behavioral Ecology and Sociobiology* 6:29–38. Nishida, T. 1989. Social interactions between resident and immigrant female chimpanzees. In Heltne and Marquardt, pp. 68–89.

176 "If we are to avoid destruction": Dyson, F. 1984. *Weapons and Hope.* New York: Harper and Row, p. 15.

177 chimpanzee populations are being decimated by humans: Teleki, G. 1989. Population status of wild chimpanzees *(Pan troglodytes)* and threats to survival. In Heltne and Marquardt, pp. 312–353.

177 "Yes, boys have": Stanley, A. 1990. Child warriors. *Time* 135(25):30–35, p. 32.

177 "Aggression is certainly part": Dyer, G. 1985. *War.* New York: Crown, p. 13.

177 "An ineluctable fact": Seabury, P., and A. Codevilla. 1989. *War Ends and Means.* New York: Basic Books, p. 6.

177 innate traits of men that accompany war: Eibl-Eibesfeldt 1989, pp. 402, 346.

178 war "is too complex": Ehrenreich, B. 1997. *Blood Rites: Origins and History of the Passions of War.* New York: Metropolitan Books, pp. 9–10, 20, 124–125, 231–232, 238, 240.

178 denials that humans possess an instinct to kill: Groebel and Hinde 1989, pp. xiii–xvi. See also Engels, F. 1984 [1964]. *The Origin of the Family, Private Property, and the State.* New York: International Publishers. Marx, K., and F. Engels. 1845 [1976]. The German ideology, part I: Feuerbach. In K. Marx and F. Engels, *Collected Words,* vol. 5, pp. 27–96. London: Lawrence and Wishert. Betzig, L. L. 1986. *Despotism and Differential Reproduction: A Darwinian View of History.* New York: Aldine, pp. 11, 19, 26–28, 66, 68.

179–180 "Weapons are truly ancient": O'Connell 1989, p. 14.

180–181 "I feel no remorse about enemy soldiers who were killed in combat": Gruenther, R. L., and D. W. Parmly. 1992. The crusade of a Green Beret. *Special Warfare* 5(1):42–55. See also *U.S. News & World Report* Staff. 1992. *Triumph Without Victory.* New York: Times Books/Random House, p. 355.

181 archconservationist David Brower: Brower, D. R. 1991. *For Earth's Sake: The Life and Times of David Brower.* Salt Lake City: Peregrine Smith, pp. 206–207.

181 "I cannot recall the least bit of thinking about having just killed a man": George, J. 1981 [1991]. *Shots Fired in Anger.* Washington, D.C.: National Rifle Association of America, p. 57.

181 "When you go into combat": Sasser, C. W., and C. Roberts. 1990. *One Shot-One Kill.* New York: Pocket Books, p. 34.

181 "[In] the intimacy of life": Caputo, P. 1977. *A Rumor of War.* New York: Holt, Rinehart and Winston, p. xv.

181–182 "It was an act of love": Manchester, W. 1980. *Goodbye Darkness: A Memoir of the Pacific War.* New York: Little, Brown, p. 391.

182 "For most men": Sherrod, D. 1987. The bonds of men: problems and possibilities in close male relationships. In H. Brod (ed.), *The Making of Masculinities: The New Men's Studies,* pp. 213–239. Boston: Allen and Unwin.

182 "Friends are described": Hammond, D., and A. Jablow. 1987. Gilgamesh and the Sundance Kid: the myth of male friendship. In Brod, pp. 241–258.

182–183 "By the second grade": Treadwell, P. 1987. Biologic influences on masculinity. In Brod, pp. 259–285, p. 268.

183 women do not bond as men do: Hrdy, S. B. 1981. *The Woman That Never Evolved.* Cambridge: Harvard University Press, p. 130.

183–184 the Israeli "experiment": Hadley, A. T. 1986. *The Straw Giant: Triumph and Failure: America's Armed Forces.* New York: Random House, p. 259.

184 "Boys were seen quarreling": Lever, J. 1976. Sex differences in the games children play. *Social Problems* 23:478–487, pp. 482–483. (Cited in F. de Waal. 1996. p. 118.)

184 women can't cope with combat or killing: Hackworth, D. H. 1991. War and the second sex. *Newsweek,* August 5:24–29.

184 "This terror is born": George 1981 [1991], p. 325.

184 "We discovered": Moore and Galloway 1992, p. xviii.

184 decorated with 110 medals: Hackworth, D. H. 1991. "We'll win, but . . ." *Newsweek,* January 21:26–31.

184–185 "Sure, I was fighting": Hackworth and Sherman 1989, pp. 111–112.

185 the "nonviolent" Semai: Dentan, R. K. 1968. *The Semai: A Nonviolent People of Malaysia.* New York: Holt, Rinehart and Winston.

185 300,000 bullets were fired: Hadley 1986, p. 189.

185 "on average": Marshall, S. L. A. 1947 [1978]. *Men Against Fire: The Problem of Battle Command in Future War.* Gloucester, Mass.: Peter Smith, p. 54.

185 more appropriate training since World War II: Grossman, D. 1995. *On Killing: The Psychological Cost of Learning to Kill in War and Society.* New York: Little, Brown, p. 181.

185 those 15 percent who fired: Hadley 1986, pp. 56, 54–55.

186 "I remember": Saitoti, T. O. 1980. Warriors of Maasailand. *Natural History* 89(8):42–55.

186 "Never in the field": Churchill, W. S. 1949. *Their Finest Hour.* Boston: Houghton Mifflin, p. 340.

186–187 reciprocal altruism: Trivers, R. L. 1971. The evolution of reciprocal altruism. *Quarterly Review of Biology* 46:35–57.

187 No biologist believes true altruism is possible as a ruling behavior, but computer scientist and psychologist Herbert A. Simon tried to prove otherwise—despite admitting that altruists themselves would become extinct. Simon says altruists can raise the fitness of their society by assisting its selfish members to breed more than average. The flaw here is that Simon is using a group selection argument, a "good-for-the-species" argument that is impossible even if altruists could raise the fitness of selfish individuals that high, which remains doubtful. But Simon "proves" otherwise by equating altruists with "docile" individuals—defined as being disposed to learn socially. In real life, selfish people—indeed, all people—are equally capable of social learning. Here Simon becomes confused. "If docility were something the individual deliberately chose," he admits, "one might even rename the accompanying altruism 'enlightened selfishness.'" (Or "reciprocal altruism," which it of course is.) But Simon claims that the only difference between altruists and selfish individuals is that altruists are "docile" social learners due to being unwittingly compelled by their genes, whereas selfish people also are "docile" but have chosen to be so. All in all, Simon's "special assumptions" add up only on his computer: Williams 1966. See also Simon, H. A. 1990. A mechanism for social selection and successful altruism. *Science* 250:1665–1668.

187 cleaner fish: Bishop, C. (ed.) 1992. *The Way Nature Works.* New York: Macmillan, pp. 125, 192.

187 vampire bats: Wilkinson, G. S. 1990. Food sharing in vampire bats. *Scientific American* 262(2):76–82.

187 female lions remain in their pride: Schaller, G. B. 1972. *The Serengeti Lion.* Chicago: University of Chicago Press.

187 lionesses defend against alien lionesses: Heinsohn, R., and C. Packer. 1995. Complex cooperative strategies in group-territorial lions. *Science* 269:1260–1262.

188 male lions travel and fight cooperatively: Bertram, B. C. R. 1973. Lion population regulation. *East African Wildlife Journal* 11:215–225.

188 infanticide of cubs: Packer, C., L. Herbst, A. E. Pusey, J. D. Bygott, J. P. Hanby, S. J. Cairns, and M. Borgerhoff Mulder. 1988. Reproductive success in lions. In T. H. Clutton-Brock (ed.), *Reproductive Success: Studies in Individual Variation in Contrasting Breeding Systems,* pp. 363–383. Chicago: University of Chicago Press, pp. 374, 371.

188 success requires cooperation: Bertram, B. C. R. 1976. Kin selection in lions and evolution. In P. P. G. Bateson and R. Hinde (eds.), *Growing Points in Ethology,* pp. 281–301. Cambridge: Cambridge University Press. See also Bertram, B. C. R. 1985. Blood relatives: kin selection in a lion pride. In D. Macdonald (ed.), *All the World's Animals: Carnivores,* pp. 26–27. New York: Torstar. Handby, J. 1982. *Lion's Share.* Boston: Houghton Mifflin. Packer, Herbst, Pusey, Bygott, Hanby, Cairns, and Borgerhoff Mulder 1988, p. 373.

188 44 percent of male "kin groups": Pusey, A. E., and C. Packer. 1983. The once and future kings. *Natural History* 92(8):54–63.

188 cheetahs: Caro, T. M. 1994. *Cheetahs of the Serengeti Plains: Group Living in an Asocial Species.* Chicago: University of Chicago Press, p. 183.

188–189 true lion kin groups and coalitions based on reciprocal altruism do not work the same: Packer, Herbst, Pusey, Bygott, Hanby, Cairns, and Borgerhoff Mulder 1988, p. 380.

189 kamikaze pilots killed 5,000 Americans: Pineau, R. (ed.). 1958 [1978]. *The Divine Wind.* New York: Bantam, p. 222.

189 "It is absolutely": McCullough, D. G. 1966. *The American Heritage Picture History of World War II.* Crown/American Heritage, p. 606.

189 2,363 kamikaze planes: Okumiya, M., J. Horikoshi, and M. Caidin. 1958 [1957]. *Zero.* New York: Ballantine, p. 254.

190 "a mere drop": Andreski, S. 1964. Origins of war. In J. D. Carthy and F. J. Ebling (eds.), *The Natural History of Aggression.* Institute of Biological Symposia, No. 13. New York: Academic Press.

190 severe hardships triggered more wars than chronic shortages: Horgan, J. 1988. Why warfare? A broad study of preindustrial societies offers some answers. *Scientific American* 259(5):20.

190 "There are significant": Homer-Dixon, T. F., J. H. Boutwell, and G. W. Rathjens. 1993. Environmental change and violent conflict. *Scientific American* 268(2):38–45.

191 hawks, doves, and bourgeois: Maynard Smith, J. 1982. *Evolution and the Theory of Games.* Cambridge: Cambridge University Press. See also Maynard Smith, J., and G. R. Price. 1973. The logic of animal conflicts. *Nature* 246:15–18.

191 "Indeed the militia": Kopel, D. B., and S. D'Andrilli. 1990. The Swiss and their guns. *American Rifleman* 138(2):38–39, 74–81. See also Kopel, D. B. 1992. *The Samurai, the Mountie, and the Cowboy: Should America Adopt the Gun Controls of Other Democracies?* Buffalo: Prometheus, pp. 278–294.

192 most animals use a bourgeois strategy: Pool, R. 1995. Putting game theory to the test. *Science* 268:1591–1593.

192 "Wars are won or lost": Seabury and Codevilla 1989, p. 8.

192 if ten tribes exist: Schmookler, A. B. 1984. *The Parable of the Tribes.* Berkeley: University of California Press, pp. 21–30.

192–193 "failed to put on a show": Chagnon, N. A. 1983. *Yanomamo: The Fierce People.* 3d ed. New York: Holt, Rinehart and Winston, p. 188.

193 17 percent of Yanomamo wives: Chagnon, N. A. 1988. Life histories, blood revenge, and warfare in a tribal population. *Science* 239:985–992.

194 cheaters and warriors: Gadgil, M. 1972. Male dimorphism as a consequence of natural selection. *American Naturalist* 106:574–580.

194 the importance of communicating one's willingness to retaliate: Schelling, T. C. 1966. *Arms and Influence.* New Haven, Conn.: Yale University Press.

194 Swiss willingness to retaliate: Kopel 1992, pp. 280–285. See also Zagare, F. C. 1987. *The Dynamics of Deterrence.* Chicago: University of Chicago Press.

194 "Most people today": Seabury and Codevilla 1989, p. 54.

195–196 "Every Dani alliance": Heider, K. 1979. *Grand Valley Dani: Peaceful Warriors.* New York: Holt, Rinehart and Winston, p. 88.

196 ideal strategy: Liddell Hart 1967, p. 338.

196 "supreme excellence": Sun Tzu. 1983. *The Art of War,* edited by J. Clavell. New York: Delacorte Press, p. 15. See also Seabury and Codevilla 1989, p. 122. Von Clausewitz 1832 [1968], p. 198.

197 "Why would anyone": Gibbons, A. 1993. Warring over women. *Science* 261:987–988.

197 "If the earliest cause": Wright 1942, p. 136.

198 Kublai Khan sired 47 sons: Polo, M. 1298. [1948]. *The Adventures of Marco Polo as Dictated in Prison to a Scribe in the Year 1298: What He Experienced and Heard During His Twenty-four Years Spent in Travel Through Asia and the Court of the Kublai Khan,* edited by R. J. Walsh. New York: John Day, p. 68.

198 men everywhere compete reproductively: Chagnon 1979. See also Betzig, L. 1988. Mating and parenting in Darwinian perspective. In L. Betzig, M. Borgerhoff Mulder, and P. Turke (eds.), *Human Reproductive Behavior: A Darwinian Perspective,* pp. 3–20. New York: Cambridge University Press. Irons, W. 1988. Reproductive behavior in humans. In Betzig, Borgerhoff Mulder, and Turke, pp. 302–314.

198 Aka pygmies: Hewlett, B. S. 1988. Sexual selection and parental investment among Aka pygmies. In Betzig, Borgerhoff Mulder, and Turke, pp. 263–276.

198 Ifaluks: Betzig, L. 1988. Redistribution: equity or exploitation? In Betzig, Borgerhoff Mulder, and Turke, pp. 49–63.

198 Yomut Turkmen: Irons, W. 1974. Nomadism as a political adaptation: the case of the Yomut Turkmen. *American Ethnologist* 1:635–658. See also Irons, W. 1979. Culture and the biological process. In Chagnon and Irons, pp. 257–272.

198 Akoa Pygmies, Ona and Yaghans, Australian Aborigines, Nootka: Coon, C. S. 1971. *The Hunting Peoples.* New York: Atlantic–Little, Brown.

200 NVA cities and supply dumps in Cambodia: Brennan, M. (ed.). 1990. *Hunter-Killer Squadron Aero-Weapons•Aero-Scouts•Aero-Rifles Vietnam 1965–1972.* Novato, Calif.: Presidio Press. See also Nolan, K. W. 1990. *Into Cambodia.* Novato, Calif.: Presidio Press.

200 Westmoreland had forbidden: Moore and Galloway 1992, pp. 308–309, 314–315, 342. See also Sheehan 1988, pp. 556, 632, 641, 650–652, 693.

200 Westmoreland's strategy: Moore and Galloway 1992, pp. 339, 345.

201–202 "Our mission": Caputo 1977, pp. xvii, 311. See also Mason, J. 1983. *Chickenhawk.* New York: Viking Press, pp. 1, 309. Hilsman, R. 1990. *American Guerrilla: My War Behind Japanese Lines.* New York: Brassey's, p. 274. Moore and Galloway 1992, pp. 339–341.

203 "We're the unwilling": Nolan, K. W. 1990.

203 "Our rule of thumb": Brennan, M. 1985. *Brennan's War: Vietnam 1965–69.* Novato, Calif.: Presidio Press, p. 1.

203 there was no plan to win the war: Arnett, P., I. McLeod, and M. MacLear. 1980. *Vietnam: The Ten Thousand Day War.* Vol. 7: *Tet!* Documentary film, Embassy Home Entertainment, Los Angeles. See also Sheehan 1988, p. 730. Downs, F. 1978 [1983]. *The Killing Zone.* New York: Berkeley, p. 208. Parker, F. C., IV. 1989. *VIETNAM Strategy for Stalemate.* New York: Paragon House, pp. 3, 142–164, 220. Hackworth and Sherman 1989, p. 594.

204 "A great nation": Parker 1989, p. 236.

Chapter Seven: Genocide

205 "It seems": Goodall, J. 1990. *Through a Window: My Thirty Years with the Chimpanzees of Gombe.* Boston: Houghton Mifflin, p. 102.

205 "Nuclear, chemical, and biological": Wills, C. 1998. *Children of Prometheus: The Accelerating Pace of Human Evolution.* Reading, Mass.: Helix/Perseus Books, p. 271.

205–209 Lloyd Paul Blanchard's Philippine experiences: Blanchard, R. 1990. Personal communications.

210 17 million people killed: Seabury, P., and A. Codevilla. 1989. *War Ends and Means.* New York: Basic Books, p. 15.

210 more than fifty significant military conflicts: Whitelaw, K. 1997. Good works, evil results. *U.S. News & World Report* 1229(20):34–35.

210 ethnic Albanians in Kosovo: Smucker, P. G. 1998. Tit for tat in the Balkans. *U.S. News & World Report* 125(3):30–34.

210 genocide in the southern Sudan: Lovgren, S. 1998. A famine made by man. *U.S. News & World Report* 125(10):38–43.

210 Indonesia is relocating its "excess" population from Java: Caufield, C. 1984. Pioneers of the outer islands. *Natural History* 93(3):22–32. See also Staff. 1992. Massacre in East Timor. *Rainforest Action Report Action Alert,* no. 68, Rainforest Action Network, San Francisco, p. 2. Staff. 1990. U.S. mining threatens Indonesian forest, University of Texas may also profit. *Rainforest Action Network Action Alert,* no. 54. Staff. 1990. Shell Oil destroying Thai rainforest and abusing villagers. *Rainforest Action Network Action Alert,* no. 48.

211 ethnocentrism and xenophobia: Eibl-Eibesfeldt, I. 1989. *Human Ethology.* New York: Aldine de Gruyter, p. 289.

211 "There is more than satire": Gelman, D. 1989. Why we all love to hate. *Newsweek,* August 28:62–64.

211 "Two motives": von Clausewitz, C. 1832 [1976]. *On War.* Princeton, N.J.: Princeton University Press, pp. 102–103.

211–212 "us" and "them" identities: Sherif, M. 1956. Experiments in group conflict. *Scientific American* 196:54–58.

212 socialization to instill patriotism: Johnson, G. R. 1986. Kin selection, socialization, and patriotism: an integrating theory. *Politics and the Life Sciences* 4(2):127–140. See also Johnson, G. R. 1987. In the name of the fatherland: an analysis of kin term usage in patriotic speech and literature. *International Political Science Review* 8(2):165–174. Johnson, G. R. 1989. The role of kin recognition mechanisms in patriotic socialization: further reflections. *Politics and the Life Sciences* 8(1):62–69.

212 "propaganda and opinion control": Wright, Q. 1942. *A Study of War.* Chicago: University of Chicago Press, pp. 1019, 1095.

212 the Eipo of Irian Jaya: Eibl-Eibesfeldt 1989, p. 403.

212 "the San lack": Konner, M. 1982. *The Tangled Wing: Biological Constraints on the Human Spirit.* New York: Holt, Rinehart and Winston, pp. 9, 204.

212 "I didn't know": Mills, J. 1986. *Underground Empire: Where Crime and Governments Embrace.* New York: Doubleday, p. 316.

213 5 million to 40 million Indians: Lord, L. 1997. How many people were here before Columbus? *U.S. News & World Report* 123(7):68–70.

213 "The weapons and tactics": Courtwright, D. T. 1996. *Violent Land: Single Men and Social Disorder from the Frontier to the Inner City.* Cambridge: Harvard University Press, p. 114.

213 Brazilian Indians' decline from 11 million to 300,000: Stern, D. 1998. New glimpse of an old world. *U.S. News & World Report* 125(3):8.

214 "As with the Indians": Courtwright 1996, p. 158.

215 "The new style of warfare": Crossette, B. 1998. Violation: an old scourge of war becomes its latest crime. *New York Times,* June 12:4-1, 4-6.

215–216 chimp infanticide and cannibalism: Bygott, J. D. 1972. Cannibalism among wild chimpanzees. *Nature* 238:410–411. See also Suzuki, A. 1971. Carnivority and cannibalism observed in forest-living chimpanzees. *Journal of the Anthropological Society of Nippon* 74:30–48. Kawanaka, K. 1981. Infanticide and cannibalism with special reference to the newly observed case in the Mahale Mountains. *African Studies Monographs* 1:69–99. Nishida, T., and K. Kawanaka. 1985. Within group cannibalism by adult male chimpanzees. *Primates* 26:274–285. Nishida, T., and M. Hiraiwa-Hasegawa. 1985. Responses to mother-son pair in the wild chimpanzee: a case report. *Primates* 26:1–13. Wolf, K., and S. T. Schulman. 1984. Males response to "stranger" females as a function of female reproductive value among chimpanzees. *American Naturalist* 123:163–174. Takasaki, H. 1985. Female life history and mating patterns among the M Group chimpanzees of the Mahale Mountains. *Primates* 26:121–129.

217 About 70 million people were killed in World War II, 16 million soldiers and the rest civilians. Roughly 10 percent of all populations died, civilian and military, from Germany eastward: roughly 35 million Russians, 6 million Jews, 5.6 million Germans, 3 million Poles, 1.6 million Yugoslavians, and 2 million other Europeans. The toll in the Far East was 11 million Chinese, 1.3 million Indonesians, 1 million Vietnamese, 2.5 million Japanese, and nearly a million more Allied soldiers: Dower, J. W. 1986. *War Without Mercy.* New York: Pantheon, pp. 295–301.

218 "All human culture": Hitler, A. 1962. *Mein Kampf.* Boston: Houghton Mifflin.

218 Jews, Slavs, Gypsies, and even Russians: O'Connell, R. L. 1989. *Of Arms and Men.* New York: Oxford University Press, p. 285.

218 Note human culture in Europe owed much to non-Aryans: Boorstein, D. J. 1983. *The Discoverers.* New York: Random House.

218 a sobering number of German scientists: Muller-Hill, B. 1988. *Murderous Science: Elimination by Scientific Selection of Jews, Gypsies, and Others, Germany 1933–45.* New York: Oxford University Press.

218 "When I came round": Hadley, A. T. 1986. *The Straw Giant: Triumph and Failure: America's Armed Forces.* New York: Random House, p. 207.

218 Hitler's SS *(Schutzstaffel)*: Parshall, G. 1995. Freeing the survivors. *U.S. News & World Report* 118(15):50–65.

219 "housecleaning of Jews": Shirer, W. L. 1960. *The Rise and Fall of the Third Reich.* New York: Simon and Schuster, p. 658.

219 Few people remember today, but Hitler's Nazi army almost won. Moreover, Hitler did not fail because he had an inferior army. The army of the Third Reich was the best on Earth. Nor did Hitler lose because he was a genocidal maniac. Rather, Hitler lost because he was a poor tactician and too egotistical to listen to his generals. Hitler made five bad decisions against their advice that cost him the conquest of Europe: (1) by pampering Hermann Goering's Luftwaffe, Hitler allowed 340,000 Allied soldiers trapped at Dunkirk in 1940 to evacuate (Churchill, W. S. 1969. The miracle of Dunkirk. In *The Reader's Digest Illustrated Story of World War II,* pp. 112–117. Pleasantville, N. Y.: Reader's Digest). (2) He withdrew from the Battle of Britain when, despite German losses (1,733 planes) exceeding British losses (915 planes), the British Air Force was nearly annihilated (Churchill, W. S. 1969. The Battle of Britain. In *The Reader's Digest Illustrated Story of World War II,* pp. 120–129). (3) He invaded Yugoslavia out of pique, then invaded Greece to help Mussolini, thus delaying by seven weeks his scheduled May 1, 1941, invasion of Russia with 3 million men, who crushed Stalin's 2.5-million-man army but were caught by the Russian winter unprepared (Shirer 1960). (4) He invaded Russia. (5) He declared war on the United States, thereby committing Germany to a furious two-front war against overwhelming Allied power. Of a U.S. population of 135 million, 16 million donned uniforms, and U.S. factories produced more weapons and war materiel than the rest of the world combined (Seabury and Codevilla 1989, p. 69.)

219 America's national effort during World War II: Millett, A. R., and P. Maslowski. 1984. *For the Common Defense: A Military History of the United States of America.* New York: Free Press, p. 408.

219 "filial piety": Dower 1986, p. 275.

219–220 war was an "eternal . . . creative": Ibid., p. 216.

220 Japanese war plans called for conquering all of Asia: Ibid., pp. 273–278.

220 "the living space of the Yamato race": Ibid., pp. 276–277.

220 evidence for racial hatred in the Pacific: Ibid., pp. 33–73.

220 Bataan Death March: Dyess, W. 1969. Death march from Bataan. In *Reader's Digest Illustrated History of World War II,* pp. 164–171.

220 Japan's treacherous attack on Pearl Harbor: Prange, G. W. 1981. *At Dawn We Slept.* New York: McGraw-Hill.

220 documented cases of massacres of noncombatants: Dower 1986, pp. 42, 45.

220 Japanese secret police: Manchester, W. 1980. *Goodbye Darkness: A Memoir of the Pacific War.* New York: Little, Brown, pp. 86, 217, 281.

220–221 "Probably in all our history": Dower 1986, p. 33.

221 Japanese Americans in detention camps: Conn, S. 1960 [1984]. The decision to evacuate the Japanese from the Pacific coast. In R. G. Kent (ed.), *Command Decisions,* pp. 125–150. Washington, D.C.: United States Army, Center of Military History.

221 public opinion polls: Dower 1986, p. 53.

221 racist "scientific theories": Ibid., pp. 118–146.

221 "You have taught": Ibid., p. 71.

221 44 percent of U.S. soldiers: Stouffer, S. A., et al. 1949. *The American Soldier.* Princeton, N.J.: Princeton University Press. (Cited in Grossman, D. 1995. *On Killing: The Psychological Cost of Learning to Kill in War and Society.* New York: Little, Brown, p. 162.)

221 Americans were preoccupied: Dower 1986, pp. 238–261.

222 GIs in the Pacific took few prisoners alive: Sledge, E. B. 1981. *With the Old Breed at Peleliu and Okinawa.* Novato, Calif.: Presidio Press.

222 no Japanese officer surrendered: Wolfert, I. 1943. *Battle for the Solomons.* Boston: Houghton Mifflin.

222 winning World War II in Europe: Ambrose, S. E. 1994. *D-Day June 6, 1944: The Climactic Battle of World War II.* New York: Simon and Schuster. See also Ambrose, S. E. 1997. *Citizen Soldiers: The U.S. Army from the Normandy Beaches to the Bulge to the Surrender of Germany, June 7, 1944–May 7, 1945.* New York: Simon and Schuster.

223 "a place where hundreds": Heminway, J. 1997. Darkest heart. *Men's Journal,* November:55–58, 104, 122–125.

223–224 Genocide in Rwanda and Zaire: Ransdell, E., I. Gilmore, L. Lief, B. B. Auster, and L. Fasulo. 1994. A descent into hell. *U.S. News & World Report* 117(5):42–46. See also Ransdell, E. 1994. Resurrecting a nation. *U.S. News & World Report* 117(7):45–46. Whitelaw, K., and S. Kiley. 1997. The wars behind the war. *U.S. News & World Report* 122(11). Whitelaw, K. 1997. Dusting Kabila for fingerprints and finding none. *U.S. News & World Report* 122(2):45. Lovgren, S., and K. Whitelaw. 1997. Mobutuism without Mobutu. *U.S. News & World Report* 123(20):50–51.

225 in 1998 Rwanda was trying to pull itself together: Pasternack, S. 1998. Personal communication, Kenya, August 1.

225 "Anyone who looked": Rhodes, R. 1986. *The Making of the Atomic Bomb.* New York: Simon and Schuster, pp. 27–28.

226 Germany's and Japan's atomic bomb programs: Rhodes, R. 1995. *Dark Sun: The Making of the Hydrogen Bomb.* New York: Simon and Schuster, pp. 40–42, 155–156, 160–161.

226 "I said I thought": Rhodes 1986, p. 442. See also Lanquette, W., and B. Szilard. 1993. *Genius in the Shadows.* New York: Scribners.

226 "[Hans] Bethe's group": Teller, E. 1987. *Better a Shield Than a Sword: Perspectives on Defense and Technology.* New York: Free Press, p. 53.

227 "Jimmy Byrnes": Rhodes 1986, p. 618.

227 "This is the biggest": Morton, L. 1960 [1984]. The decision to use the atomic bomb. In K. R. Greenfield (ed.), *Command Decisions,* pp. 493–518. Washington, D.C.: United States Army, Center of Military History, p. 499.

227 "Now I am become Death": Rhodes 1995, p. 471.

227 "Little Boy" destroyed Hiroshima: Rhodes 1986, pp. 703, 711.

227–228 "The mushroom itself": Ibid., p. 711.

228 both atomic bombs accounted for a mere 3 percent: Okumiya, M., J. Horikoshi, and M. Caidin. 1956 [1966]. *Zero.* New York: Ballantine, p. 6.

228 Japan's final army size: Ibid., p. 286.

228 Fermi and Teller and the H-bomb: Rhodes 1995, p. 248.

229 top science adviser Vannevar Bush convinced Truman to approve the H-bomb: Bush, V. 1949. *Modern Arms and Free Men.* New York: Simon and Schuster, p. 128.

229 Truman encouraged Edward Teller and Stanislaw Ulam to design an H-bomb: Broad, W. J. 1982. Rewriting the history of the H-bomb. *Science* 218:769–772.

229 Teller worried about the H-bomb igniting the earth's atmosphere: Rhodes 1995, pp. 254, 402.

229 "Mike" explodes Elugelab: Ibid., pp. 484–511.

229 lithium-deuteride-fueled successors to "Mike": Ibid., p. 525.

229 thermonuclear research, spying, and the Soviet H-bomb: Ibid., pp. 82, 87, 98–145, 153–155, 168–176, 182, 193, 202, 206–209, 217–218, 257, 259–260, 411–414, 421–422, 427–428, 524, 569. See also Williams, R. C. 1987. *Klaus Fuchs: Atom Spy.* Cambridge: Harvard University Press.

229–230 "This [H-bomb]": Rhodes 1995, p. 529.

230 the Americans and Soviets cut their megatonnages: Auster, B. B. 1991. The arms race in reverse. *U.S. News & World Report* 111(17):54.

230 can we ever really trust the other side?: Nincic, M. 1986. Can the U.S. trust the U.S.S.R.? *Scientific American* 254(4):33–41. See also Drell, S., and M. Eimer. 1987. Policy forum: verification and arms control. *Science* 235:406–414.

230 we cannot rid ourselves of nuclear weapons; they are here to stay: Allison, G. T., A. Carnesdale, and J. Nye (eds.). 1985. *Hawks, Doves, and Owls.* New York: Norton.

230 U.S. Senate bans biowarfare development: Crawford, M. H. 1989. Banning the biological bomb. *Science* 246:1385.

230–231 "A world in which": Koshland, D. E., Jr. 1990. Research policy and the peace dividend. *Science* 247:1165. See also Galloway, J. L., and B. B. Auster. 1994. The most dangerous place on Earth. *U.S. News & World Report* 116(24):40–56. Nye, J. S., Jr. 1992. New approaches to nuclear proliferation policy. *Science* 256:1293–1297.

231 "every nation": Rhodes 1995, p. 162.

231 terrorist attacks on U.S. embassies in Tanzania and Kenya, numbers of international terrorist acts: Cooperman, A. 1998. Terror strikes again. *U.S. News & World Report* 125(7):8–17.

232 terrorist Osama bin Laden: Auster, B. B. 1998. The recruiter for hate. *U.S. News & World Report* 125(8):48–50. See also Diamond, J. 1998. War on terror. *Arizona Daily Sun,* August 21:1, 12.

232 terrorism is not new: Long, D. E. 1990. *The Anatomy of Terrorism.* New York: Free Press.

232 The RAND Corporation analyzed 63 major terrorist incidents and found terrorism works: Kupperman, R. H., and D. M. Trent. 1990. Terrorism: threat, reality, and response. In N. A. Weiner, M. A. Zahn, and R. Sagi (eds.), *Violence: Patterns, Causes, Public Policy.* New York: Harcourt Brace Jovanovich, p. 197.

232 organizations train and offer terrorists for hire: Melman, Y. 1986. *The Master Terrorist: The True Story of Abu-Nidal.* New York: Avon.

PART THREE: THE ANTIDOTE

235 "We must, however, acknowledge": Darwin, C. 1871. *The Descent of Man and Selection in Relation to Sex.* New York: Modern Library, p. 920.

Chapter Eight: Who, Me?

236–237 Robert Lezelle Courtney murders Cynthia Volpe: Brownlee, J. R. Personal communication. See also Walters, R. 1993. Niece traces history of family mental disorders. *Bakersfield Californian,* August 22, p. A-6.

238 every nineteen seconds another violent crime: Federal Bureau of Investigation (FBI). 1999. *Crime in the United States.* Washington, D.C.: U.S. Department of Justice, p. 4.

238–239 the dilemma of the commons: Hardin, G. 1968. The tragedy of the commons. *Science* 162:1243–1248.

240 Prisoner's Dilemma: Dawkins, R. 1989. *The Selfish Gene.* New ed. New York: Oxford University Press.

240 nasty to nice strategies: Axlerod, R. T., and W. D. Hamilton. 1981. The evolution of cooperation. *Science* 211:1390–1396. See also Axlerod, R. T., and D. Dion. 1989. The further evolution of cooperation. *Science* 246:1385–1390.

241 more complicated games: Ridley, M. 1996. *The Origins of Virtue: Human Instincts and the Evolution of Cooperation.* New York: Viking, pp. 54–82.

241–242 a single factor made all the difference in Linnda Caporeal's experiments: Bower, B. 1990. Getting out from number one: selfishness may not dominate human behavior. *Science News* 137(17):266–267.

242 our genes program us to be selfish: Dawkins 1989, p. 3.

242 when the "shadow of the future": Axlerod and Dion 1989.

242 Kitcher and Batali enter a tit-for-tat program with the tactic to "opt out": Flam, F. 1994. The artifice of cooperation. *Science* 265:868–869.

242 the budget of the United States: Shuman, H. E. 1988. *Politics and the Budget.* New York: Prentice-Hall.

243 $5 trillion debt: Rainie, H. 1992. State of the Union. *U.S. News & World Report* 111(27):36–39.

243 "The rapaciousness": Shuman 1988, p. 73.

243–244 "Punishment allows": Boyd, R., and P. J. Richerson. 1989. Punishment allows the evolution of reciprocity (or anything else) in sizable groups. Paper presented at the first annual meeting of the Human Behavior and Evolution Society, Northwestern University, Evanston, Ill., August 25–27.

244 religious cults: Galanter, M. 1989. *Cults: Faith, Healing, and Coercion.* New York: Oxford University Press.

244 the FBI at Waco, Texas: LaPierre, W. 1994. *Guns, Crime, and Freedom.* Washington, D.C.: Regnery, pp. 179–200; Rainie, H., J. Popkin, D. McGraw, B. Duffy, T. Gest, J. A. Tooley, and D. Bowermaster. 1993. Armageddon in Waco, the final days of David Koresh. *U.S. News & World Report* 114(17):24–34.

244 when moralist strategy is too weak, vigilantes arise: Wilson, J. Q., and R. J. Herrnstein. 1985. *Crime and Human Nature.* New York: Simon and Schuster, p. 506.

244 five hundred U.S. vigilante groups: Kopel, D. B. 1992. *The Samurai, the Mountie, and the Cowboy: Should America Adopt the Gun Controls of Other Democracies?* Buffalo: Prometheus, p. 324.

245 The most revered philosophers and prophets in history have known that, of all options, cooperation works best. "In dealing with others," advised Lao-tzu twenty-five centuries ago, "be gentle and kind": Lao-tzu. 1972. *Tao Te Ching,* translated by Gia-Fu Feng and Jane English. New York: Vintage, p. 8. See also Psalms 37:27. Smart, N., and R. D. Hecht (eds.). 1982. *Sacred Texts of the World: An Anthology.* New York: Crossroad, pp. 14, 154, 315. Matthew 7:7, 22:39.

245 "the social instincts": Darwin 1871, p. 495.

245–246 de Waal's moral chimps: de Waal, F. 1996. *Good Natured: The Origins of Right* *and Wrong in Humans and Other Animals.* Cambridge: Harvard University Press, pp. 157–161, 217.

246 humans putatively can only represent to ourselves what goes on in other people's minds due to our highly specialized brains: Gibbons, A. 1993. Empathy and brain evolution. *Science* 259:1250–1251.

247 the Swiss enjoy the greatest freedom from crime: Kopel, D. B., and S. D'Andrilli. 1990. The Swiss and their guns. *American Rifleman* 138(2):38–39, 74–81.

247 rescuing the last victim: Salholz, E. 1989. Bracing for the big one. *Newsweek,* October 30:28–32. See also Magnuson, E. 1989. Earthquake. *Time* 134(18):30–40.

247 arrests for looting dropped to 25 percent: Magnuson 1989.

247 the human psyche evolved: Alexander, R. D. 1987. *The Biology of Moral Systems.* New York: Aldine de Gruyter, pp. 261–262.

247 we are gene machines: Dawkins, R. 1976. *The Selfish Gene.* New York: Oxford University Press. See also Dawkins, R. 1982. *The Extended Phenotype.* San Francisco: W. H. Freeman.

248 some genes are selfish: Bull, J. J., I. J. Molineux, and J. H. Werren. 1992. Selfish genes. *Science* 256:65.

248 "To say that we are evolved": Alexander 1987, p. 40.

248 "Police investigating alleged rapes, and attorneys defending accused rapists," notes Donald Symons, "often argue that a given act of intercourse was not really rape because the victim did not physically resist, yet police generally advocate nonresistance—as the most adaptive strategy—for victims of violent, or potentially violent crimes": Symons, D. 1979. *The Evolution of Human Sexuality.* New York: Oxford University Press, p. 277.

248 Consider this victim's report: "After I managed to loosen the ropes with which I was tied up, I went to my neighbor's and immediately called for the police. They didn't arrive for more than an hour, and when they did arrive, they were very rude and insensitive. Despite my bruises and my excited condition, the first police officer who arrived, asked me 'Lady, what makes you think you were raped?'": Herrington, L. H. 1982. *President's Task Force on Victims of Crime.* Washington, D.C.: White House, p. 58.

248 many courts also fail to identify: Herrington 1982, pp. 72–82. See also Epstein, J., and S. Langenbahn. 1994. *The Criminal Justice Response to Rape.* Washington, D.C.: U.S. Department of Justice, Office of Justice Programs, p. 46.

248 55 percent of suspects arrested for rape: Reaves, B. 1990. *Felony Defendants in Large Urban Counties, 1988.* Washington, D.C.: Bureau of Justice Statistics Executive Summary.

248–249 "In most states": Swasey, E. J. 1992. NRA woman's voice. *American Rifleman* 140(10):20. See also Schorer, J. 1990. Special series on Nancy Ziegenmeyer's rape case. *Des Moines Register* special reprint.

249 each rapist in prison: Langan, P. A. 1991. America's soaring prison populations. *Science* 251:1568–1573. See also Dillingham, S. D. 1991. *Violent Crime in the United States.* Washington, D.C.: U.S. Department of Justice, Bureau of Labor Statistics, p. 2.

249 the 300 percent increase in prisoners in jail: Swasey, E. J. 1994. The real cause of violent crime. *American Rifleman* 142(7):40–41.

249 The following professional advice—and license for irresponsibility—was given to victims of rape by authors Caren Adams and Jennifer Fay in answer to the question, "Should I report my rape to the police or not?": "Although others may make you feel so, it is not your responsibility to stop a rapist. You did not create him; he is not your responsibility. If you report it to the police, he may be stopped, but he may not be": Adams, C., and J. Fay. 1989. *Free of the Shadows: Recovering from Sexual Violence.* Oakland, Calif.: New Harbinger, p. 202.

249 only four felons go to prison: Swasey 1994, The real cause.

249 capital punishment for murder: Daly, M., and M. Wilson. 1988. *Homicide*. New York: Aldine de Gruyter, p. 226.

249 to work, punishment must be certain and swift: Wilson and Herrnstein, pp. 44–45, 49–56, 61–62. See also Reiss, A. J., Jr., and J. A. Roth (eds.). 1993. *Understanding and Preventing Violence*. Washington, D.C.: National Academy Press, pp. 291–294.

250 about $100 billion: Maguire, K., and A. L. Pastore (eds.). 1997. *Sourcebook of Criminal Justice Statistics, 1996*. Washington, D.C.: U.S. Department of Justice, p. 3.

250 70 percent of the world's lawyers: Gergen, D. 1991. America's legal mess. *U.S. News & World Report* 111(8):72.

250 42,361,840 crimes: Maguire and Pastore 1997, p. 224.

250 twelve-year-old child: Kopel 1992, pp. 375–376.

250 three-quarters of serious crimes: Baker, J. R. 1991. War for your guns. *American Rifleman* 139(9):36–39.

250 of the 4.3 million people serving: Swasey, E. 1994. NRA woman's voice. *American Rifleman* 142(3):16.

250 "I'm disappointed": Timnick, L. 1989. Jury votes death for Night Stalker Ramirez. *Los Angeles Times,* October 5:1, 36.

250–251 Without linking pain to prison: Robbins, A. 1992. *Awaken the Giant Within*. New York: Summit Books, pp. 511–512. See also Gest, T., J. Seter, D. Friedman, and K. Whitelaw. 1995. Crime and punishment. *U.S. News & World Report* 119(1):24–26.

251 "Whatever factors contribute": Wilson and Herrnstein 1985, pp. 24, 42–43, 63, 493, 496, 495, 506, 507.

251–252 analysis of murder rates: Witkin, G. 1998. The crime bust. *U.S. News & World Report* 124(20):28–37.

252 fourteen people are murdered: Jones, S. 1994. A typical crime day in America. *American Rifleman* 142(6):26.

252 only 19 percent of Americans: Maguire and Pastore 1996, pp. 203, 128, 129, 131, 133, 142, 151, 153.

252–253 85 percent of Americans: Ibid., p. 173. See also Forst, B. E., and J. C. Hernon. 1985. *The Criminal Justice Response to Victim Harm*. Washington, D.C.: U.S. Department of Justice, National Institute of Justice Research in Brief, p. 5.

253 Americans' preferred solutions to violent crime: Maguire and Pastore 1996, pp. 165, 167, 173, 176, 178.

253 by 1997, 75 percent of Americans: Maguire and Pastore 1997, pp. 159, 161, 165.

253 a 1997 look: Brownlee, S., D. McGraw, and J. Vest. 1997. The place for vengeance. *U.S. News & World Report* 122(23):24–32.

253 24 percent of Americans blame lack of parental discipline: Maguire and Pastore 1996.

253 by April 30, 1996: Maguire and Pastore 1997, pp. 568, 561, 558.

253–254 "some form of strong social control": Alder, F. 1983. *Nations Not Obsessed with Crime*. Littleton, Colo.: Fred B. Rothman. (Cited in Kopel 1992, pp. 410–411.)

254 by 1991 only half of American kids: Bernstein, A. 1994. Outlook family nuclear fallout. *U.S. News & World Report* 117(10):25. See also Fuchs, V. R., and D. M. Reklis. 1992. America's children: economic perspectives and policy options. *Science* 255:41–46.

254 Israeli rescue raid on Entebbe: Stevenson, W. 1976. *90 Minutes at Entebbe*. New York: Bantam.

254–255 antiterrorist reprisals against Osama bin Laden: Newman, R. J. 1998. America fights back. *U.S. News & World Report* 125(8):38–46.

255–256 "the combined measure": Sampson, R. J., S. W. Raudenbush, and F. Earis. 1997. Neighborhoods and crime: a multilevel study of collective efficacy. *Science* 277:918–924.

257 "Changing an organization": Robbins 1992, p. 24.

257 "The prime factor": Roosevelt, T. 1957. The fight goes on. In H. Hagedorn (ed.), *The Theodore Roosevelt Treasury,* pp. 313–317. New York: Putnam.

INDEX

ABOUT THE AUTHOR

Michael P. Ghiglieri received his Ph.D. from the University of California at Davis, and is currently Associate Professor of Anthropology at the University of Northern Arizona. He is the author of *East of the Mountains of the Moon: Chimpanzee Society in the African Rain Forest*, *The Chimpanzees of Kibale Forest: A Field Study of Ecology and Social Structure*, and *Canyon*.